Observability

Advances in Design and Control

SIAM's Advances in Design and Control series consists of texts and monographs dealing with all areas of design and control and their applications. Topics of interest include shape optimization, multidisciplinary design, trajectory optimization, feedback, and optimal control. The series focuses on the mathematical and computational aspects of engineering design and control that are usable in a wide variety of scientific and engineering disciplines.

Editor-in-Chief
Ralph C. Smith, North Carolina State University

Editorial Board

Stephen L. Campbell, North Carolina State University
Michel C. Delfour, University of Montreal
Fariba Fahroo, Air Force Office of Scientific Research
J. William Helton, University of California, San Diego
Birgit Jacob, University of Wuppertal
Kirsten Morris, University of Waterloo
Jennifer Mueller, Colorado State University
Michael Ross, Naval Postgraduate School
John Singler, Missouri University of Science and Technology
Stefan Volkwein, Universität Konstanz

Series Volumes

Martinelli, Agostino, *Observability: A New Theory Based on the Group of Invariance*

Betts, John T., *Practical Methods for Optimal Control Using Nonlinear Programming, Third Edition*

Rodrigues, Luis, Samadi, Behzad, and Moarref, Miad, *Piecewise Affine Control: Continuous-Time, Sampled-Data, and Networked Systems*

Ferrara, A., Incremona, G. P., and Cucuzzella, C., *Advanced and Optimization Based Sliding Mode Control: Theory and Applications*

Morelli, A. and Smith, M., *Passive Network Synthesis: An Approach to Classification*

Özbay, Hitay, Gümüşsoy, Suat, Kashima, Kenji, and Yamamoto, Yutaka, *Frequency Domain Techniques for $\mathcal{H}_\infty$ Control of Distributed Parameter Systems*

Khalil, Hassan K., *High-Gain Observers in Nonlinear Feedback Control*

Bauso, Dario, *Game Theory with Engineering Applications*

Corless, M., King, C., Shorten, R., and Wirth, F., *AIMD Dynamics and Distributed Resource Allocation*

Walker, Shawn W., *The Shapes of Things: A Practical Guide to Differential Geometry and the Shape Derivative*

Michiels, Wim and Niculescu, Silviu-Iulian, *Stability, Control, and Computation for Time-Delay Systems: An Eigenvalue-Based Approach, Second Edition*

Narang-Siddarth, Anshu and Valasek, John, *Nonlinear Time Scale Systems in Standard and Nonstandard Forms: Analysis and Control*

Bekiaris-Liberis, Nikolaos and Krstic, Miroslav, *Nonlinear Control under Nonconstant Delays*

Osmolovskii, Nikolai P. and Maurer, Helmut, *Applications to Regular and Bang-Bang Control: Second-Order Necessary and Sufficient Optimality Conditions in Calculus of Variations and Optimal Control*

Biegler, Lorenz T., Campbell, Stephen L., and Mehrmann, Volker, eds., *Control and Optimization with Differential-Algebraic Constraints*

Delfour, M. C. and Zolésio, J.-P., *Shapes and Geometries: Metrics, Analysis, Differential Calculus, and Optimization, Second Edition*

Hovakimyan, Naira and Cao, Chengyu, *$\mathcal{L}_1$ Adaptive Control Theory: Guaranteed Robustness with Fast Adaptation*

Speyer, Jason L. and Jacobson, David H., *Primer on Optimal Control Theory*

Betts, John T., *Practical Methods for Optimal Control and Estimation Using Nonlinear Programming, Second Edition*

Shima, Tal and Rasmussen, Steven, eds., *UAV Cooperative Decision and Control: Challenges and Practical Approaches*

Speyer, Jason L. and Chung, Walter H., *Stochastic Processes, Estimation, and Control*

Krstic, Miroslav and Smyshlyaev, Andrey, *Boundary Control of PDEs: A Course on Backstepping Designs*

Ito, Kazufumi and Kunisch, Karl, *Lagrange Multiplier Approach to Variational Problems and Applications*

Xue, Dingyü, Chen, YangQuan, and Atherton, Derek P., *Linear Feedback Control: Analysis and Design with MATLAB*

Hanson, Floyd B., *Applied Stochastic Processes and Control for Jump-Diffusions: Modeling, Analysis, and Computation*

Michiels, Wim and Niculescu, Silviu-Iulian, *Stability and Stabilization of Time-Delay Systems: An Eigenvalue-Based Approach*

Ioannou, Petros and Fidan, Barış, *Adaptive Control Tutorial*

Bhaya, Amit and Kaszkurewicz, Eugenius, *Control Perspectives on Numerical Algorithms and Matrix Problems*

Robinett III, Rush D., Wilson, David G., Eisler, G. Richard, and Hurtado, John E., *Applied Dynamic Programming for Optimization of Dynamical Systems*

Huang, J., *Nonlinear Output Regulation: Theory and Applications*

Haslinger, J. and Mäkinen, R. A. E., *Introduction to Shape Optimization: Theory, Approximation, and Computation*

Antoulas, Athanasios C., *Approximation of Large-Scale Dynamical Systems*

Gunzburger, Max D., *Perspectives in Flow Control and Optimization*

Delfour, M. C. and Zolésio, J.-P., *Shapes and Geometries: Analysis, Differential Calculus, and Optimization*

Betts, John T., *Practical Methods for Optimal Control Using Nonlinear Programming*

El Ghaoui, Laurent and Niculescu, Silviu-Iulian, eds., *Advances in Linear Matrix Inequality Methods in Control*

Helton, J. William and James, Matthew R., *Extending $\mathcal{H}^\infty$ Control to Nonlinear Systems: Control of Nonlinear Systems to Achieve Performance Objectives*

Observability
A New Theory Based on the Group of Invariance

Agostino Martinelli
Inria Grenoble - Rhône-Alpes Research Centre
Montbonnot-Saint-Martin, France

siam.
Society for Industrial and Applied Mathematics
Philadelphia

Copyright © 2020 by the Society for Industrial and Applied Mathematics

10 9 8 7 6 5 4 3 2 1

All rights reserved. Printed in the United States of America. No part of this book may be reproduced, stored, or transmitted in any manner without the written permission of the publisher. For information, write to the Society for Industrial and Applied Mathematics, 3600 Market Street, 6th Floor, Philadelphia, PA 19104-2688 USA.

No warranties, express or implied, are made by the publisher, authors, and their employers that the programs contained in this volume are free of error. They should not be relied on as the sole basis to solve a problem whose incorrect solution could result in injury to person or property. If the programs are employed in such a manner, it is at the user's own risk and the publisher, authors, and their employers disclaim all liability for such misuse.

Trademarked names may be used in this book without the inclusion of a trademark symbol. These names are used in an editorial context only; no infringement of trademark is intended.

MATLAB is a registered trademark of The MathWorks, Inc. For MATLAB product information, please contact The MathWorks, Inc., 3 Apple Hill Drive, Natick, MA 01760-2098 USA, 508-647-7000, Fax: 508-647-7001, info@mathworks.com, www.mathworks.com.

Publications Director	Kivmars H. Bowling
Executive Editor	Elizabeth Greenspan
Developmental Editor	Mellisa Pascale
Managing Editor	Kelly Thomas
Production Editor	David Riegelhaupt
Copy Editor	Claudine Dugan
Production Manager	Donna Witzleben
Production Coordinator	Cally A. Shrader
Compositor	Cheryl Hufnagle
Graphic Designer	Doug Smock

Library of Congress Cataloging-in-Publication Data
Names: Martinelli, Agostino, author.
Title: Observability : a new theory based on the group of invariance / Agostino Martinelli.
Description: Philadelphia : Society for Industrial and Applied Mathematics, 2020. | Series: Advances in design and control ; 37 | Includes bibliographical references and index. | Summary: "This book is about nonlinear observability. It provides a modern theory of observability based on a new paradigm borrowed from theoretical physics. Notably, it presents the first general analytic solution of a very complex open problem: the nonlinear Unknown Input Observability (nonlinear UIO)"-- Provided by publisher.
Identifiers: LCCN 2020016945 (print) | LCCN 2020016946 (ebook) | ISBN 9781611976243 (paperback) | ISBN 9781611976250 (ebook)
Subjects: LCSH: Observers (Control theory) | Symmetry (Mathematics)
Classification: LCC QA402.3 .M3426 2020 (print) | LCC QA402.3 (ebook) | DDC 515/.642--dc23
LC record available at *https://lccn.loc.gov/2020016945*
LC ebook record available at *https://lccn.loc.gov/2020016946*

siam is a registered trademark.

To my parents.

My mother, who fully transmitted to me her immense love for nature.

My father, one of the most talented scientists I have ever met.

⌘

Contents

List of Algorithms		ix
Preface		xi
1	**Introduction**	**1**
	1.1 Elementary example of an input-output system	1
	1.2 Observability	2
	1.3 Main features of a complete theory of observability	7
	1.4 The twofold role of time	12
	1.5 Unknown input observability	12
	1.6 Nonlinear observability for time-variant systems	16
I	**Reminders on Tensors and Lie Groups**	**19**
2	**Manifolds, Tensors, and Lie Groups**	**21**
	2.1 Manifolds	21
	2.2 Tensors	25
	2.3 Distributions and codistributions	36
	2.4 Lie groups	45
	2.5 Tensors associated with a group of transformations	53
II	**Nonlinear Observability**	**55**
3	**Group of Invariance of Observability**	**57**
	3.1 The chronostate and the chronospace	58
	3.2 Group of invariance in the absence of unknown inputs	61
	3.3 Group of invariance in the presence of unknown inputs	63
4	**Theory of Nonlinear Observability in the Absence of Unknown Inputs**	**67**
	4.1 Theory based on a constructive approach	68
	4.2 Observability rank condition for the simplified system	68
	4.3 Observability rank condition in the general case	74
	4.4 Observable function	82
	4.5 Theory based on the standard approach	85
	4.6 Unobservability and continuous symmetries	87
	4.7 Extension of the observability rank condition to time-variant systems	90

5	**Applications: Observability Analysis for Systems in the Absence of Unknown Inputs**	**95**
	5.1 The unicycle	96
	5.2 Vehicles moving on parallel lines	97
	5.3 Simultaneous odometry and camera calibration	98
	5.4 Simultaneous odometry and camera calibration in the case of circular trajectories	101
	5.5 Visual inertial sensor fusion with calibrated sensors	105
	5.6 Visual inertial sensor fusion with uncalibrated sensors	111
	5.7 Visual inertial sensor fusion in the cooperative case	117
	5.8 Visual inertial sensor fusion with virtual point features	124
III	**Nonlinear Unknown Input Observability**	**133**
6	**General Concepts on Nonlinear Unknown Input Observability**	**135**
	6.1 A constructive definition of observability	136
	6.2 Basic properties to obtain the observable codistribution	138
	6.3 A partial tool to investigate the observability properties in the presence of unknown inputs	142
	6.4 Theory based on the standard approach	148
7	**Unknown Input Observability for Driftless Systems with a Single Unknown Input**	**153**
	7.1 Extension of the observability rank condition for the simplified systems	153
	7.2 Applications	158
	7.3 Analytic derivations	175
8	**Unknown Input Observability for the General Case**	**187**
	8.1 Extension of the observability rank condition for the general case	187
	8.2 Applications	196
	8.3 Analytic derivations	218
	8.4 Extension to time-variant systems	234
A	**Proof of Theorem 4.5**	**237**
B	**Reminders on Quaternions and Rotations**	**243**
C	**Canonic Form with Respect to the Unknown Inputs**	**247**
	C.1 System canonization in the case of a single unknown input	248
	C.2 System canonization in the general case	249
Bibliography		**257**
Index		**261**

List of Algorithms

Algorithm 2.1	Smallest involutive distribution containing span$\{f_1,\ldots,f_d\}$	38
Algorithm 2.2	Smallest integrable codistribution containing span$\{\frac{\partial}{\partial x}h_1,\ldots,\frac{\partial}{\partial x}h_k\}$ and invariant under span$\{f_1,\ldots,f_d\}$	40
Algorithm 4.1	Observable codistribution for the simplified system in the absence of unknown inputs	71
Algorithm 4.2	Observable codistribution in the absence of unknown inputs	77
Algorithm 4.3	Observable codistribution in the chronospace (in the absence of unknown inputs)	91
Algorithm 4.4	Observable codistribution for time-variant systems (in the absence of unknown inputs)	91
Algorithm 6.1	Observable codistribution in the augmented space that includes the unknown inputs	143
Algorithm 7.1	Vectors ${}^i\phi_k$ for driftless systems with a single unknown input	154
Algorithm 7.2	Observable codistribution for driftless systems with a single unknown input	155
Algorithm 7.3	Fusion of Algorithm 7.1 and Algorithm 7.2	156
Algorithm 7.4	Observable codistribution in the augmented space for driftless systems with $m_w = m_u = 1$	176
Algorithm 8.1	Vectors ${}^i\phi_k^{\alpha_1,\ldots,\alpha_k}$ in the general case with unknown inputs	193
Algorithm 8.2	Observable codistribution in the general case with unknown inputs	193
Algorithm 8.3	Fusion of Algorithm 8.1 and Algorithm 8.2	194
Algorithm 8.4	Observable codistribution in the chronospace (in the presence of unknown inputs)	235
Algorithm 8.5	Observable codistribution for time-variant systems (in the presence of unknown inputs)	236
Algorithm C.1	Canonization for driftless systems with a single unknown input	249
Algorithm C.2	Canonization in the general case	252

Preface

This book is the result of almost three decades of research in very different scientific and technological domains: robotics, computer vision, Brownian dynamics, astrophysics/cosmology, and neuroscience. From one side, my interest in observability originates from very practical exigencies. Specifically, for a long time, I studied the possibility of using monocular vision and inertial sensors to make a drone able to fly autonomously in challenging real-world scenarios, such as remote inspection and search and rescue after natural disasters. This made it necessary to develop sophisticated and robust estimation strategies able to work in real time. In this respect, I was forced to learn more and more on the theory of nonlinear observability, i.e., in order to design suitable methodologies for very complex estimation problems. From another side, my background and long experience in theoretical physics allowed me to look at the problem of observability from a completely different perspective, and to adopt methods that are original and, I believe, innovative with respect to the existing literature in control theory.

Very roughly speaking, and from a very practical point view, we can think of observability as follows. Observability refers to the state that characterizes a system (e.g., if the system is an aerial drone, its state, under suitable conditions, can be its position and its orientation). A system is also characterized by one or more inputs, which drive its dynamics and one or more outputs (e.g., for the drone, the inputs could be the speeds of its rotators and the outputs the ones provided by an on-board monocular camera and/or a GPS). Very roughly speaking, a state is observable if the knowledge of the system inputs and outputs, during a given time interval, allows us its determination.

The observability rank condition [7, 19], introduced in the 1970s by Hermann and Krener [19], is a simple systematic procedure that allows us to give an answer to the previous fundamental question, i.e., whether the state is observable or not. It is based on very simple systematic computation (differentiation and matrix rank determination) on the functions that describe the system (i.e., the functions that express its dynamics in terms of the inputs and the functions that express its outputs in terms of the components of the state). The observability rank condition can deal with any system, independently of its complexity and type of nonlinearity. It works automatically.

On the other hand, the derivations of this and similar results in nonlinear observability do not capture/exploit more profound and fundamental features that are intimately related to the concept of observability. This results in two important limitations:

- These derivations, although simple and based on elementary mathematics, can be sometimes burdensome with the risk of easily losing the meaning of the results and losing the meaning of their assumptions.

- More complex observability problems (e.g., the unknown input observability problem[1] to which this book provides the complete analytic solution) remained unsolved for half a century.

The key to overcoming the two above limitations consists in building a new theory that accounts for the *group of invariance that is inherent to the concept of observability*. This is the typical manner in which the research in physics has always proceeded. To this regard, I wish to emphasize that the derivation of the basic equations of any physics theory (e.g., general relativity Yang–Mills, quantum chromodynamics) starts precisely from the characterization of the group of invariance of the theory. Making manifest the group of transformations under which a given problem is invariant is the fundamental step to capture the essence of the problem.[2] Maybe the most important example is the Standard Model of particle physics that provides a full description of the electromagnetic, weak, and strong interactions. This model is invariant under several symmetries simultaneously. They are basically the following two: the global Poincaré symmetry (that consists of the familiar translational symmetry, rotational symmetry, and the inertial reference frame invariance central to the theory of special relativity) and the local $SU(3) \times SU(2) \times U(1)$ gauge symmetry, which is an internal symmetry that essentially defines the Standard Model.

One of the major novelties introduced by this book is the characterization of the group of invariance of observability and, regarding the case of unknown inputs, the characterization of a subgroup that will be called the *Simultaneous Unknown Input-Output transformations' group* ($\mathcal{SUIO}$).

First of all, the group of invariance of observability allowed me to identify simplified systems for which obtaining the solution is much easier than in the general case. On the other hand, these systems are still complex enough to be representative of the general case.

I wish to emphasize that, in many scientific/technological domains, a complex problem is solved by first solving some simplified versions of the general formulation. The identification of these simplified versions is a key step and deserves a profound and intelligent investigation. A naive idea is to take the linear case as a simplified version of the general problem. In many cases, this choice has no rational foundation and becomes a useless exercise. From science we learn that the problems that can be analytically solved are characterized by a set of symmetries. This fact holds in classical mechanics, in quantum mechanics, in field theory, in general relativity, etc. The problems which are defined as *integrable* are very few, and they are characterized by one or more continuous symmetries (e.g., in mechanics they are the gyroscope, the two-bodies problem, the harmonic oscillator, and the pendulum). Without providing a quantitative discussion of this fundamental aspect, I wish to mention Noether's theorem that establishes the link between symmetries and conservation laws, which are crucial to obtain an analytic solution. Noether's theorem is a very general result, and it exists in classical and quantum mechanics and also in field theory.

In a very broad sense, for a scientific/technological problem characterized by a group of invariance, a reasonable manner to determine simplified versions of the problem is to understand under which conditions this group of invariance becomes an Abelian group. As we will see in section 2.4, an Abelian group is very simple and can be represented by scalars (see Schur's

---

[1] The problem of unknown input observability is the same problem of observability when some of the inputs are unknown. For instance, in the case of our drone, its dynamics could be driven also by the wind, which is in general unknown. The wind is a disturbance and acts on the dynamics as an (unknown) input.

[2] Even from a philosophical point of view, we could say that the comprehension of a phenomenon is the identification of the link between this phenomenon with other apparently very different phenomena, i.e., the set of transformations that translates this phenomenon in other different phenomena (e.g., Newton's apple with the trajectory of a planet with the space-time distortion in a black hole).

lemma in section 2.4.2). This means that, to deal with these simplified problems, we do not require the use of tensors with respect to the group of invariance (section 2.5). On the other hand, these simplified problems are still representative of the original problem since they are still characterized by the same group of invariance.

This book provides several novelties with respect to the existing literature in control theory. Specifically, the reader will learn the following:

- The solution of two open problems in control theory (the book provides separately the solution and the derivation), which are

    - the extension of the observability rank condition to nonlinear systems driven by also unknown inputs and
    - the extension of the observability rank condition to nonlinear, time-variant systems (both in the presence and in the absence of unknown inputs).

- A new and more palatable derivation of the existing results in nonlinear observability.

- A new manner of approaching scientific and technological problems, borrowed from theoretical physics (Chapter 2 summarizes in a very intuitive and quick manner the basic mathematics, which includes tensorial calculus).

- A new manner of dealing with the variable *time* in system theory, which is obtained by introducing a new framework, which will be called the *chronospace*.

In addition, the book provides many examples, most of them from robotics and autonomous navigation. Some of them are deliberately elementary, and they are chosen for educational purposes. On the other hand, the book also provides very complex examples. Most of them are in the framework of vision-aided inertial navigation for aerial vehicles. In particular, for these systems, the book provides all the derivations needed to separate the observable part of the system from the one that is unobservable. This analysis has a fundamental practical importance since it allows us to obtain the basic equations necessary for the implementation of any estimation scheme (e.g., filter based, optimization based, etc.) or even to obtain a closed-form solution of the problem (when it exists). In addition, the book also provides very complex examples that can have applications in neuroscience.[3]

Finally, I hope this book is also an opportunity for the control and information theory communities to borrow basic mathematics, tricks, and types of reasoning from theoretical physics and potentially revisit many aspects of control and information theory.

---

[3] Interestingly, I obtained the analytic solution of the unknown input observability problem, pushed by the curiosity of solving one of these examples.

Chapter 1

Introduction

This book deals with *input-output systems* and is about *observability*. In order to illustrate these two concepts, and to highlight the scope of the book and in particular its novelties with respect to the existing books, we start by referring to a specific and elementary example of an input-output system.

1.1 ▪ Elementary example of an input-output system

We consider a vehicle that moves on a plane. By introducing on this plane a global frame, we can characterize the position and orientation of the vehicle by the three parameters x_v, y_v, and θ_v with $x_v \in \mathbb{R}$, $y_v \in \mathbb{R}$, and $\theta_v \in [-\pi, \pi)$ (see Figure 1.1).

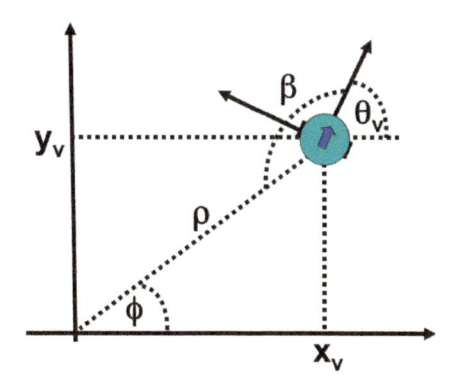

Figure 1.1. *Example of an input-output system: a vehicle that moves on a plane.*

Under the assumption of the unicycle constraint, these parameters satisfy the following dynamics equations (unicycle dynamics):

$$\begin{bmatrix} \dot{x}_v &= v\cos\theta_v, \\ \dot{y}_v &= v\sin\theta_v, \\ \dot{\theta}_v &= \omega, \end{bmatrix} \quad (1.1)$$

where v and ω are the linear and the rotational vehicle speeds, respectively.

Let us assume that a landmark is placed at the origin of the global frame. In addition, the vehicle is equipped with a sensor that is able to perceive the landmark and that simultaneously provides its distance and bearing angle in the local frame (i.e., ρ, β in Figure 1.1). Both ρ and β can be expressed in terms of the vehicle position and orientation. We have

$$\begin{aligned} \rho &= \sqrt{x_v^2 + y_v^2}, \\ \beta &= \pi - \theta_v + \operatorname{atan2}(y_v,\ x_v). \end{aligned} \quad (1.2)$$

We say that this system is an input-output system characterized by the state $[x_v,\ y_v,\ \theta_v]^T$. The inputs are v, ω, which are two functions of time that drive the dynamics of the state according to the equations in (1.1). The outputs are the two quantities ρ and β, which can be expressed in terms of the components of the state, according to the equations in (1.2).

This system is a special case of systems characterized by a state x that satisfy the dynamics

$$\begin{cases} \dot{x} &= f^0(x) + \sum_{k=1}^m f^k(x) u_k, \\ y &= h(x). \end{cases} \quad (1.3)$$

In particular the following hold:

- x is the state and $x = [x_v,\ y_v,\ \theta_v]^T$.

- $f^0(x)$, called the drift, in our specific example vanishes (i.e., $f^0(x) = [0,\ 0,\ 0]^T$).

- m, the number of system inputs, is 2. Specifically, $u_1 = v$, $u_2 = \omega$ and $f^1(x) = [\cos\theta_v,\ \sin\theta_v,\ 0]^T$ and $f^2(x) = [0,\ 0,\ 1]^T$.

- y is the output and has two components, i.e., $y = h(x) = [\sqrt{x_v^2 + y_v^2},\ \pi - \theta_v + \operatorname{atan2}(y_v,\ x_v)]^T$.

The first aspect that we wish to emphasize is that the state is a point that belongs to an object that is very similar to (but is not) a Euclidean space. For our specific example, this object is $\mathbb{R}^2 \times \mathcal{S}^1$. This is a special case of a differentiable manifold of dimension 3. In the case of a vehicle that moves in the entire 3D space (e.g., a drone), the position and orientation are characterized by 6 parameters (3 for the position and 3 for the orientation). The corresponding manifold has dimension 6, and it is $\mathbb{R}^3 \times \mathcal{S}^3$.

A differentiable manifold is locally similar to a Euclidean space near each point. This allows us to introduce basic operations (e.g., the Lie derivative). However, there are also important differences that need attention. For instance, in general, vectors do not belong to the manifold. In Chapter 2, we remind the reader of the concept of manifold and tensor (the latter generalizes the concept of vector) and basic operations with them.

1.2 ▪ Observability

The basic problem that the theory of observability investigates is to understand whether an input-output system contains the necessary information to perform the estimation of the state. The information comes from the data delivered by one or more sensors that provide measurements on the system inputs and outputs. A quantitative manner to formulate this problem consists in checking whether, by using the knowledge of the system inputs and outputs during a given time interval, we can uniquely determine the initial state, starting from a completely lost situation and by having the possibility to choose any time assignment for the inputs (provided that some regularity conditions are satisfied). We want to provide an answer to this question, for the elementary

1.2. Observability

system introduced in the previous section, by following intuitive geometric reasoning. This will allow us to better understand the concept of observability and in particular the main features that a complete theory of observability must account for, namely

- the role of the inputs,
- the link between observability and continuous symmetries,
- the concept of observable function,
- the group of invariance of the theory,
- the twofold role of time, and
- the assumptions of the theory.

Specifically, we analyze separately the following systems, which are characterized by the same dynamics given in (1.1) and that differ for the output:

- System 1: The output is the distance of the origin ($y = \rho$).
- System 2: The output is the bearing angle of the origin in the local frame of the vehicle ($y = \beta$).
- System 3: The outputs are the distance and the bearing angle of the origin in the local frame of the vehicle ($y = [\rho, \ \beta]^T$).
- System 4: The outputs are the Cartesian coordinates of the origin in the local frame of the vehicle ($y = [\rho \cos \beta, \ \rho \sin \beta]^T$).
- System 5: The output is the bearing of the vehicle in the global frame ($y = \phi$).

System 1

The system is characterized by the dynamics in (1.1) and the following output:

$$y = \rho = \sqrt{x_v^2 + y_v^2}. \tag{1.4}$$

We want to check whether we have the necessary information to uniquely reconstruct the initial vehicle configuration by knowing the inputs and outputs during a given time interval.

When at the initial time the distance ρ from the origin is available, the vehicle position must belong to a circumference (see Figure 1.2). Additionally, any orientation is possible. As soon as the vehicle moves in accordance with the inputs $v(t)$ and $\omega(t)$, it is possible to find one trajectory starting from each point on the circumference that provides the same distance from the origin, at any time. This is obtained by suitably choosing the initial vehicle orientation (e.g., in Figure 1.2, the two indicated trajectories provide the same distance from the origin, at any time). Therefore, the dimension of the *unobservable* region is 1. In particular, we introduce the following transformation:

$$\begin{aligned} x_v &\to x_v' = \cos \tau \ x_v - \sin \tau \ y_v, \\ y_v &\to y_v' = \sin \tau \ x_v + \cos \tau \ y_v, \\ \theta_v &\to \theta_v' = \theta_v + \tau, \end{aligned} \tag{1.5}$$

where $\tau \in [-\pi, \ \pi)$ is the parameter that defines the transformation. The system inputs and output at any time are compatible with all the trajectories that differ because the initial state was transformed as above.

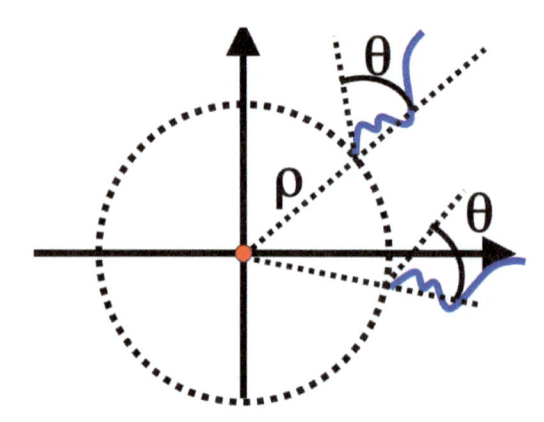

Figure 1.2. *System 1: the two trajectories are compatible with the same system inputs and outputs (the outputs consist of the distances from the origin).*

System 2

Now, instead of a range sensor, the vehicle is equipped with a bearing sensor (e.g., a camera), which provides the bearing angle of the origin in its own frame. Therefore, our system has the following output (see Figure 1.1):

$$y = \beta = \pi - \theta_v + \operatorname{atan2}(y_v, x_v) = \pi - \theta_v + \phi. \tag{1.6}$$

To check whether the vehicle configuration $[x_v, y_v, \theta_v]^T$ is observable, we have to prove that it is possible to uniquely reconstruct the initial vehicle configuration by knowing the inputs and the outputs in a given time interval. When at the initial time the bearing angle β of the origin is available, the vehicle can be everywhere in the plane, but, for each position, only one orientation provides the right bearing β. In Figure 1.3, all three positions A, B, and C are compatible with the observation β, provided that the vehicle orientation satisfies (1.6). In particular, the orientation is the same for A and B but not for C.

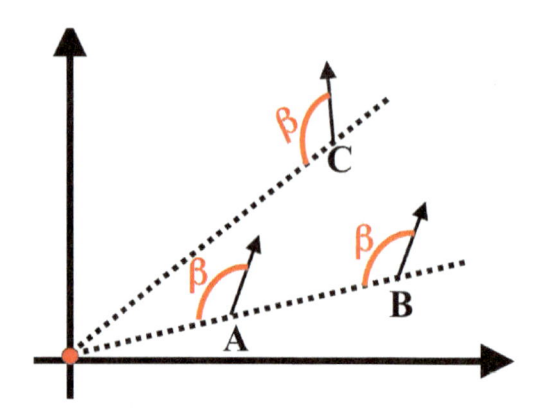

Figure 1.3. *System 2: the three initial vehicle configurations are compatible with the same initial observation (β).* © 2012 IEEE. Reprinted, with permission, from A. Martinelli, Vision and IMU Data Fusion: Closed-Form Solutions for Attitude, Speed, Absolute Scale and Bias Determination, Transaction on Robotics, Volume 28 (2012), Issue 1 (February), pp. 44–60.

1.2. Observability

Let us suppose that the vehicle moves according to the inputs $v(t)$ and $\omega(t)$. With the exception of the special motion consisting of a line passing by the origin, by only performing a further bearing observation it is possible to distinguish all the points belonging to the same line passing by the origin. In Figure 1.4(a) the two initial positions in A and B do not reproduce the same observations ($\beta_A \neq \beta_B$). On the other hand, all the initial positions whose distance from the origin is the same cannot be distinguished independently of the chosen trajectory. In Figure 1.4(b), the two indicated trajectories provide the same bearing observations, at any time. Therefore, the dimension of the *unobservable* region is 1. Note that this region coincides with the unobservable region of the previous system, and it is characterized by the transformation in (1.5).

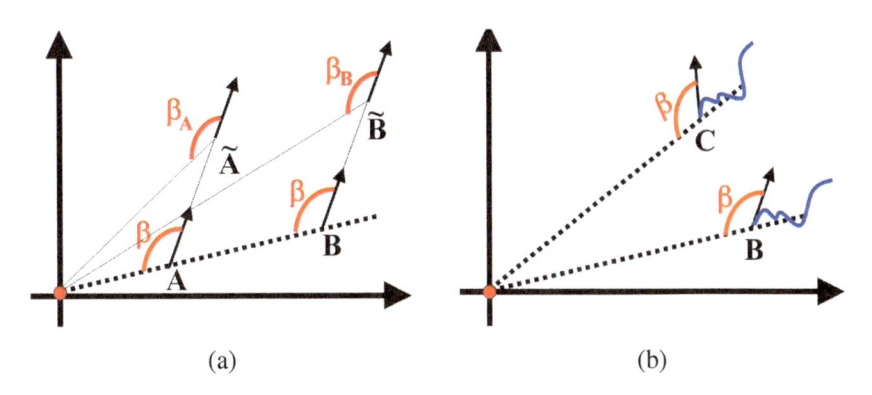

Figure 1.4. *In* (a) *the two initial positions (A and B) do not reproduce the same observations ($\beta_A \neq \beta_B$). In* (b) *the two indicated trajectories provide the same bearing observations at every time.* © 2011 *IEEE. Reprinted, with permission, from* [32].

System 3

The system is characterized by (1.1) and (1.2). This system has both the outputs that characterize, separately, the previous two systems. Since both these systems have exactly the same unobservable region, this region remains unobservable when both the outputs are available. In other words, the system inputs and outputs at any time are compatible with all the trajectories that differ because the initial state was transformed in accordance with (1.5).

System 4

Having the Cartesian coordinates of the origin, i.e., $\rho \cos \beta$ and $\rho \sin \beta$, is equivalent to having the polar coordinates ρ and β. Hence, this case coincides with the previous one.

System 5

Now, instead of having a bearing sensor on board, the bearing sensor is fixed at the origin and provides the bearing angle of the vehicle in the global frame. Therefore, our system has the following output (see Figure 1.1):

$$y = \phi. \tag{1.7}$$

As in the previous cases, we start by considering the first bearing observation ϕ_1. Let us refer to Figure 1.5. All the initial vehicle positions standing on the infinite line from the origin and passing through A_1 agree with this measurement. Additionally, any orientation is possible. If the vehicle moves along a straight line, with the exception of the radial motion passing through A_1, B_1, C_1, as soon as a second measurement is acquired (ϕ_2) we still obtain infinite initial states that agree with all the measurements. Specifically, all the positions standing on the segment $O - C_1$ agree with all the measurements, provided that the orientation is such that the distance traveled by the vehicle between the two bearing measurements is the same. In particular, the three trajectories shown in Figure 1.5 agree with all the input measurements and the two bearing measurements ϕ_1 and ϕ_2. Note that $\overline{A_1A_2} = \overline{B_1B_2} = \overline{C_1C_2}$. Now let us suppose that the vehicle still moves straight. In order to agree with the input measurements, we have $\overline{A_2A_3} = \overline{B_2B_3} = \overline{C_2C_3}$. On the other hand, a third bearing measurement performed after this last displacement would provide a different value for all three trajectories. In particular, we remark that only the trajectory $B_1 - B_2 - B_3$ simultaneously agrees with the input measurements and the three observations ϕ_1, ϕ_2, and ϕ_3, and we conclude that in this case the entire configuration is observable.

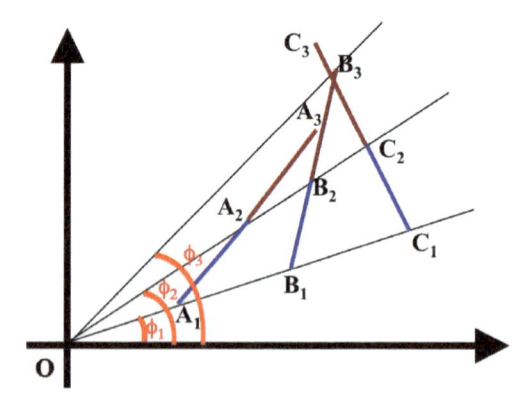

Figure 1.5. *System 5: only the trajectory $B_1 - B_2 - B_3$ simultaneously agrees with the inputs and the output at three distinct times ϕ_1, ϕ_2, and ϕ_3. The other two trajectories agree with the inputs and the output at the first two times ϕ_1, ϕ_2.*

Let us suppose now that the sensor at the origin provides the bearing angle modulo π (instead of 2π). In other words,

$$y = \phi \bmod \pi.$$

In this case, independently of the input time assignment, the inputs and the output at any time are compatible with the two trajectories that differ because the initial state was transformed by a rotation of π, i.e., according to the equations below (see also Figure 1.6):

$$\begin{aligned} x_v &\to x'_v = \cos \pi \, x_v - \sin \pi \, y_v = -x_v, \\ y_v &\to y'_v = \sin \pi \, x_v + \cos \pi \, y_v = -y_v, \\ \theta_v &\to \theta'_v = \theta_v + \pi. \end{aligned}$$

For this case, we cannot uniquely reconstruct the initial state from the knowledge of the system inputs and outputs in a given time interval. However, we only have two possible initial states.

1.3. Main features of a complete theory of observability

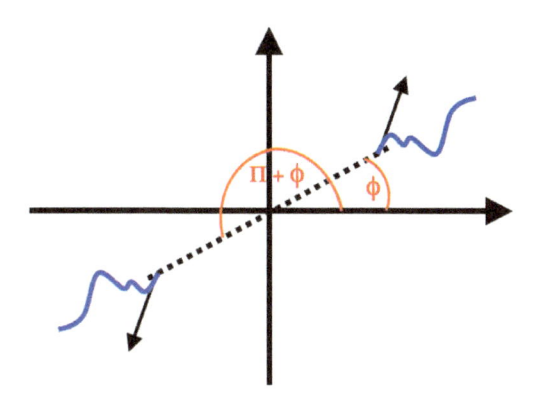

Figure 1.6. *The two trajectories agree with the same inputs and outputs at any time.*

In addition, and more importantly, these states are not "close" and could be distinguished by exploiting an extra rough knowledge provided by a further source of information.

We conclude this section by remarking that the considered systems are elementary. This allowed us to obtain the observability properties by following simple intuitive procedures. On the other hand, for very complex systems (e.g., characterized by a high dimensions state), it would be desirable to have an analytic test to check the state observability by simply following an automatic procedure. This automatic procedure exists: it is the observability rank condition [7, 19]. We analytically derive this criterion in Chapter 4. However, this criterion can only be adopted in the case when all the inputs are known. In addition, this criterion cannot be applied to time-variant systems.[4] The extensions of the observability rank condition to the two above more general systems are among the novelties of this book. The introduction of the criterion that holds when some of the system inputs are unknown requires a new theory of observability that accounts for several important aspects. We provide a qualitative discussion of these aspects in the next three sections. Accounting for them quantitatively is the matter of this book.

1.3 ▪ Main features of a complete theory of observability

From the previous analysis we can draw some statements about the features that must characterize a complete theory of observability. We discuss part of them in the following subsections, but postpone the discussion of another fundamental aspect (the twofold role of time) to section 1.4. Additionally, regarding the role of the inputs (in particular the case when they are unknown and act as disturbances), we postpone further remarks to section 1.5.

1.3.1 ▪ Definition of observability

In discussing the observability properties of our systems, we checked the possibility of reconstructing the initial state by using the knowledge of the inputs and the outputs during a given time interval. The definition of observability follows this idea. As illustrated in the last example, there are cases where we cannot uniquely determine the initial state, but it suffices to add extra

---
[4]A time-variant system is a system whose behavior changes with time. In particular, the system will respond differently to the same input at different times.

rough information to make it possible. To deal with this important aspect, we will introduce a property that is weaker than observability, which will be called *weak* observability.

Once we have defined the concept of observability (and weak observability), the fundamental problem is to obtain the analytic test to automatically check the observability of a general system. In other words, the problem is to obtain the procedure that allows us to check the observability by performing automatic computation on the functions that appear in (1.3). In the case of only known inputs, this analytic test was derived in the 1970s: it is the observability rank condition. In Chapter 4, we provide a derivation of this analytic criterion. In particular, we show that the observability rank condition is a simple consequence of the inverse function theorem.[5] In section 4.5, we also provide the concept of *indistinguishability* and we discuss its link with observability.

1.3.2 ▪ The role of the inputs

In all the examples discussed in the previous section, the observability properties increase by applying nonvanishing inputs. In particular, we were free to choose any time assignment for the inputs, provided that several regularity conditions are satisfied (we discuss this issue in section 1.3.6). As we will see, in the definition of observability, this aspect is explicitly outlined by stating that the existence of a single set of input functions $(u_1(t), \ldots, u_m(t))$ that make the state observable suffices to define the state as observable.

1.3.3 ▪ The link between observability and continuous symmetries

In all the systems where we found that the state is not observable (and even not weak observable), we found that the unobservable region is defined by a continuous transformation. This is always the case. There is an intimate relation between the observability (or better the unobservability) and the concept of *continuous symmetry*. In our case, the continuous symmetry was defined by (1.5), and it is a rotation about the vertical axis. The concept of continuous symmetry and its link with observability is introduced in Chapter 4 (section 4.6).

1.3.4 ▪ The concept of observable function

When the state is not observable, there are functions of the state which are observable. This was the case in the first four systems analyzed in the previous section. For those systems, we know that the unobservable region has dimension 1. Since the dimension of the state is 3, we have $3 - 1 = 2$ independent functions of the components of the state that are observable (i.e., the value that they take at the initial state can be reconstructed by using the inputs and the outputs in a given time interval). These functions must be invariant with respect to the transform given in (1.5). It is immediate to verify that the following two functions are invariant under this transform:

$$\begin{aligned} \rho &= \rho(x_v, y_v, \theta_v) = \sqrt{x_v^2 + y_v^2}, \\ \theta &= \theta(x_v, y_v, \theta_v) = \theta_v - \arctan 2(y_v, x_v). \end{aligned} \quad (1.8)$$

We formalize the concept of observable function in Chapter 4.

---

[5] Note that the inverse function theorem provides a sufficient and "almost" a necessary condition for a function to be invertible. The same also holds for the observability rank condition.

1.3.5 ▪ The group of invariance of the theory

In section 1.3.3, we discussed the link between observability and continuous symmetry. In particular, when the state that characterizes a given system is unobservable, there exists at least one transformation on the initial state that cannot be perceived by the analysis of the inputs and outputs (e.g., the rotation around the vertical axis defined by (1.5)). By "group of invariance of observability" we mean a completely different concept. We are not referring to a specific system, and we are not considering transformations of the initial state. This group of transformations is intrinsic to the concept of observability itself. In order to fully characterize this group of invariance, we need to deal with the variable time in a new (and atypical in control theory) manner. In particular, as will be discussed in section 1.4, in general, time acts on an input-output system in two distinct manners. This requires the use of a new framework, the *chronospace*, that will be introduced in section 3.1. In this section, we only give an overview of the group of invariance of observability without taking into account the twofold role of time. Later, in section 1.4, we intuitively discuss the twofold role of time. Finally, in Chapter 3, we introduce the new framework to properly deal with time and we fully and quantitatively characterize the group of invariance of observability.

In what follows, we provide the transformations under which observability must be invariant.

State coordinates transformations' group

Obviously, the result of an observability analysis must be independent of the coordinates that we adopt to define the state. For instance, for the vehicle discussed in our examples, we must be free to adopt polar coordinates to define its position (i.e., ρ and ϕ instead of x_v and y_v). The choice of the coordinates that define the state can be crucial to simplify the analysis, but the result must be independent. This means that the basic equations of the theory of observability must be written in terms of tensors. Tensors are the extension of scalars, vectors, and covectors (scalars are tensors of rank 0, and vectors and covectors are tensors of rank 1). They will be introduced in Chapter 2. In a very rough manner, we can say that they are defined with respect to a group of transformations and they transform in a given manner with respect to these transformations. In this case, the group of transformations includes all the possible changes of coordinates that define the state. Note that the state does not belong, in general, to a Euclidean space but to a manifold. Vectors and covectors do not belong to a manifold (with the exception of some special cases). More in general, tensors do not belong to a manifold. For this reason, we need to introduce the concepts of tangent space and the dual, which is the cotangent space. They will be introduced in Chapter 2. For instance, for our vehicle, $[x_v,\ y_v,\ \theta_v]^T$ is not a vector. However, as we will see in Chapter 2, $[\dot{x}_v,\ \dot{y}_v,\ \dot{\theta}_v]^T$ is. In particular, it belongs to the tangent space and it transforms in the right manner by changing the coordinates, e.g., by expressing the vehicle position in polar coordinates. Note that, even when the state belongs to a Euclidean space (in which case the manifold coincides with its tangent space at any point), the state is not a vector. This is because we are building a theory invariant with respect to any coordinates' change and not only linear coordinates' changes (e.g., as we mentioned, the theory must be invariant under the coordinates' change from Cartesian to polar coordinates). See section 2.2 for the definition of vectors (and, more in general, tensors).

It is fundamental to express the results of the theory in terms of tensorial equations. This guarantees its invariance.

Output transformations' group

As we saw in the discussion of system 4, expressing the output in terms of the polar coordinates (ρ, β), or in terms of the Cartesian coordinates ($\rho \cos \beta$, $\rho \sin \beta$), is equivalent. More in general, by denoting with $h(x) = [h_1(x), \ldots, h_p(x)]^T$, the results of the observability analysis must be invariant under the transformation

$$h_1 \to h_1' = h_1'(h_1, \ldots, h_p), \quad \ldots, \quad h_p \to h_p' = h_p'(h_1, \ldots, h_p), \tag{1.9}$$

provided that the Jacobian of the previous transformation, i.e., the following $p \times p$ matrix,

$$\frac{\partial h'}{\partial h} = \begin{bmatrix} \frac{\partial h_1'}{\partial h_1} & \frac{\partial h_1'}{\partial h_2} & \cdots & \frac{\partial h_1'}{\partial h_p} \\ \frac{\partial h_2'}{\partial h_1} & \frac{\partial h_2'}{\partial h_2} & \cdots & \frac{\partial h_2'}{\partial h_p} \\ \cdots & \cdots & \cdots & \cdots \\ \frac{\partial h_p'}{\partial h_1} & \frac{\partial h_p'}{\partial h_2} & \cdots & \frac{\partial h_p'}{\partial h_p} \end{bmatrix},$$

is nonsingular. Note that this group of transformations is independent of a state coordinate change. In other words, the output transformations' group is a new group of transformations under which the theory is invariant and we must account for this.

Input transformations' group

We can also operate a change in the inputs without altering the information on the initial state. Let us refer to our general system defined by (1.3), and let us introduce the following linear transformation:

$$u_k \to u_k' = \sum_{j=1}^{m} M_k^j u_j, \quad k = 1, \ldots, m, \tag{1.10}$$

where M is a nonsingular $m \times m$ matrix. In order to maintain the same observability properties, we need to make unaffected the time derivative of the state, i.e., $\dot{x}$. Hence, we need to simultaneously operate the following transformation:

$$f^k \to f'^k = \sum_{j=1}^{m} N_j^k f^j, \quad k = 1, \ldots, m, \tag{1.11}$$

where the matrix N is the inverse of M, namely

$$\sum_{j=1}^{M} M_l^j N_j^k = \delta_l^k,$$

where δ is the Kronecker delta, defined as follows:

$$\delta_l^k = \begin{bmatrix} 0, & l \neq k, \\ 1, & l = k. \end{bmatrix}$$

Note that this group of transformations is independent of the previous two (i.e., the state coordinate transformations' group and the output transformations' group). In other words, the input transformations' group is a new group of transformations under which the theory is invariant and we must account for this.

1.3. Main features of a complete theory of observability

Note that, in the above equations, we adopted sometimes an upper index (e.g., f^k) and sometimes a lower index (e.g., u_k). This is consistent with the tensor notation. In Chapter 2, we introduce this notation. We also introduce the Einstein notation that is very common in theoretical physics and significantly helps to achieve notational brevity.

The basic equations of a complete theory of observability must account for the invariance of the theory with respect to the previous three groups of transformations. Therefore, these equations must be tensorial equations. Note that we have three distinct types of tensors. For instance, we have the following:

- The quantity f^k is a vector with respect to the state coordinate transformations' group, a scalar with respect to the output transformations' group, and a vector (or better one of the m components of a vector) with respect to the input transformations' group.

- The quantity u_k is a scalar with respect to the first two transformations' groups and a covector (or better one of the m components of a covector) with respect to the last one.

- The quantity $\dot{x}$ is a vector with respect to the first transformations' group and a scalar with respect to the last two transformations' groups.

Making manifest the invariance with respect to these transformations' groups will be fundamental to obtain the analytic criterion to check the state observability, in particular in the presence of unknown inputs. Hence, our theory adopts different types of tensors.

The coexistence of tensors of different types is not new. In particular, in theoretical physics, we have several examples. The most important is the Standard Model of particle physics that provides a full description of the electromagnetic, weak, and strong interactions.[6]

1.3.6 ▪ The assumptions of the theory

In discussing our examples, we always assumed that all the measurements are noiseless and provided continuously in time (i.e., at infinite frequency). These are common assumptions in the theory of observability. Note that a theory that does not introduce any assumption on the noise would be trivial. No state would be observable. As a result, a theory that accounts for the noise must also introduce some assumptions to characterize the noise. This is certainly possible but makes the theory applicable to very specific cases. Assuming no noise is a strong assumption. On the other hand, it allows us to obtain simple results that can be very useful in real situations. Similar remarks hold for the frequency of the measurements (e.g., contemplating a finite frequency without any assumption on the state dynamics would bring us again to a trivial theory: no state would be observable).

In this book, we assume that, in most cases, the functions are analytic with respect to their arguments. This assumption could be motivated by remarking that any "real" function (i.e., that characterizes a real phenomenon) can be approximated by an analytic function. This fact, together with the above assumption of noiseless and continuous-time measurements, makes in some sense futile the investigation of the cases when the functions are nonanalytic. However, we wish to mention an important exception regarding the interaction between a known input and an unknown input. We basically assume more regular conditions on the latter. This means that, in the presence of both, we will be authorized to set the known input as a discontinuous function of time.

---

[6]In this model, two kinds of tensors coexist. They refer to the following two groups of invariance: the global Poincaré symmetry and the local $SU(3) \times SU(2) \times U(1)$ gauge symmetry.

1.4 • The twofold role of time

The discussion provided in this section is fundamental to obtain a complete characterization of the group of invariance of observability. In particular, time intervenes separately in the output transformations' group and in the input transformations' group.

Let us start by referring to our simple example, and let us suppose now that the dynamics of our system are also driven by an external force. For instance, the vehicle is a boat in the presence of a current. The dynamics equations become

$$\left[\begin{array}{ll} \dot{x}_v & = v_x + v\cos\theta_v, \\ \dot{y}_v & = v_y + v\sin\theta_v, \\ \dot{\theta}_v & = \omega, \end{array} \right. \quad (1.12)$$

where $[v_x, v_y]^T$ is the speed of the current. We assume that it is known. This system is still a special case of the system characterized by (1.3). With respect to the dynamics in (1.1), we also have a drift. Specifically, $f^0(x) = [v_x, v_y, 0]^T$. We write the above equations by using differentials. We have

$$\left[\begin{array}{ll} dx_v & = v_x dt + ds\cos\theta_v, \\ dy_v & = v_y dt + ds\sin\theta_v, \\ d\theta_v & = d\alpha, \end{array} \right. \quad (1.13)$$

where $ds = vdt$ and $d\alpha = \omega dt$. Additionally, we suppose that the system is characterized by a given output (for instance, the one given in 1.2).

We can assert that the dynamics in (1.13) are driven by three independent inputs that are dt, ds, and $d\alpha$. In particular, we remark the following fundamental aspect. Time acts in two completely different manners. From one side, time is an index (in this case a continuous index) that synchronizes all the system inputs and outputs. We will call it the *chronological time* (or simply time, when it is not ambiguous). From another side, time acts as a system input (with the particularity that it is ineluctably assigned). We will call this time the *ordinary time* (or simply time, when it is not ambiguous). Note that this second role of time only occurs in systems characterized by a nonvanishing drift (i.e., the term f^0 in (1.3) has at least a nonvanishing component). In Chapter 3, we introduce a new framework, the *chronospace*, where these two roles of time are separated. Note that, when the inputs and the outputs are measured by sensors, these sensors provide their measurements together with the time when each measurement has occurred. This means that our real system is also equipped with an additional sensor that is the clock (i.e., a sensor that measures the chronological time). By introducing our new framework, we will show that the measurements provided by this clock are the measurements of a new system output.

Let us go back to the group of invariance discussed in section 1.3.5. In the new framework we must extend the transformation defined by (1.9) to account for the presence of a new output, which is the chronological time. Additionally, we must extend the transformation defined by (1.10)–(1.11) to account for the fact that time is a system input. This makes manifest the twofold role of time since it comes in two independent groups of transformations: the output transformations' group and the input transformations' group. This will be done in Chapter 3.

1.5 • Unknown input observability

The problem of unknown input observability is the same problem of observability when some of the inputs are unknown. For instance, in the case of a drone, its dynamics could be driven also by the wind, which is in general unknown. The wind is a disturbance and acts on the dynamics as an (unknown) input.

1.5. Unknown input observability

A general characterization of an input-output system, when some of the inputs are unknown, is given by the following equations:

$$\begin{cases} \dot{x} = g^0(x) + \sum_{k=1}^{m_u} f^k(x)u_k + \sum_{j=1}^{m_w} g^j(x)w_j, \\ y = h(x), \end{cases} \quad (1.14)$$

where g^0 is the drift, m_u is the number of inputs which are known, and m_w is the number of inputs which are unknown. This generalizes the system defined by (1.3), which is obtained by setting $m_u = m$ and $m_w = 0$.

Note that, in the presence of unknown inputs, we prefer to denote the drift by g^0 instead of f^0. An unknown input differs from a known input not only because it is unknown but also because it cannot be assigned. In this sense, when in section 1.4 we discussed the input role of time, we must remark that this input behaves as a known input because it is known but it behaves as an unknown input because it cannot be assigned. This is the reason why, in the above equation, the drift was denoted by g^0 instead of f^0.

1.5.1 ▪ An example of an unknown input system

Let us go back to our elementary system, and let us investigate the observability properties of systems 1–5 discussed in section 1.2 but when one of the two inputs is unknown. Specifically, we will investigate the case when v is unknown. For those systems where the output is an angle (systems 2 and 5, i.e., $y = \beta$ and $y = \phi$, respectively), we certainly lose a degree of freedom, which corresponds to the absolute scale. In other words, the unobservable region includes the states that are obtained by transforming the initial state as follows:

$$\begin{aligned} x_v &\to x'_v = \lambda\, x_v, \\ y_v &\to y'_v = \lambda\, y_v, \\ \theta_v &\to \theta'_v = \theta_v, \end{aligned}$$

where λ is the continuous parameter that characterizes the transform. This transformation alone defines the unobservable region of system 5. In the case of system 2, we also have the transformation in (1.5).

Let us consider the case when the output is $y = \rho$. We know that, when both the inputs are known, all the independent observable functions are the ones given in (1.8). Hence, we can redefine the state as $[\rho,\ \theta]^T$. It can be easily verified that the new state satisfies the following equations:

$$\begin{bmatrix} \dot{\rho} = v\cos\theta, \\ \dot{\theta} = \omega - \frac{v}{\rho}\sin\theta, \end{bmatrix} \quad y = \rho, \quad (1.15)$$

which express the link between the new state $[\rho,\ \theta]^T$ and the system inputs (v, ω) and the output (ρ). We want to understand whether this reduced state is still observable when v becomes unknown. We proceed with a simple heuristic procedure. In particular, we show that we can retrieve the initial value of the state by setting the angular speed as follows:

$$\omega(t) = \begin{bmatrix} 0 & = t \le 0, \\ \omega_0 & = t > 0, \end{bmatrix}$$

with $\omega_0 > 0$. Since ρ is an output, and since it is provided continuously in time, also $\dot{\rho}$ and $\ddot{\rho}$ are available at any time. From (1.15) we obtain

$$\ddot{\rho} = \dot{v}\cos\theta - v\dot{\theta}\sin\theta = \dot{v}\cos\theta - v\sin\theta\left(\omega - \frac{v}{\rho}\sin\theta\right).$$

By computing the above quantity in 0^+ and 0^- we obtain

$$\ddot{\rho}(0^+) = \dot{v}_0 \cos\theta_0 - v_0 \sin\theta_0 \left(\omega_0 - \tfrac{v_0}{\rho_0}\sin\theta_0\right),$$
$$\ddot{\rho}(0^-) = \dot{v}_0 \cos\theta_0 + \tfrac{v_0^2}{\rho_0}\sin^2\theta_0,$$

where ρ_0, θ_0, v_0, and $\dot{v}_0$ are the values of the respective quantities computed at $t=0$ (note that we assumed v analytic). By subtracting the two above equations we obtain

$$\ddot{\rho}(0^-) - \ddot{\rho}(0^+) = v_0 \sin\theta_0\, \omega_0. \tag{1.16}$$

This equation, together with the first equation in (1.15) computed at $t=0$, i.e.,

$$\dot{\rho}_0 = v_0 \cos\theta_0, \tag{1.17}$$

provides a system of two independent equations in v_0 and θ_0 and, consequently, it allows their determination.

1.5.2 ▪ The role of the inputs in the case of unknown inputs

From these examples, we remark that the fact that a given input becomes unknown reduces in general the observability properties. This is the case of systems 2 and 5 when the linear speed becomes unknown. However, in some cases, the unobservable region remains the same (e.g., this is the case of system 1 when v becomes unknown). What we can certainly say is that the fact that a given input becomes unknown cannot increase the observability properties or, equivalently, the unobservable region cannot reduce. This is because the continuous symmetries that define the unobservable region cannot be broken by the fact that a given input becomes unknown.

On the other hand, the observability properties in general increase by adding a new degree of freedom to the dynamics even if the corresponding input is unknown. If in system 1 we set the linear speed to zero, even by knowing this fact, the unobservable region has dimension 2 (see the discussion in section 1.2 when we only exploit the output at the initial time). In contrast, by setting this linear speed to a given nonzero value, even by not knowing this value and even if it is time dependent (but it is an analytic function of time), the dimension of the unobservable region reduces to 1. In other words, the presence of the input, even if it is unknown, is able to break part of the symmetries that characterize the unobservability of the system in the absence of the input.

1.5.3 ▪ Extension of the observability rank condition

Obtaining the general analytic procedure to check the state observability for systems characterized by (1.14) is very complex. Note that this problem was an open problem in control theory for half a century. It was introduced and first investigated in [2] at the end of the 1960s. Specifically, [2] investigated the observability of linear, time-invariant dynamical systems when some of the input functions are unknown and obtained an expression for the observability subspace. A huge effort has since been devoted to design observers for both linear and nonlinear systems in the presence of unknown inputs, in many cases in the context of fault diagnosis, e.g., [1, 5, 8, 9, 12, 15, 16, 17, 18, 20, 27, 45, 47]. In some of the above works, interesting conditions for the existence of an unknown input observer were introduced. On the other hand, these conditions have the following limitations:

1.5. Unknown input observability

- They refer to a restricted class of systems since they are often characterized by linearity (or some specific type of nonlinearity) with respect to the state in some of the functions that characterize the dynamics and/or the system outputs.[7] No condition holds without any restriction on the nonlinearity with respect to the state in the aforementioned functions.

- They cannot be implemented automatically, i.e., by following a systematic procedure (e.g., by the usage of a simple code that adopts a symbolic computation tool, as in the case of the observability rank condition).

- They do not characterize the system observability since they only check the existence of an unknown input observer that belongs to a specific class of observers.

This book provides the complete analytic solution to the unknown input observability problem (which fully overcomes all the above limitations). The key to achieving this fundamental result is in building a new theory that accounts for the group of invariance discussed in section 1.3.5. Note that this is also very useful to obtain the criterion in the case without unknown inputs (i.e., for the systems characterized by (1.3)). First of all, the group of invariance of observability allows us to identify simplified systems for which obtaining the analytic test is much easier than in the general case. On the other hand, these systems are still complex enough to be representative of the general case.

In a very broad sense, for a scientific/technological problem characterized by a group of invariance, a reasonable manner to determine simplified versions of the problem is to understand under which conditions this group of invariance becomes an Abelian group. As we will see in section 2.4, an Abelian group is very simple and can be represented by scalars (see Schur's lemma in section 2.4.2). This means that, to deal with these simplified problems, we do not require the use of tensors with respect to the group of invariance (section 2.5). On the other hand, these simplified problems are still representative of the original problem since they are still characterized by the same group of invariance.[8]

Regarding the systems without unknown inputs (i.e., characterized by (1.3)), it is possible to show (see Chapter 3) that the input transformations' group becomes Abelian for two distinct classes of systems. In particular, they satisfy the following two conditions, respectively:

- The dynamics are driftless (i.e., f^0 vanishes) and driven by a single input (i.e., $m = 1$).

- There are no inputs (i.e., $m = 0$), and the dynamics are uniquely driven by a drift.

In addition, the output transformations' group becomes Abelian when the system has a single output ($h(x)$ is a scalar). We refer to the two above classes of systems, both characterized by a single output, as the *simplified* systems.

For the simplified systems, a complete observability theory does not require the introduction of tensors with respect to these two (Abelian) groups of transformations and the derivation of the observability rank condition becomes trivial. In other words, to obtain the analytic solution of the observability problem (i.e., the observability rank condition for the systems characterized by (1.3)) it is very convenient to analyze first the driftless case with a single input and a single output or the case without inputs, with a drift and a single output (instead of analyzing the linear case).

---

[7] These functions are the functions that appear in (1.14), i.e., $g^0(x), g^1(x), \ldots, g^{m_w}(x), f^1(x), \ldots, f^{m_u}(x), h(x)$.

[8] In particular, all the key quantities that characterize the analytic solution are scalars and in the general, non-Abelian case, they become multi-index tensors.

For the systems characterized by (1.14), the situation is more complex. We found that it is very convenient to introduce a new group of transformations that basically merges together the output transformations' group and the input transformations' group and, concerning the inputs, only includes the unknown inputs and the ordinary time. In other words, the known inputs will be scalars with respect to this transformations' group. We call this new group of transformations the *Simultaneous Unknown Input-Output transformations' group* ($\mathcal{SUIO}$). Obviously, the input transformations' group and the output transformations' group are independent and the input transformations' group also includes the known inputs. Hence, by proceeding in this manner, we lose generality. However, we gain simplicity. In practice, our theory will use two distinct types of tensors: tensors with respect to the state coordinate transformations' group and tensors with respect to $\mathcal{SUIO}$. Note that this group is built in the new framework that accounts for the twofold role of time. This will be done in Chapter 3.

The $\mathcal{SUIO}$ drives us towards the analytic solution of the observability problem (i.e., the derivation of the extension of the observability rank condition in the presence of unknown inputs) by providing the suitable class of simplified systems to be investigated at first. As we will see in Chapter 3, the $\mathcal{SUIO}$ becomes an Abelian group for the simplified systems defined by the following conditions:

- The dynamics are driftless (i.e., g^0 vanishes).

- The dynamics are driven by several known inputs and a single unknown input (i.e., $m_w = 1$).

- The system has a single output ($h(x)$ is a scalar).

In practice, to deal with these simplified systems, we do not need to introduce the new type of tensors, and the only tensors of the theory are the ones with respect to the state coordinates transformations' group. In other words, we have the same tensors that characterize the theory of observability without unknown inputs, for which the derivation of the observability rank condition is trivial. In Chapter 7, we deal with these simplified systems and we provide the extension of the observability rank condition for them. This solution was first introduced in [38].

Finally, in Chapter 8, we consider the general case (i.e, characterized by (1.14)). To deal with this case we need to introduce a new class of tensors, which are the tensors with respect to the $\mathcal{SUIO}$. This will make it possible to identify all the key quantities that govern the observability in the presence of unknown inputs. Note that the fundamental quantity will be a tensor of rank 3 with respect to $\mathcal{SUIO}$.

Note that, in contrast with the complexity of its analytic derivation, the complexity of the overall analytic criterion is comparable to the complexity of the standard observability rank condition. Given any nonlinear system characterized by any type of nonlinearity, driven by both known and unknown inputs, the state observability is obtained automatically, i.e., without human intervention (e.g., by the usage of a very simple code that uses symbolic computation). This is a fundamental practical (and unexpected) advantage.

1.6 • Nonlinear observability for time-variant systems

The introduction of the chronospace in Chapter 3 will allow us to easily provide the solution of a further important open problem in control theory, that is, the extension of the observability rank condition to nonlinear time-variant systems. These systems differ from the ones characterized by (1.3) (or by (1.14) in the presence of unknown inputs) because they explicitly depend on time.

1.6. Nonlinear observability for time-variant systems

In other words, all the functions that characterize their dynamics and/or their outputs explicitly depend on time. A typical example of a time-variant system is an aircraft. For this system, there are two main factors that make it time-variant: decreasing weight due to consumption of fuel, and the different configuration of control surfaces during takeoff, cruising, and landing.

In the absence of unknown inputs, these systems are characterized by the equations below:

$$\begin{cases} \dot{x} &= f^0(x,\, t) + \sum_{i=1}^{m} f^i(x,\, t) u_i, \\ y &= [h_1(x,\, t), \ldots, h_p(x,\, t)]^T. \end{cases} \quad (1.18)$$

The extension of the observability rank condition to these systems will be given in section 4.7. Note that this extension was an open problem in control theory. The solution was only obtained in the simple linear case.

In the presence of also unknown inputs, the systems will be characterized by the equations below:

$$\begin{cases} \dot{x} &= g^0(x,\, t) + \sum_{i=1}^{m_u} f^i(x,\, t) u_i + \sum_{j=1}^{m_w} g^j(x,\, t) w_j, \\ y &= [h_1(x,\, t), \ldots, h_p(x,\, t)]^T. \end{cases} \quad (1.19)$$

The extension of the observability rank condition to these systems will be given in section 8.4.

Part I
Reminders on Tensors and Lie Groups

Chapter 2
Manifolds, Tensors, and Lie Groups

In this chapter we remind the reader of several basic concepts on manifolds, tensors, and group theory. We only remind the reader the basic concepts and properties that will be used in the next chapters. The interested reader can find more details in other books (e.g., [46]).

2.1 ▪ Manifolds

A manifold is an object very similar to the Euclidean space. In particular, it is characterized by a given dimension n and, locally, is similar to $\mathbb{R}^n$ near each point.

As mentioned in Chapter 1, in this book we will consider input-output systems characterized by a state that belongs to a manifold. We define the concept of manifold in this section.

We start by reminding the reader of some general properties of $\mathbb{R}^n$, in particular the property of *continuity* and the Hausdorff criterion. For this scope, we require the concept of open set.

Let us consider a given point $x = [x^1, \ldots, x^n]^T \in \mathbb{R}^n$. An *open ball* of x of radius $r(>0)$ (or *open neighborhood* or more simply *neighborhood*) is the set $B(x,r)$ defined as follows:

$$B(x,r) = \left\{ y \in \mathbb{R}^n : \sqrt{\sum_{i=1}^n (x^i - y^i)^2} < r \right\}.$$

A set $U \subset \mathbb{R}^n$ is *open* if, for any $x \in U$, there exists an open ball $B(x,r)$ such that $B(x,r) \subset U$. Note that an open ball is an open set.

From this concept, we can easily formalize the concept of *continuous space*. Intuitively, we require that there are points arbitrarily close to any given point. A typical example of a noncontinuous space is a lattice. To characterize a continuum space we use the Hausdorff criterion: *any two distinct points of a continuum space have open balls that do not intersect.*

2.1.1 ▪ Topological spaces

In $\mathbb{R}^n$, the union of open sets is still an open set (this even holds for infinite open sets). Also the intersection of open sets is an open set. However, this is true for a finite number of open sets. The intersection of an infinite number of open sets could not be an open set. Starting from this general property, which regards $\mathbb{R}^n$, we introduce a topological space as follows. We consider a generic set T. In addition, we consider a collection of subsets of T, say $U = \{U_i\}$, and we call them open sets. We say that the couple (T, U) formed by the set and the collection of subsets is a topological space if it satisfies the following two properties:

1. The union of any collection of subsets of U is an open set.
2. The intersection of any finite collection of subsets of U is an open set.

Obviously, the couple $(\mathbb{R}^n, \Gamma)$, with Γ the collection of all the open subsets of $\mathbb{R}^n$, is a topological space. On the other hand, $\mathbb{R}^n$ is characterized by many other properties that are not necessarily required to define a topological space. In particular, a topological space is not necessarily a metric space. In other words, a topological space does not require the notion of distance. Our characterization only requires the notion of open sets.

2.1.2 ▪ Map among topological spaces

Given two topological spaces, T_1 and T_2, a map from T_1 to T_2 is a rule that associates with an element $x \in T_1$ a unique element $y = f(x) \in T_2$. Note that a map provides a unique $f(x)$ for every x, but not necessarily a unique x for every $f(x)$. A trivial example is provided by the case when $T_1 = T_2 = \mathbb{R}$, and we consider the function $f(x) = x^2$. For a given $f(x) > 0$, we always have two values of x. When the reverse also holds, i.e., there exists a unique x for every $f(x)$, the map is a *one-to-one correspondence*.

Image and inverse image

Given two topological spaces T_1 and T_2 and a map $f : T_1 \to T_2$, for a given set $A \subset T_1$ we call the *image* of A, the following set:

$$B = f(A) = \{y \in T_2 : \exists x \in A, \ y = f(x)\}.$$

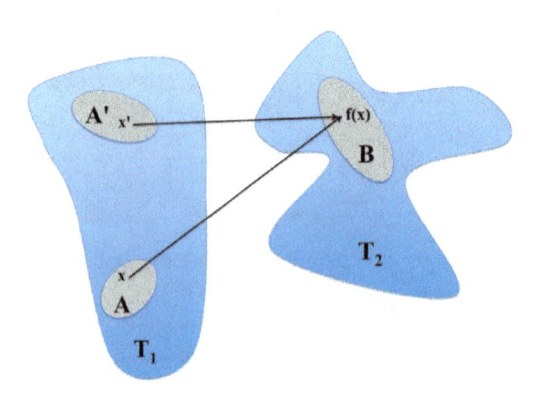

Figure 2.1. $B \subset T_2$ *is the image of* $A \subset T_1$.

Similarly, we define the *inverse image* of a subset $B \subset T_2$ as follows:

$$C = \{x \in T_1 : \exists y \in B, \ y = f(x)\}.$$

Since in general the map f is not one-to-one, the inverse image of $B = f(A)$ does not necessarily coincide with A (e.g., in Figure 2.1, the inverse map of B is $A \cup A'$).

Continuous map

Given two topological spaces T_1 and T_2 and a map $f : T_1 \to T_2$, f is continuous if the image of any open set $A \subset T_1$ is an open set of T_2. In particular, f is continuous at $x \in T_1$ if any open set of T_2 that contains $f(x)$ contains the image of an open set of T_1. This definition generalizes the familiar notion of continuous function introduced in the elementary calculus. The function $f : \mathbb{R} \to \mathbb{R}$ is continuous at x if, $\forall \epsilon > 0, \exists \delta > 0$ such that $|f(x') - f(x)| < \epsilon$ when $|x' - x| < \delta$. Note that the set in $\mathbb{R}$ defined by $|x' - x| < \delta$ is an open set. The definition introduced in the elementary calculus agrees with the above definition for metric spaces. The above definition does not require one to introduce a metric.

2.1.3 ▪ Manifolds and differential manifolds

We are now ready to define the concept of manifold. It is characterized by a given integer n, which is its dimension. We define a manifold as follows.

Definition 2.1 (Manifold). *A manifold $\mathcal{M}$ of dimension n is a topological space, which satisfies the Hausdorff criterion, and such that each point of $\mathcal{M}$ has an open neighborhood which has a continuous one-to-one map onto an open set of $\mathbb{R}^n$.*

Figure 2.2 shows an example of a set that is not a manifold. In particular, there does not exist any continuous one-to-one map between any open set which includes the point P and any open set of $\mathbb{R}$ (note the importance of requiring the continuity of the map to define a manifold).

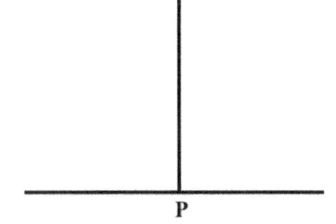

Figure 2.2. *A typical example of a set that is not a manifold.*

The concept of manifold is fundamental to define a coordinate system.

We wish to emphasize that the definition of manifold does not involve the whole of $\mathcal{M}$. Indeed, we do not want to restrict the global topology of $\mathcal{M}$. Additionally, at this stage, we only require the map to be a continuous one-to-one correspondence. We have not yet introduced any geometrical notion as distances, angles, etc. We only require that the local topology of $\mathcal{M}$ is the same as that of $\mathbb{R}^n$. A manifold is a space with this topology.

Chart and atlas

Given a manifold $\mathcal{M}$, we define a chart and an atlas as follows:

- A *chart* is a couple $(\mathcal{U}, \phi)$, where $\mathcal{U} \subset \mathcal{M}$ is an open set and $\phi : \mathcal{U} \to U$ is a one-to-one continuous map, with U an open set of $\mathbb{R}^n$. Given a point $P \in \mathcal{U}$, we call $(x^1, \ldots, x^n) = \phi(P) \in \mathbb{R}^n$ the coordinates of P in the chart $(\mathcal{U}, \phi)$.

- An *atlas* is a collection of charts $(\mathcal{U}_k, \phi_k)$ such that $\bigcup_k \mathcal{U}_k = \mathcal{M}$.

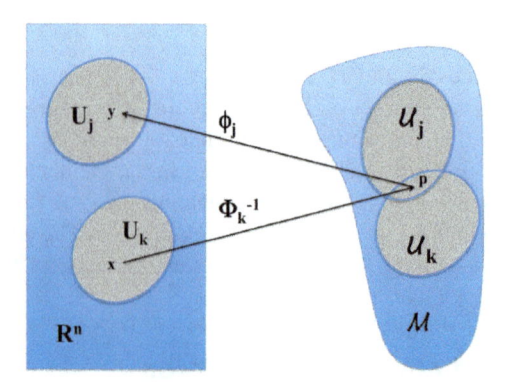

Figure 2.3. *Two charts for the same nonempty intersection set $\mathcal{U}_k \cap \mathcal{U}_j$.*

For a given atlas, let us suppose that the intersection $\mathcal{U}_k \cap \mathcal{U}_j$ is nonempty. The composition of the maps $\phi_j \circ \phi_k^{-1}$ is a map $\mathbb{R}^n \to \mathbb{R}^n$, that is (see also Figure 2.3),

$$\begin{bmatrix} y^1 &= y^1(x^1, \ldots, x^n), \\ \ldots \\ y^n &= y^n(x^1, \ldots, x^n). \end{bmatrix} \quad (2.1)$$

When the partial derivatives of the above functions up to the k order exist and are continuous functions, the two charts $(\mathcal{U}_j, \phi_j)$ and $(\mathcal{U}_k, \phi_k)$ are said to be $\mathcal{C}^k$ related. If it is possible to construct an atlas such that every chart is $\mathcal{C}^k$ related to every other one it overlaps with, then the manifold is said to be a $\mathcal{C}^k$ manifold. If $k = 1$, it is called a differentiable manifold. The notion of differentiable manifold is crucial because it allows us to add basic objects (i.e., one can define the notion of tangent space and tensors) and basic operations (e.g., Lie derivatives).

Equation (2.1) can be regarded as a coordinates' change. In practice, when we need to change the coordinates, instead of defining an atlas, we simply introduce a set of equations as in (2.1). They represent a change of coordinates when the Jacobian of the transformation is nonsingular, in other words, when the matrix

$$\frac{\partial y}{\partial x} = \begin{bmatrix} \frac{\partial y^1}{\partial x^1} & \frac{\partial y^1}{\partial x^2} & \cdots & \frac{\partial y^1}{\partial x^n} \\ \frac{\partial y^2}{\partial x^1} & \frac{\partial y^2}{\partial x^2} & \cdots & \frac{\partial y^2}{\partial x^n} \\ \cdots & \cdots & \cdots & \cdots \\ \frac{\partial y^n}{\partial x^1} & \frac{\partial y^n}{\partial x^2} & \cdots & \frac{\partial y^n}{\partial x^n} \end{bmatrix}$$

has a nonvanishing determinant. Indeed, because of the inverse function theorem, the equations in (2.1) can be locally inverted.

2-sphere

We conclude this section by providing a simple example of a typical differentiable manifold of dimension $n = 2$: the 2-sphere, often denoted by S^2. This set can be defined as a surface in $\mathbb{R}^3$ as follows:

$$S^2 = \left\{ (\xi^1, \xi^2, \xi^3) \in \mathbb{R}^3 \,:\, (\xi^1)^2 + (\xi^2)^2 + (\xi^3)^2 = R^2 \right\}$$

2.2. Tensors

for a given $R > 0$. Let us build an atlas. We need to find continuous one-to-one maps between open sets of S^2 and open sets of $\mathbb{R}^2$. We can take advantage of the spherical coordinates:

$$\left[\begin{array}{ll} \xi^1 & = R\cos\phi\sin\psi, \\ \xi^2 & = R\cos\phi\cos\psi, \\ \xi^3 & = R\sin\phi. \end{array} \right. \tag{2.2}$$

We note that the entire S^2 is obtained by considering any $\phi \in \left[-\frac{\pi}{2}, \frac{\pi}{2}\right]$ and $\psi \in [-\pi, \pi]$. On the other hand, the set in $\mathbb{R}^2$ defined by $x^1 = \phi$ and $x^2 = \psi$ for $-\frac{\pi}{2} \leq x^1 \leq \frac{\pi}{2}$ and $-\pi \leq x^2 \leq \pi$ is not an open set of $\mathbb{R}^2$. In addition, there are points where the map in (2.2) is not one-to-one. Specifically, the north pole of S^2 is obtained by all the values $x^1 = \frac{\pi}{2}$, $x^2 \in [-\pi, \pi]$ and all the points of S^2 obtained for $x^2 = -\pi$ can be also obtained by setting $x^2 = \pi$. To avoid these problems, we need to restrict the map in (2.2) to the open set of $\mathbb{R}^2$:

$$-\frac{\pi}{2} < x^1 < \frac{\pi}{2}, \qquad -\pi < x^2 < \pi.$$

However, the two poles and the semicircle $\psi = \pi$ are left out. The only possibility is to introduce a second chart, again in spherical coordinates but rotated in such a way that the line $\psi = 0$ would coincide with the equator of the old system. Then every point of the sphere would be covered by one of the two charts, and in principle one should be able to find the coordinate transformation for the overlapping region. Hence, the atlas consists of two charts.

2.2 ▪ Tensors

2.2.1 ▪ Formal definition of vectors and covectors

Let us consider a differentiable manifold $\mathcal{M}$ of dimension n and a given coordinates' change $x \to y$ defined as follows:

$$\left[\begin{array}{ll} y^1 & = y^1(x^1, \ldots, x^n), \\ \ldots \\ y^n & = y^n(x^1, \ldots, x^n). \end{array} \right. \tag{2.3}$$

We know that the Jacobian

$$\frac{\partial y}{\partial x} = \begin{bmatrix} \frac{\partial y^1}{\partial x^1} & \frac{\partial y^1}{\partial x^2} & \cdots & \frac{\partial y^1}{\partial x^n} \\ \frac{\partial y^2}{\partial x^1} & \frac{\partial y^2}{\partial x^2} & \cdots & \frac{\partial y^2}{\partial x^n} \\ \cdots & \cdots & \cdots & \cdots \\ \frac{\partial y^n}{\partial x^1} & \frac{\partial y^n}{\partial x^2} & \cdots & \frac{\partial y^n}{\partial x^n} \end{bmatrix}$$

is a nonsingular matrix.

A *contravariant* vector (or simply vector) is an object that consists of n components, whose values depend on the chosen coordinates. In particular, by denoting these components with

$$\xi^1, \xi^2, \ldots, \xi^n$$

in the x coordinates, the values of these components in the y coordinates become

$$\zeta^i = \sum_{j=1}^n \frac{\partial y^i}{\partial x^j} \xi^j, \qquad i = 1, \ldots, n. \tag{2.4}$$

Note that we only use the above expression (which provides the transformation of the components of the vector under a coordinates' change) in order to define a vector.

The reader can verify that, in the 3D space, the velocity of an object is a vector according to the above definition. In particular, the reader can verify that the velocity satisfies (2.4) when we consider in (2.3) the Cartesian coordinates for x^i and the spheric coordinates for y^i. By using this coordinates' change, the reader can also verify that the position of an object is not a vector.

We also define a dual object that is a *covariant* vector (or simply a covector). It is still an object that consists of n components, whose values depend on the chosen coordinates. In particular, by denoting these components with

$$\sigma_1, \sigma_2, \ldots, \sigma_n$$

in the x coordinates, the values of these components in the y coordinates become

$$\tau_i = \sum_{j=1}^{n} \frac{\partial x^j}{\partial y^i} \sigma_j, \qquad i = 1, \ldots, n. \tag{2.5}$$

In other words, the components of a covector transform with the inverse and transpose matrix that defines the transformation of the components of a vector.

The reader can verify that the gradient of a scalar field is a covector according to the above definition.

It is a common convention to use upper indices for the components of a vector and lower indices for the components of a covector.

2.2.2 ▪ Einstein notation

This notation is very common in theoretical physics to achieve conciseness in the equations. It implies summation over a set of indexed terms in a given equation. According to this notation, when an index appears twice and is not otherwise defined, it implies summation of that term over all the values of the index. This index is called a dummy index since any symbol can replace it without changing the meaning of the expression, provided that it does not collide with index symbols in the same equation. In theoretical physics, when the dummy index is Latin, the sum is from 1 to 3 (i.e., all the spatial dimensions). When it is Greek, the sum is from 0 to 3 (i.e., all the dimensions of the space-time). Also in this book Latin indices take positive values, $1, 2, \ldots,$ and Greek indices also take the value 0. In particular, when the state that characterizes a given input-output system belongs to a manifold of dimension n, Latin indices take the values $1, 2, \ldots, n$. In section 3.1, we introduce the chronospace, which is a new manifold with dimension $n+1$. In this context, the index of a given tensor can also take the value 0 and Greek indices will be adopted when they can take all the values $0, 1, 2, \ldots, n$. Finally, in section 2.5 we extend the concept of tensor. The new tensors will always be characterized by a given number of indices. When these indices will take positive integer values, they will be denoted by Latin letters; when they can also take the zero value, they will be Greek.

According to this notation, (2.4) and (2.5) read as follows:

$$\zeta^i = \frac{\partial y^i}{\partial x^j} \xi^j, \qquad \tau_i = \frac{\partial x^j}{\partial y^i} \sigma_j.$$

We also have

$$\frac{\partial y^i}{\partial x^j} \frac{\partial x^j}{\partial y^k} = \delta^i_k.$$

2.2. Tensors

In all the above equations, the index j is a dummy index and is summed over $1, \ldots, n$. In addition, since we did not specify the value of the indices i and k, it means that the above expressions hold for any $i = 1, \ldots, n$ and $k = 1, \ldots, n$.

2.2.3 ▪ Tangent and cotangent space

In general, vectors do not belong to a manifold, but, for each point P of a given manifold $\mathcal{M}$, we can build a space that locally resembles the manifold and that is a vector space with the same dimension of $\mathcal{M}$. It is usually denoted by T_P. For instance, in the example provided at the end of section 2.1, we cannot define a vector as an arrow on the sphere. However, we can introduce a new space that is the plane tangent to the sphere at each point. On this plane, a vector can be defined. For more general manifolds, it is not easy to visualize T_P. In this section, we provide a definition of T_P for a generic differential manifold.

Curve

We start by introducing the concept of curve on a given differentiable manifold. We define a curve γ as a one-parameter correspondence between the open set $(-1, 1) \subset \mathbb{R}$ and $\mathcal{M}$:

$$\gamma(t) = \begin{bmatrix} \gamma^1(t) \\ \cdots \\ \gamma^n(t) \end{bmatrix}.$$

We assume that the above functions are at least of class $\mathcal{C}^1$.

Directional derivative

We now consider a given point $P \in \mathcal{M}$ and we consider a curve $\gamma(t)$ that passes by this point. Without loss of generality, we assume that $\gamma(0) = P$. We also consider a function

$$f(x) = f(x_1, \ldots, x_n) : \mathcal{M} \to \mathbb{R}.$$

We define the directional derivative of the function f along γ as follows:

$$\Delta_\gamma f = \left. \frac{df(\gamma(t))}{dt} \right|_{t=0} = \left. \frac{d\gamma^i}{dt} \right|_{t=0} \frac{\partial f}{\partial x^i}.$$

Since f is any function, we can consider the operator Δ_γ as follows:

$$\Delta_\gamma = \left. \frac{d\gamma^i}{dt} \right|_{t=0} \frac{\partial}{\partial x^i} = \left. \frac{d\gamma^i}{dt} \right|_{t=0} \partial_i,$$

where, for notational brevity, we denote $\partial_i = \frac{\partial}{\partial x^i}$.

Tangent space

It is possible to show that the directional derivatives form a vector space of dimension n. In order to prove this, we first need to prove that the space of the directional derivatives is closed with respect to the sum and the product by a real value. In other words, we must prove the following:

- The sum of two directional derivatives is a directional derivative.
- The product of a directional derivative by a real value is still a directional derivative.

Let us prove the former (we leave the proof of the latter as an exercise). We consider two curves $\alpha(r)$ and $\beta(s)$. We must prove that
$$D_\alpha + \Delta_\beta$$
is a directional derivative. In other words, we need to prove that there exists a curve $\gamma(t)$ such that $\Delta_\gamma = D_\alpha + \Delta_\beta$, i.e.,
$$\left.\frac{d\gamma^i}{dt}\right|_{t=0} = \left.\frac{d\alpha^i}{dr}\right|_{r=0} + \left.\frac{d\beta^i}{ds}\right|_{s=0}.$$

We can find the one-to-one correspondences that relate r to t and s to t. We denote them by $t(r)$ and $t(s)$, respectively. We denote the inverse of the above correspondences with $r(t)$ and $s(t)$. We have the above equality by setting
$$\gamma(t) = \left.\frac{dt}{dr}\right|_{r=0} \alpha(r(t)) + \left.\frac{dt}{ds}\right|_{s=0} \beta(s(t)).$$

To complete the proof that the directional derivatives form a vector space, we need to exhibit the existence of the zero and the existence of the opposite of any vector of the space. The zero is simply obtained by setting $\gamma(t) = P$ for any $t \in (-1, 1)$. The existence of the opposite of the directional derivative defined by a given curve $\gamma(t)$ is obtained by changing the sign of the parametrization that defines the curve (i.e., $t \to -t$).

We denote by T_P the space of the directional derivatives in P. First of all, we remark that this space has the same dimension n of the manifold. Indeed, any vector can be expressed as a linear combination of the vectors $\partial_i = \frac{\partial}{\partial x^i}$. Let us consider a given vector $V \in T_P$. We have
$$V = \xi^i \frac{\partial}{\partial x^i}.$$

We can also express the same V in terms of another basis of T_P. We have
$$V = \zeta^i \frac{\partial}{\partial y^i}.$$

We want to determine the expressions of the new components ζ^i in terms of ξ^i and the change of the basis. First of all, we have
$$\frac{\partial}{\partial y^i} = \frac{\partial x^k}{\partial y^i} \frac{\partial}{\partial x^k}.$$

Therefore,
$$V = \xi^i \frac{\partial}{\partial x^i} = \zeta^i \frac{\partial}{\partial y^i} = \zeta^i \frac{\partial x^k}{\partial y^i} \frac{\partial}{\partial x^k}$$

and, as a result,
$$\zeta^i = \frac{\partial y^i}{\partial x^j} \xi^j,$$

which agrees with (2.4). In other words, the components of V are the components of a vector according to our formal definition of vectors provided in section 2.2.1.

1-forms and the cotangent space

Now we introduce another vector space that is the dual of T_P. It is the space of the 1-forms. We define a 1-form as follows. Given the vector space T_P, a 1-form is a linear function

$$\omega : T_P \to \mathbb{R}.$$

We build the space of the 1-forms by defining the sum of two 1-forms, ω_1 and ω_2, as

$$(\omega_1 + \omega_2)(V) = \omega_1(V) + \omega_2(V) \qquad \forall V \in T_P$$

and the product of a 1-form ω by a real value c as

$$(c\omega)(V) = c\,\omega(V) \qquad \forall V \in T_P.$$

It is easy to prove that the space of all the 1-forms is a vector space. We define n specific 1-forms. For a given vector $V \in T_P$, with $V = \xi^i \partial_i$, we define

$$dx^i(V) = \xi^i. \tag{2.6}$$

It is evident that dx^i is a linear function $T_P \to \mathbb{R}$. Hence, it is a 1-form. We remark that any 1-form ω can be expressed in terms of the n dx^i. We have, for any $V \in T_P$,

$$\omega(V) = \omega(\xi^i \partial_i) = \xi^i \omega(\partial_i) = dx^i(V)\omega(\partial_i).$$

Note that each ∂_i is an element of T_P, and consequently $\omega(\partial_i) = \sigma_i \in \mathbb{R}$. Since the above equality holds for any $V \in T_P$ we have

$$\omega = \sigma_i dx^i.$$

This proves that the 1-forms $dx^1, \ldots, dx^n$ constitute a basis of the space of the 1-forms. $\sigma_1, \ldots, \sigma_n$ are the components of ω in this basis. This means that the space of the 1-forms has the same dimension of the manifold (n). We denote this space by T_P^*.

We conclude by computing how the components of a 1-form transform by changing the coordinates. We have

$$dx^i(V) = \xi^i \to dy^i(V) = \zeta^i = \frac{\partial y^i}{\partial x^k}\xi^k = \frac{\partial y^i}{\partial x^k}dx^k(V).$$

Since this holds for any $V \in T_P$, we have

$$dx^i \to dy^i = \frac{\partial y^i}{\partial x^k}dx^k.$$

Given a generic $\omega = \sigma_i dx^i$, we denote by τ_i the components of ω in the basis $dy^1, \ldots, dy^n$. We have

$$\omega = \sigma_i dx^i = \tau_i dy^i = \tau_i \frac{\partial y^i}{\partial x^k}dx^k$$

and, as a result,

$$\tau_i = \frac{\partial x^j}{\partial y^i}\sigma_j,$$

which agrees with (2.5). In other words, the components of ω are the components of a covector according to our formal definition of covectors provided in section 2.2.1.

2.2.4 • Tensors of rank larger than 1

Formal definition

We provide the formal definition of general tensors by using the same approach adopted in section 2.2.1 to define vectors and covectors. We start by saying that vectors and covectors are tensors of rank 1. Specifically, we say that vectors are tensors of type $(0, 1)$ and covectors are tensors of type $(1, 0)$. A tensor of rank $q + p$ and in particular of type (q, p) consists of $n^{(q+p)}$ components, where n is the dimension of the manifold where it is defined. We denote its components as follows:
$$T^{j_1,\ldots,j_p}_{i_1,\ldots,i_q}.$$
The value of these components depends on the chosen coordinates on the manifold. In particular, let us consider the same coordinates' change $x \to y$ defined by (2.3). By denoting by
$$S^{j_1,\ldots,j_p}_{i_1,\ldots,i_q}$$
the components of the same tensor in the new coordinates, we have
$$S^{j_1,\ldots,j_p}_{i_1,\ldots,i_q} = \frac{\partial y^{j_1}}{\partial x^{l_1}} \cdots \frac{\partial y^{j_p}}{\partial x^{l_p}} \frac{\partial x^{k_1}}{\partial y^{i_1}} \cdots \frac{\partial x^{k_q}}{\partial y^{i_q}} T^{l_1,\ldots,l_p}_{k_1,\ldots,k_q}. \tag{2.7}$$
It is immediate to obtain for $p = 1$, $q = 0$ and for $p = 0$, $q = 1$ that
$$S^j = \frac{\partial y^j}{\partial x^l} T^l, \qquad S_i = \frac{\partial x^k}{\partial y^i} T_k,$$
which agree with (2.4) and (2.5), respectively. As for vectors and covectors, given a manifold $\mathcal{M}$ and a point $P \in \mathcal{M}$, tensors do not belong to the manifold. A tensor of type (q, p) is defined in P and belongs to the following space:
$$\underbrace{T_P \times \cdots \times T_P}_{p \text{ times}} \times \underbrace{T_P^* \times \cdots \times T_P^*}_{q \text{ times}}.$$

Contraction

Given a tensor of type (a, b),
$$M^{j_1,\ldots,j_b}_{i_1,\ldots,i_a},$$
we can set one of the lower indices (e.g., i_1) equal to one of the upper indices (e.g., j_1). In other words, we set $i_1 = j_1 = k$. By summing on all the values of k, we obtain a new tensor of smaller rank, in particular of type $(a - 1, b - 1)$. Specifically, we have (in the Einstein notation)
$$P^{j_2,\ldots,j_b}_{i_2,\ldots,i_a} = M^{k,j_2,\ldots,j_b}_{k,i_2,\ldots,i_a}.$$
It is immediate to verify that the components of P transform as indicated in (2.7), in particular with $q = a - 1$ and $p = b - 1$. We call this operation *contraction*. In particular, we say that we contracted the first upper index with the first lower index. We can contract more indices simultaneously, and we always obtain smaller rank tensors. In particular, if $a = b$ and we contract all the indices, we obtain a scalar (that is a tensor of rank 0, i.e., a tensor of type $(0, 0)$):
$$S = M^{i_1,i_2,\ldots,i_a}_{i_1,i_2,\ldots,i_a}.$$

2.2. Tensors

Note that, in the above equality, it is clear that we are using the Einstein notation and all the indices are dummy indices.

The contraction can be performed also on different tensors that appear in a given product. We always obtain a tensor, provided that we contract upper indices with lower indices. Let us consider the case of two tensors: M, defined as above (i.e., of type (a, b)), and a tensor of type (c, d):

$$N^{j_1,\ldots,j_d}_{i_1,\ldots,i_c}.$$

We can easily build a tensor of type $(a+c, b+d)$ as follows:

$$P^{j_1,\ldots,j_b,l_1,\ldots,l_d}_{i_1,\ldots,i_a,k_1,\ldots,k_c} = M^{j_1,\ldots,j_b}_{i_1,\ldots,i_a} N^{l_1,\ldots,l_d}_{k_1,\ldots,k_c}.$$

It is immediate to verify that the components of P transform as indicated in (2.7), in particular with $q = a + c$ and $p = b + d$. If in the product on the right-hand side of the above equation one or more lower (upper) indices of M are contracted with one or more upper (lower) indices of N, we obtain a tensor whose type is given by the remaining indices. For instance, suppose we contract the first lower index of M, i.e., i_1, with the last upper index of N, i.e., l_d. We have

$$P^{j_1,\ldots,j_b,l_1,\ldots,l_{d-1}}_{i_2,\ldots,i_a,k_1,\ldots,k_c} = M^{j_1,\ldots,j_b}_{i,i_2,\ldots,i_a} N^{l_1,\ldots,l_{d-1},i}_{k_1,\ldots,k_c}.$$

It is immediate to verify that the components of P transform as indicated in (2.7), in particular with $q = a + c - 1$ and $p = b + d - 1$.

Tensor fields

As we mentioned above, tensors are defined at a given point $P \in \mathcal{M}$. Given an open set $\mathcal{U} \subset \mathcal{M}$, we define the vector spaces

$$T_\mathcal{U} = \bigcup_{P \in \mathcal{U}} T_P, \qquad T^*_\mathcal{U} = \bigcup_{P \in \mathcal{U}} T^*_P.$$

Starting from them we can also define, for any couple (q, p), the spaces

$$\underbrace{T_\mathcal{U} \times \cdots \times T_\mathcal{U}}_{p \text{ times}} \times \underbrace{T^*_\mathcal{U} \times \cdots \times T^*_\mathcal{U}}_{q \text{ times}}.$$

A tensor field associates to every point $P \in \mathcal{U}$ a tensor. We always assume that the dependence on P is smooth. When it is not ambiguous, with a tensor we actually intend a tensor field.

Examples

We can find many examples of tensors in physics. We provide several examples in mechanics. Note that, in mechanics, the manifold is the Euclidean space $\mathbb{R}^3$ and we only consider orthonormal reference frames. Hence, we only consider coordinates' changes characterized by an orthonormal matrix. Therefore, vectors and covectors transform in the same manner (the inverse and transpose of an orthonormal matrix is the matrix itself) and we only use lower indices.

The speed and the acceleration of a mass point are vectors, i.e., tensors of rank 1. For a rigid body, the angular speed is also a tensor of rank 1.[9] A typical example of a tensor of rank 2 is the

---

[9]Interestingly, this only holds in 3D. In general, the angular speed is an antisymmetric tensor of rank 2. In 3D, the antisymmetric tensors of rank 2 are isomorphic with vectors ($\frac{d(d-1)}{2} = d$ only for $d = 3$).

moment of inertia. For a rigid body, the rotational energy is given by

$$E_{rot} = \frac{1}{2}\sum_r m(r)[|r|^2|\omega|^2 - (\omega \cdot r)^2],$$

where $\omega = [\omega_1, \omega_2, \omega_3]^T$ is the angular speed, and $r = [r_1, r_2, r_3]^T$ is the position of the point of mass m with respect to a local frame attached to the rigid body at its center of mass. We can write the expression above as

$$E_{rot} = \frac{1}{2}\sum_r m(r)[|r|^2\delta_{ij} - r_i r_j]\omega_i \omega_j = \frac{1}{2}\mathcal{I}_{ij}\omega_i \omega_j,$$

which is the contraction of the tensor $\mathcal{I}_{ij}$ with the angular speed, twice. Since E_{rot} must be a scalar, $\mathcal{I}_{ij}$ is a tensor of rank 2. This is precisely the moment of inertia:

$$\mathcal{I}_{ij} = \sum_r m(r)[|r|^2\delta_{ij} - r_i r_j].$$

The theory of general relativity largely makes use of tensors. In this case, the theory is defined on a differential manifold that does not coincide with a Euclidean space. Hence, we need to distinguish upper indices from lower indices. Specifically, the theory is defined on a pseudo-Riemannian manifold. As for a Riemannian manifold, its properties are entirely described by the Riemann curvature tensor, which is a tensor of type (3, 1). Einstein's equations, which provide the link between the geometric properties of the space-time and the local energy and momentum within that space, involve scalar fields and tensor fields of type (2, 0), namely

- the metric tensor g, which defines the distance $ds^2 = g_{\alpha\beta}dx^\alpha dx^\beta$;
- the stress-energy tensor, which is of type (2, 0);
- the Ricci tensor, which is obtained from the Riemann curvature tensor by contracting the upper index with the second lower index; and
- the trace of the Ricci curvature tensor with respect to the metric, which is a scalar.

As a last example, we wish to mention that the theory of electromagnetism can be written very concisely in terms of a tensor of type (2, 0), which is the electromagnetic tensor

$$F_{\mu\nu} = \partial_\mu A_\nu - \partial_\nu A_\mu,$$

where ∂ is the four-gradient and A the four-potential:

$$A^0 = V, \quad [A^1, A^2, A^3]^T = \vec{A},$$

in which V is the electric potential, and $\vec{A}$ is the vector magnetic potential.

2.2.5 ▪ Lie derivative

There are several important operations that can be defined with tensors. In this book, we only introduce the Lie derivative.

2.2. Tensors

We start by remarking that, given a tensor field T, its derivative $\frac{\partial T}{\partial x^k}$ is not a tensor, with the exception of the case when T is a scalar (i.e., a tensor of type $(0, 0)$). Indeed, for a scalar Φ, for the coordinates' change $x \to y$ given by (2.3), the derivative $\partial_k \Phi$ transforms as

$$\frac{\partial}{\partial x^k}\Phi \to \frac{\partial}{\partial y^k}\Phi = \frac{\partial x^j}{\partial y^k}\frac{\partial}{\partial x^j}\Phi,$$

which is a covector, or a tensor of type $(1, 0)$, according to the definition given in (2.5) (or in (2.7) with $q = 1$ and $p = 0$). For a vector field T^j, the derivative $\frac{\partial T^j}{\partial x^k}$ is not a tensor of type $(1, 1)$. Indeed,

$$T^j \to \frac{\partial y^j}{\partial x^a}T^a, \qquad \frac{\partial}{\partial x^k} \to \frac{\partial x^b}{\partial y^k}\frac{\partial}{\partial x^b},$$

and, consequently,

$$\frac{\partial T^j}{\partial x^k} \to \frac{\partial x^b}{\partial y^k}\frac{\partial}{\partial x^b}\left(\frac{\partial y^j}{\partial x^a}T^a\right) = \frac{\partial x^b}{\partial y^k}\frac{\partial y^j}{\partial x^a}\frac{\partial T_a}{\partial x^b} + \frac{\partial x^b}{\partial y^k}\frac{\partial^2 y^j}{\partial x^a \partial x^b}T_a.$$

The last term makes the transformation different from the one of a tensor of type $(1, 1)$. For tensors of higher rank, the situation becomes even worse: there are more and more terms that make the transformation different from the one that characterizes a tensor. This fact is not surprising. Tensors do not belong to the manifold but to the tangent space. Comparing the same tensor field at two different points of the manifold has no intrinsic meaning. The values that the field takes at the two points belong to two distinct spaces. In order to avoid this inconvenience, we basically have two options:

- We introduce a *connection* able to connect tangent spaces, so it permits tensor fields to be differentiated as if they were functions on the manifold with values in a fixed tangent space.

- We introduce a vector field that defines a local diffeomorphism and consider the variation of the tensor field established by this diffeomorphism.

The first option allows introducing the *covariant derivative*. When applied to a tensor of type (q, p), it returns a tensor of type $(q+1, p)$. In this book, we do not introduce this operation. The second option is used to introduce the *Lie derivative*. When applied to a given tensor, it returns a tensor of the same type. We introduce this operation in this section.

We start by providing the abstract definition. Given a tensor field of type (q, p),

$$T^{j_1,\ldots,j_p}_{i_1,\ldots,i_q},$$

and a vector field V, with components V^k, the Lie derivative of T along V is

$$\mathcal{L}_V T^{j_1,\ldots,j_p}_{i_1,\ldots,i_q} = V^s \frac{\partial T^{j_1,\ldots,j_p}_{i_1,\ldots,i_q}}{\partial x^s} + T^{j_1,\ldots,j_p}_{k,i_2,\ldots,i_q}\frac{\partial V^k}{\partial x^{i_1}} + \cdots + T^{j_1,\ldots,j_p}_{i_1,\ldots,i_{q-1},k}\frac{\partial V^k}{\partial x^{i_q}} \qquad (2.8)$$

$$- T^{l,j_2,\ldots,j_p}_{i_1,\ldots,i_q}\frac{\partial V^{j_1}}{\partial x^l} - \cdots - T^{j_1,\ldots,j_{p-1},l}_{i_1,\ldots,i_q}\frac{\partial V^{j_p}}{\partial x^l}.$$

By a direct computation, it is possible to prove that $\mathcal{L}_V T^{j_1,\ldots,j_p}_{i_1,\ldots,i_q}$ is a tensor of type (q, p). Note that the single addends in the above expression are not tensors. However, their sum is able to cancel out all the terms that do not transform as tensors.

We introduce now the geometric idea behind the above definition.

The local diffeomorphism

Given the vector field V defined on an open set of the manifold $\mathcal{M}$, we can consider the following set of n differential equations:

$$\dot{x}^i(t) = V^i(x^1(t), \ldots, x^n(t)), \qquad i = 1, \ldots, n,$$

where $(x^1, \ldots, x^n)$ are the coordinates on the considered open set. Let us consider a given point with coordinates $(x_0^1, \ldots, x_0^n)$. We can compute the solution of the above differential equations at a given t by setting the following initial conditions:

$$x^i(0) = x_0^i.$$

We denote the solution as follows:

$$F_t^i(x_0^1, \ldots, x_0^n) = x^i(t).$$

This can be regarded as an application of the considered open set in itself, i.e.,

$$F_t : \quad (x_0^1, \ldots, x_0^n) \to (x^1(t), \ldots, x^n(t)),$$

defined by the continuous parameter t. For small t, we have

$$x^i(t) = x_0^i + t\, V^i(x_0^1, \ldots, x_0^n) + o(t^2).$$

The Jacobian of the inverse of F_t is

$$\frac{\partial x_0^i}{\partial x^j} = \delta_j^i - t\frac{\partial V^i}{\partial x^j} + o(t^2). \tag{2.9}$$

Let us consider now the tensor field of type (q, p):

$$T^{j_1,\ldots,j_p}_{i_1,\ldots,i_q}.$$

We consider the change of coordinates

$$x^i(t) \to x_0^i.$$

When we transport the tensor $T^{j_1,\ldots,j_p}_{i_1,\ldots,i_q}$ from $(x^1(t), \ldots, x^n(t))$ to $(x_0^1, \ldots, x_0^n)$, we obtain (see (2.7))

$$(F_t T)^{j_1,\ldots,j_p}_{i_1,\ldots,i_q} = \frac{\partial x_0^{j_1}}{\partial x^{l_1}} \cdots \frac{\partial x_0^{j_p}}{\partial x^{l_p}} \frac{\partial x^{k_1}}{\partial x_0^{i_1}} \cdots \frac{\partial x^{k_q}}{\partial x_0^{i_q}} T^{l_1,\ldots,l_p}_{k_1,\ldots,k_q}. \tag{2.10}$$

Geometric definition of the Lie derivative

We define the Lie derivative of the tensor field $T^{j_1,\ldots,j_p}_{i_1,\ldots,i_q}$ along the vector field V^j as follows:

$$\mathcal{L}_V T^{j_1,\ldots,j_p}_{i_1,\ldots,i_q} = \frac{d}{dt}(F_t T)^{j_1,\ldots,j_p}_{i_1,\ldots,i_q}\bigg|_{t=0}.$$

To compute this derivative, we use the expression given in (2.10), where, for all the Jacobians, we substitute the expression in (2.9) and we take the first order in t. We finally obtain the expression provided in (2.8).

Lie derivatives of tensors of ranks 0 and 1

We use the definition given in (2.8) to obtain the expression of the Lie derivative of scalars, vectors and covectors. Let us start with scalars. We have $T = T(x^1, \ldots, x^n)$. We need to use (2.8) for $q = p = 0$. We obtain

$$\mathcal{L}_V T = V^s \frac{\partial T}{\partial x^s}. \tag{2.11}$$

We can also write the above expression in a vector format. We have

$$\mathcal{L}_V T = \frac{\partial T}{\partial x} \cdot V.$$

We can apply the Lie derivative operator on the scalar field T multiple times: the result will be a scalar field. In this book, we adopt the following notation:

$$\mathcal{L}_V^j T = \underbrace{\mathcal{L}_V \cdots \mathcal{L}_V}_{j \text{ times}} T. \tag{2.12}$$

Finally, if we have several vector fields, $V_1, \ldots, V_k$, we adopt the following notation:

$$\mathcal{L}^j_{V_{i_1} V_{i_2} \cdots V_{i_j}} T = \mathcal{L}_{V_{i_j}} \mathcal{L}_{V_{i_{j-1}}} \cdots \mathcal{L}_{V_{i_1}} T, \tag{2.13}$$

where the indices $i_1, i_2, \ldots, i_j$ take values among $1, 2, \ldots, k$.

Let us study now the case of vectors. We have $T = [T^1(x^1, \ldots, x^n), \ldots, T^n(x^1, \ldots, x^n)]^T$. We need to use (2.8) for $q = 0$ and $p = 1$. We obtain

$$\mathcal{L}_V T^j = V^s \frac{\partial T^j}{\partial x^s} - T^l \frac{\partial V^j}{\partial x^l}. \tag{2.14}$$

This is still a vector. In particular, the above expression provides its jth component. We can also read the above expression as the product of the $n \times n$ matrix Jacobian $\frac{\partial T}{\partial x}$ times the vector V minus the $n \times n$ matrix Jacobian $\frac{\partial V}{\partial x}$ times the vector T. In other words, we can write the above expression in a matrix format as follows:[10]

$$\mathcal{L}_V T = \frac{\partial T}{\partial x} V - \frac{\partial V}{\partial x} T.$$

It is also very common to call this operation the Lie brackets of V and T, and it is denoted as follows:

$$\mathcal{L}_V T = [V, T]. \tag{2.15}$$

In this book, we often use this notation.

By a direct computation, it is possible to prove the following fundamental properties:

- Given two vector fields T and S and two scalar fields Φ and Ψ, we have

$$[\Phi T, \ \Psi S] = \Phi \Psi [T, S] - \Psi T \mathcal{L}_S \Phi + \Phi S \mathcal{L}_T \Psi. \tag{2.16}$$

- Given two vector fields T and S and a scalar field Φ, we have

$$\mathcal{L}_{[T, S]} \Phi = \mathcal{L}_T \mathcal{L}_S \Phi - \mathcal{L}_S \mathcal{L}_T \Phi. \tag{2.17}$$

---

[10] In this book, when we use a matrix format, we write vectors as columns and covectors as rows.

In this book, we will often use the above equalities.

We conclude by considering the case of covectors. We have

$$T = [T_1(x^1, \ldots, x^n), \ldots, T_n(x^1, \ldots, x^n)].$$

We need to use (2.8) for $q = 1$ and $p = 0$. We obtain

$$\mathcal{L}_V T_i = V^s \frac{\partial T_i}{\partial x^s} + T_k \frac{\partial V^k}{\partial x^i}. \tag{2.18}$$

A special case occurs when the covector is the differential of a given scalar field, in other words when $T_i = \partial_i \Phi$ or, in a matrix format, $T = \frac{\partial \Phi}{\partial x}$. By a direct computation, it is possible to prove the fundamental property

$$\mathcal{L}_V \partial_i \Phi = \partial_i \mathcal{L}_V \Phi \tag{2.19}$$

or, in a matrix format,

$$\mathcal{L}_V \frac{\partial}{\partial x} \Phi = \frac{\partial}{\partial x} \mathcal{L}_V \Phi.$$

In other words, the operations $\mathcal{L}_V$ and $\frac{\partial}{\partial x}$ are commutative when they are applied to a scalar field.

2.3 • Distributions and codistributions

This section is devoted to introducing two objects (the distribution and the codistribution) that play a fundamental role in the theory of nonlinear observability and, more broadly, in nonlinear control theory. These objects are defined on an open set of a manifold and are the vector spaces generated by a set of tensor fields of rank 1. In particular, a distribution is the span of tensors of type $(0, 1)$ (i.e., vector fields), while a codistribution is the span of tensors of type $(1, 0)$ (i.e., covector fields).

2.3.1 • Distribution

Let us consider a set of d vector fields defined on a given open set $U \subset \mathcal{M}$, where $\mathcal{M}$ is a differentiable manifold of dimension n. Let us denote them by $f_1(x), \ldots, f_d(x)$. Note that each f_i is a vector field. In particular, the index i does not indicate the ith component of this vector field (actually, this ambiguity does not exist because the index would have been an upper index). We introduce in any case the following notation, which is useful to disambiguate other situations (e.g., the case of covectors). If we want to specify the jth component ($j = 1, \ldots, n$) of the ith vector field ($i = 1, \ldots, d$), we write

$$(f_i)^j.$$

In other words, when we have a set of tensors and we use a further index to identify one of these tensors, we introduce the parentheses as above to specify a given component of this tensor. For instance, if we have d tensors, $T_1, \ldots, T_d$, of type (q, p), to indicate a given component of the tensor T_i, we write

$$(T_i)^{j_1, \ldots, j_p}_{i_1, \ldots, i_q}.$$

As usual, we assume that $f_1(x), \ldots, f_d(x)$ are analytic functions of x. A distribution is the vector space

$$\Delta(x) = \text{span}\{f_1(x), \ldots, f_d(x)\}.$$

2.3. Distributions and codistributions

For a given x, a distribution is a vector space of a given dimension $d \leq n$. This vector space is a subspace of $\mathbb{R}^n$. The dimension of this vector space may vary by varying x. A point $x_0 \in U$ is a *regular* point for Δ if there exists an open ball B_0 of x_0 such that the dimension of $\Delta(x)$ is the same $\forall x \in B_0$. Similarly, a distribution is *nonsingular* on a given open set B_0 if its dimension is the same $\forall x \in B_0$. In the following, we always refer to open sets where the considered distribution is nonsingular.

As for vector spaces, we define the sum of two distributions, Δ_1 and Δ_2, as the span of the generators of both the distributions. This is still a distribution. We simply denote it by $\Delta = \Delta_1 \oplus \Delta_2$.

Let us consider a distribution of dimension d. Let us denote by $f_1(x), \ldots, f_d(x)$ its generators. Hence, any $f(x) \in \Delta(x)$ can be expressed in terms of $f_1(x), \ldots, f_d(x)$:

$$f(x) = \sum_{i=1}^{d} c_i(x) f_i(x),$$

where $c_i(x)$ are scalar functions.

Involutive distribution

The distribution Δ is *involutive* if, for any pair $f, g \in \Delta$, we have

$$[f, g] \in \Delta,$$

where the above square brackets denote the Lie brackets defined in (2.15). Let us denote by $f_1, \ldots, f_d$ a set of generators of Δ. It is immediate to realize that, to check whether Δ is involutive, it suffices to check that $[f_i, f_j] \in \Delta$ for any pair $i, j = 1, \ldots, d$. Indeed, we have $f(x) = \sum_{i=1}^{d} c_i(x) f_i(x)$ and $g(x) = \sum_{i=1}^{d} d_i(x) f_i(x)$. Hence, by using (2.16),

$$[f, g] = \sum_{i=1}^{d} \sum_{j=1}^{d} [c_i f_i, d_j f_j] = \sum_{i,j=1}^{d} \left(c_i d_j [f_i, f_j] - d_j f_j \mathcal{L}_{f_j} c_i + c_i f_j \mathcal{L}_{f_i} d_j \right),$$

which belongs to Δ if $[f_i, f_j] \in \Delta$ for any pair $i, j = 1, \ldots, d$.

For a given distribution Δ and a given vector field g, we denote by $[\Delta, g]$ the distribution generated by all the vector fields $[f, g]$ $\forall f \in \Delta$. We wish to find a set of generators of the distribution $\Delta \oplus [\Delta, g]$. We show that these can be obtained by selecting a set of independent vector fields among

$$f_1, \ldots, f_d, [f_1, g], \ldots, [f_d, g].$$

Obviously, the span of the previous vectors includes Δ. We must prove that they also generate $[\Delta, g]$. In other words, we must prove that, for any $f \in \Delta$, the vector $[f, g]$ is generated by $f_1, \ldots, f_d, [f_1, g], \ldots, [f_d, g]$. We have $f = \sum_{i=1}^{d} c_i(x) f_i$. Hence,

$$[f, g] = \sum_{i=1}^{d} [c_i f_i, g] = \sum_{i=1}^{d} c_i [f_i, g] - \sum_{i=1}^{d} f_i \mathcal{L}_g c_i,$$

which belongs to the span of $f_1, \ldots, f_d, [f_1, g], \ldots, [f_d, g]$.

We can extend this concept for a set of vector fields $g_1, \ldots, g_k$. We denote by Δ_g the distribution generated by these vector fields. We define $[\Delta, \Delta_g]$ as the distribution generated by all the vector fields $[f, g] \, \forall f \in \Delta$ and $\forall g \in \Delta_g$. By proceeding as in the case of a single vector field ($k = 1$), we obtain that the distribution $\Delta \oplus \Delta_g \oplus [\Delta, \Delta_g]$ is generated by a set of independent vector fields among

$$f_1, \ldots, f_d, g_1, \ldots, g_k, [f_1, g_1], \ldots, [f_d, g_1], \ldots, [f_1, g_k], \ldots, [f_d, g_k].$$

We proved the following result.

Proposition 2.2. *Given the two sets of vector fields $f_1, \ldots, f_d$ and $g_1, \ldots, g_k$, the distribution $\Delta \oplus \Delta_g \oplus [\Delta, \Delta_g]$ is generated by a set of independent vector fields among*

$$f_1, \ldots, f_d, g_1, \ldots, g_k, [f_1, g_1], \ldots, [f_d, g_1], \ldots, [f_1, g_k], \ldots, [f_d, g_k].$$

The smallest involutive distribution that contains a given distribution

By using the above property, we derive a simple algorithm to build the smallest involutive distribution that contains a given distribution. In other words, starting from a given set of vector fields $f_1, \ldots, f_d$, we provide a simple algorithm to build the smallest involutive distribution that contains span$\{f_1, \ldots, f_d\}$. In particular, we show that this algorithm requires at most $n - 1$ iterations.

The algorithm is the following.

ALGORITHM 2.1. Smallest involutive distribution containing span$\{f_1, \ldots, f_d\}$.

    Set $\Delta = \text{span}\{f_1, \ldots, f_d\}$
    Set $s = d$
    Set $\Delta = \Delta \oplus [\Delta, \Delta]$
    while $\dim(\Delta) > s$ **do**
        $s = \dim(\Delta)$
        Set $\Delta = \Delta \oplus [\Delta, \Delta]$
    end while

Since the dimension of Δ cannot exceed n, the algorithm converges in at most $n - 1$ steps. It is immediate to realize that Δ is involutive. Indeed, after the convergence, $[\Delta, \Delta] \subset \Delta$, meaning that $[f, g] \in \Delta \, \forall f, g \in \Delta$. By construction, span$\{f_1, \ldots, f_d\} \subset \Delta$. Finally, let us prove that Δ is the smallest involutive distribution that contains span$\{f_1, \ldots, f_d\}$. At a given iteration of the algorithm, the dimension of Δ is s. Hence,

$$\Delta = \text{span}\{f_1, \ldots, f_s\}.$$

As it has been proved above, we can build $\Delta \oplus [\Delta, \Delta]$ by only computing the vector fields

$$[f_i, f_j], \quad i, j = 1, \ldots, s,$$

and by selecting the largest number of independent vectors among $f_1, \ldots, f_s$ and the above vectors. The distribution $\Delta \oplus [\Delta, \Delta]$ is precisely the span of these independent vectors. On the other hand, all the above vectors must be included in any involutive distribution that contains span$\{f_1, \ldots, f_d\}$. In other words, at each iteration, we only include in the set of generators vectors that belong to any involutive distribution that contains span$\{f_1, \ldots, f_d\}$.

2.3. Distributions and codistributions

2.3.2 ▪ Codistribution

Exactly as for the distributions, a codistribution is the span of a set of k covector fields. We have

$$\Omega(x) = \text{span}\{\omega_1(x), \ldots, \omega_k(x)\}.$$

For a given x, a codistribution is a vector space of a given dimension $k \leq n$. This dimension may vary by varying x. Exactly as in the case of distributions, we define the concepts of regularity and nonsingularity and, in the following, we always refer to open sets where the considered codistribution is nonsingular.

As for the distributions, we define the sum of two codistributions, Ω_1 and Ω_2, the span of the generators of both the codistributions. This is still a codistribution. We simply denote it by $\Omega = \Omega_1 \oplus \Omega_2$.

Let us consider a codistribution of dimension k. Let us denote by $\omega_1(x), \ldots, \omega_k(x)$ its generators. Hence, any $\omega(x) \in \Omega(x)$ can be expressed in terms of $\omega_1(x), \ldots, \omega_k(x)$:

$$\omega(x) = \sum_{i=1}^{k} c_i(x)\omega_i(x),$$

where $c_i(x)$ are scalar functions.

Integrable codistribution

A codistribution is *integrable* if it can be generated by covectors which are the gradients of scalar functions. In this book, we always work with integrable codistributions. An integrable codistribution of dimension k is defined by a set of scalar functions, $h_1, \ldots, h_k$, whose gradients are independent. We have

$$\Omega(x) = \text{span}\left\{\frac{\partial}{\partial x}h_1(x), \ldots, \frac{\partial}{\partial x}h_k(x)\right\}.$$

Let us consider an integrable codistribution Ω, and let us suppose that it is generated by the gradients of the scalar functions $h_1, \ldots, h_k$. We often refer to these scalar functions as the *generators* of Ω. Note that a given covector that belongs to an integrable codistribution is not necessarily the gradient of a scalar function. Let us denote by f a given vector field. We denote by $\mathcal{L}_f \Omega$ the set of all the covectors which are the Lie derivatives along f of the covectors in Ω. In other words,

$$\mu \in \mathcal{L}_f\Omega \text{ iff } \exists \omega \in \Omega \text{ s.t. } \mu = \mathcal{L}_f\omega.$$

We prove that the codistribution $\Omega \oplus \mathcal{L}_f\Omega$ is still an integrable codistribution and it is generated by the gradients of $h_1, \ldots, h_k, \mathcal{L}_f h_1, \ldots, \mathcal{L}_f h_k$. As we will see, this is a consequence of the fact that, for any scalar h, we have $\mathcal{L}_f \frac{\partial}{\partial x} h = \frac{\partial}{\partial x} \mathcal{L}_f h$ (see 2.19).

Let us denote by μ a given covector of $\mathcal{L}_f\Omega$. We have $\mu = \mathcal{L}_f\omega$, with $\omega = \sum_{i=1}^{k} c_i(x)\frac{\partial}{\partial x}h_i$. Hence,

$$\mu = \mathcal{L}_f\omega = \sum_{i=1}^{k}\left(\mathcal{L}_f c_i \frac{\partial}{\partial x}h_i + c_i \mathcal{L}_f \frac{\partial}{\partial x}h_i\right) = \sum_{i=1}^{k}\left(\mathcal{L}_f c_i \frac{\partial}{\partial x}h_i + c_i \frac{\partial}{\partial x}\mathcal{L}_f h_i\right).$$

We can extend this concept for a set of vector fields $f_1, \ldots, f_d$. We obtain that the codistribution $\Omega \oplus \mathcal{L}_{f_1}\Omega \oplus \cdots \oplus \mathcal{L}_{f_d}\Omega$ is still an integrable codistribution and it is generated by the gradients of $h_1, \ldots, h_k, \mathcal{L}_{f_1}h_1, \ldots, \mathcal{L}_{f_1}h_k, \ldots, \mathcal{L}_{f_d}h_1, \ldots, \mathcal{L}_{f_d}h_k$.

We proved the following result.

Proposition 2.3. *Given the integrable codistribution $\Omega = \text{span}\left\{\frac{\partial}{\partial x}h_1(x), \ldots, \frac{\partial}{\partial x}h_k(x)\right\}$ and a set of vector fields $f_1, \ldots, f_d$, the codistribution $\Omega \oplus \mathcal{L}_{f_1}\Omega \oplus \cdots \oplus \mathcal{L}_{f_d}\Omega$ is still an integrable codistribution and it is generated by the gradients of the following functions:*

$$h_1, \ldots, h_k, \mathcal{L}_{f_1}h_1, \ldots, \mathcal{L}_{f_1}h_k, \ldots, \mathcal{L}_{f_d}h_1, \ldots, \mathcal{L}_{f_d}h_k.$$

Invariant codistributions

Given an integrable codistribution Ω, we say that it is *invariant* under the vector field f if $\mathcal{L}_f\Omega \subset \Omega$.

We derive a simple algorithm to build the smallest codistribution that contains $\text{span}\{\frac{\partial}{\partial x}h_1, \ldots, \frac{\partial}{\partial x}h_k\}$ and that is invariant under a set of vector fields $f_1, \ldots, f_d$. The algorithm is the following.

ALGORITHM 2.2. Smallest integrable codistribution containing $\text{span}\{\frac{\partial}{\partial x}h_1, \ldots, \frac{\partial}{\partial x}h_k\}$ and invariant under $\text{span}\{f_1, \ldots, f_d\}$.

Set $\Omega = \text{span}\left\{\frac{\partial}{\partial x}h_1, \ldots, \frac{\partial}{\partial x}h_k\right\}$
Set $s = k$
Set $\Omega = \Omega \oplus \bigoplus_{i=1}^{d} \mathcal{L}_{f_i}\Omega$
while $\dim(\Omega) > s$ **do**
　$s = \dim(\Omega)$
　Set $\Omega = \Omega \oplus \bigoplus_{i=1}^{d} \mathcal{L}_{f_i}\Omega$
end while

where $\bigoplus_{i=1}^{d} \mathcal{L}_{f_i}\Omega \triangleq \mathcal{L}_{f_1}\Omega \oplus \mathcal{L}_{f_2}\Omega \oplus \cdots \oplus \mathcal{L}_{f_d}\Omega$.

Since the dimension of Ω cannot exceed n, the algorithm converges in at most $n - 1$ steps. It is immediate to realize that Ω is invariant under $f_1, \ldots, f_d$. Indeed, after the convergence, $\bigoplus_{i=1}^{d} \mathcal{L}_{f_i}\Omega \subset \Omega$. By construction, $\text{span}\{\frac{\partial}{\partial x}h_1, \ldots, \frac{\partial}{\partial x}h_k\} \subset \Omega$. Finally, let us prove that Ω is the smallest codistribution that contains $\text{span}\{\frac{\partial}{\partial x}h_1, \ldots, \frac{\partial}{\partial x}h_k\}$. At a given iteration of the algorithm, the dimension of Ω is s. Hence, it is the span of the gradients of s scalar functions, $l^1, \ldots, l^s$, i.e.,

$$\Omega = \text{span}\left\{\frac{\partial}{\partial x}l^1, \ldots, \frac{\partial}{\partial x}l^s\right\}.$$

In particular, $s \geq k$ and we can set $l^1 = h_1, \ldots, l^k = h_k$. As it has been proved above, we can build $\Omega \oplus \bigoplus_{i=1}^{d} \mathcal{L}_{f_i}\Omega$ by only computing the scalars

$$l^i, \quad \mathcal{L}_{f_j}l^i, \qquad i = 1, \ldots, s, \quad j = 1, \ldots, d,$$

and by selecting the largest number of functions whose gradients are independent. The codistribution $\Omega \oplus \bigoplus_{i=1}^{d} \mathcal{L}_{f_i}\Omega$ is generated by these functions. On the other hand, the gradients of all the above functions must be included in any codistribution that contains $\text{span}\{\frac{\partial}{\partial x}l^1, \ldots, \frac{\partial}{\partial x}l^s\}$ and

2.3. Distributions and codistributions

that is invariant under $f_1, \ldots, f_d$. In other words, at each iteration, we only include in the set of generators functions whose gradients belong to any codistribution that contains span$\{\frac{\partial}{\partial x}l^1, \ldots, \frac{\partial}{\partial x}l^s\}$ and that is invariant under $f_1, \ldots, f_d$.

Let us consider an integrable codistribution generated by $h_1, \ldots, h_k$. Let us consider a given $x_0 \in \mathcal{M}$ where the codistribution is nonsingular. We assume that we do not know x_0 but only that the values of $h_1(x_0), \ldots, h_k(x_0)$ are available. We prove a very important result, which is a consequence of the inverse function theorem. Even if we do not know x_0, we can compute the value of $h(x_0)$ for any scalar function h whose gradient belongs to the considered codistribution.

Theorem 2.4. *Let us denote by Ω the codistribution generated by $h_1, \ldots, h_k$. If x_0 is a regular point for Ω and $\frac{\partial}{\partial x}h \in \Omega$, then, by knowing the values of the generators at x_0, we can compute $h(x_0)$.*

Proof. We know that the gradients of $h_1(x), \ldots, h_k(x)$ are independent on an open set that includes x_0. Hence, there exists a local coordinate change such that in the new coordinates, the first k are $h_1(x), \ldots, h_k(x)$ and the last $n - k$ coordinates are $n - k$ coordinates among $x_1, \ldots, x_n$. Without loss of generality, we assume that they are the last $n - k$ (this is obtained by reordering the variables $x_1, \ldots, x_n$). Hence, our local change is

$$(x_1, \ldots, x_n) \to (h_1, \ldots, h_k, x_{k+1}, \ldots, x_n).$$

Note that this coordinate change is possible thanks to the inverse function theorem that guarantees the local invertibility of the above transformation. In particular, the fact that x_0 is a regular point for our codistribution is a sufficient condition for the existence of such local coordinates. Since $\frac{\partial}{\partial x}h$ belongs to the codistribution generated by $h_1, \ldots, h_k$, the function h can be uniquely expressed in terms of $h_1, \ldots, h_k$. In other words,

$$h(x) = F(h_1(x), \ldots, h_k(x))$$

and $h(x_0)$ can be obtained by only knowing $h_1(x_0), \ldots, h_k(x_0)$. ◂

Let us go back to Algorithm 2.2. An immediate consequence of the previous theorem is the following result.

Corollary 2.5. *Let us denote by Ω the smallest codistribution invariant under $f_1, \ldots, f_d$ and that contains span$\{\frac{\partial}{\partial x}h_1, \ldots, \frac{\partial}{\partial x}h_k\}$, and let us denote by s its dimension at a given x_0, which is a regular point for Ω. By knowing the values of the generators at x_0, we can compute the value of any order Lie derivatives of $h_1, \ldots, h_k$ along any vector field $f_1, \ldots, f_d$.*

In other words, we can obtain the value that any function

$$\mathcal{L}_{f_{i_1}} \mathcal{L}_{f_{i_2}} \cdots \mathcal{L}_{f_{i_N}} h_i \quad \forall N, \quad i_1, i_2, \ldots, i_N = 1, \ldots, d, \quad i = 1, \ldots, k,$$

takes at x_0 without the need of knowing x_0 but only the values of the generators at x_0.

2.3.3 ▪ Frobenius theorem

Given a distribution Δ, if its dimension is $d < n$ on a given open set, it means that, on this open set, we can find $n - d$ independent covectors which are orthogonal to Δ. Let us denote these covectors by $\omega_1, \ldots, \omega_{n-d}$. We have

$$\omega_i \cdot f = 0, \quad i = 1, \ldots, n - d \quad \forall f \in \Delta.$$

It is immediate to prove that any covector that is in the span of $\omega_1, \ldots, \omega_{n-d}$ is orthogonal to Δ. Therefore, we define the codistribution $\Omega = \text{span}\{\omega_1, \ldots, \omega_{n-d}\}$ and we say that Ω is the *orthogonal codistribution*.

In a similar manner, given a codistribution Ω, whose dimension is $k < n$ on a given open set, we can build a distribution Δ of dimension $n - k$, such that

$$\omega \cdot f = 0 \quad \forall f \in \Delta, \; \forall \omega \in \Omega.$$

Δ is the *orthogonal distribution*.

We have the following fundamental result.

Theorem 2.6 (Frobenius Theorem). *Let Ω be a nonsingular codistribution, and let us denote by Δ its orthogonal distribution. Δ is involutive if and only if Ω is integrable.*

The proof of this theorem is here reported only for completeness. It is unnecessary for the comprehension of the rest of this book.

Proof. Let us start by proving that if Ω is integrable, Δ is involutive. By assumption, there exist k scalar functions, $h_1, \ldots, h_k$, such that

$$\Omega = \text{span}\left\{\frac{\partial}{\partial x}h_1, \ldots, \frac{\partial}{\partial x}h_k\right\}.$$

Let us consider two generic vector fields, f, g, that belong to Δ. In other words, for any $j = 1, \ldots, k$, we have

$$f^i \frac{\partial}{\partial x_i} h_j = 0.$$

On the other hand,

$$f^i \frac{\partial}{\partial x_i} h_j = \mathcal{L}_f h_j.$$

Hence we know that $\mathcal{L}_f h_j = 0$. The same holds for g, i.e., $\mathcal{L}_g h_j = 0$. We need to prove that Δ is involutive. Hence, we need to prove that

$$\mathcal{L}_{[f,\,g]} h_j = 0, \quad j = 1, \ldots, k.$$

By using the equality given by (2.17) we obtain

$$\mathcal{L}_{[f,\,g]} h_j = \mathcal{L}_f \mathcal{L}_g h_j - \mathcal{L}_g \mathcal{L}_f h_j = \mathcal{L}_f 0 - \mathcal{L}_g 0 = 0.$$

Let us prove the reverse. Let us denote by $f_1, \ldots, f_d$ a set of generators of Δ, i.e.,

$$\Delta = \text{span}\{f_1, \ldots, f_d\}.$$

We need to find $n - d$ scalar fields, $h_1, \ldots, h_{n-d}$, such that their gradients are independent and orthogonal to all the above generators of Δ, namely

$$\frac{\partial h_j}{\partial x_i}(f_k)^i = 0, \quad j = 1, \ldots, n-d, \; k = 1, \ldots, d. \tag{2.20}$$

This is a system of d first-order partial differential equations.

2.3. Distributions and codistributions

Let us consider a given x_0, where Δ is nonsingular. For a given vector field f, we denote by $\phi_y^f(x_0)$ the solution of the differential equation

$$\dot{x} = f(x), \quad x(0) = x_0$$

at $t = y$. Let us denote by $f_{d+1}, \ldots, f_n$ a set of $n - d$ vector fields such that the dimension of the distribution

$$\text{span}\{f_1, \ldots, f_d, f_{d+1}, \ldots, f_n\}$$

is equal to n. We consider the following map:

$$\Phi: \mathcal{U} \to \mathcal{M},$$
$$(y_1, \ldots, y_n) \to \phi_{y_1}^{f_1} \circ \cdots \circ \phi_{y_n}^{f_n}(x_0),$$

where the symbol $\circ$ denotes the composition and $\mathcal{U}$ includes points whose norm is smaller than a given $\epsilon > 0$. We consider the Jacobian

$$\frac{\partial \Phi}{\partial y}.$$

Let us study its properties. Its ith column is

$$\frac{\partial \Phi}{\partial y_i} = \tilde{\phi}_{y_1}^{f_1} \cdots \tilde{\phi}_{y_{i-1}}^{f_{i-1}} \frac{\partial}{\partial y_i} \phi_{y_i}^{f_i} \circ \cdots \circ \phi_{y_n}^{f_n}(x_0), \tag{2.21}$$

where the tilde is used to denote the Jacobian, i.e.,

$$\tilde{\phi} = \frac{\partial \phi}{\partial x}.$$

On the other hand,

$$\frac{\partial}{\partial y_i} \phi_{y_i}^{f_i} \circ \cdots \circ \phi_{y_n}^{f_n}(x_0) = f_i(\phi_{y_i}^{f_i} \circ \cdots \circ \phi_{y_n}^{f_n}(x_0)),$$

but $\phi_{y_i}^{f_i} \circ \cdots \circ \phi_{y_n}^{f_n}(x_0) = \phi_{-y_{i-1}}^{f_{i-1}} \circ \cdots \circ \phi_{-y_1}^{f_1}(\Phi(y))$. Therefore,

$$\frac{\partial}{\partial y_i} \phi_{y_i}^{f_i} \circ \cdots \circ \phi_{y_n}^{f_n}(x_0) = f_i(\phi_{-y_{i-1}}^{f_{i-1}} \circ \cdots \circ \phi_{-y_1}^{f_1}(\Phi(y))).$$

By substituting this in (2.21) we obtain

$$\frac{\partial \Phi}{\partial y_i} = \tilde{\phi}_{y_1}^{f_1} \cdots \tilde{\phi}_{y_{i-1}}^{f_{i-1}} f_i(\phi_{-y_{i-1}}^{f_{i-1}} \circ \cdots \circ \phi_{-y_1}^{f_1}(\Phi(y))). \tag{2.22}$$

Since $\Phi(0) = x_0$, we have

$$\left.\frac{\partial \Phi}{\partial y_i}\right|_{y=0} = f_i(x_0).$$

Since these vector fields are independent, the Jacobian $\frac{\partial \Phi}{\partial y}$ is full rank at x_0 and the map Φ can be locally inverted. We have

$$\Phi^{-1}(x) = \begin{bmatrix} h_1(x) \\ \cdots \\ h_n(x) \end{bmatrix}.$$

We claim that the last $n - d$ of these functions are independent solutions of (2.20). We start by remarking that we have

$$\left.\frac{\partial \Phi^{-1}}{\partial x}\right|_{x=\Phi(y)} \frac{\partial \Phi}{\partial y} = I,$$

where I is the identity matrix. Hence, the last $n - d$ lines of $\left.\frac{\partial \Phi^{-1}}{\partial x}\right|_{x=\Phi(y)}$, which are precisely the gradients $\frac{\partial h_j}{\partial x}$, $j = d+1, \ldots, n$, are orthogonal to the first d columns of $\frac{\partial \Phi}{\partial y}$. As a result, it suffices to prove that the first d columns of $\frac{\partial \Phi}{\partial y}$ form a basis of Δ at any $x = \Phi(y)$. From (2.22) we have

$$\frac{\partial \Phi}{\partial y_i} = \widetilde{\phi}_{y_1}^{f_1} \cdots \widetilde{\phi}_{y_{i-1}}^{f_{i-1}} f_i(\phi_{-y_{i-1}}^{f_{i-1}} \circ \cdots \circ \phi_{-y_1}^{f_1}(x)).$$

We must prove that, for any two vector fields in Δ, f and g, the following vector also belongs to Δ:

$$\widetilde{\phi}_t^f g \phi_{-t}^f(x).$$

Since $g \in \Delta$, it suffices to prove that, for any $i = 1, \ldots, d$,

$$\widetilde{\phi}_t^f f_i \phi_{-t}^f(x) \in \Delta.$$

Let us consider the vectors:

$$F_i(t) = \widetilde{\phi}_{-t}^f f_i \circ \phi_t^f(x), \qquad i = 1, \ldots, d.$$

By differentiating the identity $\widetilde{\phi}_{-t}^f \widetilde{\phi}_t^f = I$ with respect to t and by exchanging the order of $\frac{d}{dt}$ and $\frac{\partial}{\partial x}$, we have

$$\frac{d}{dt} \widetilde{\phi}_{-t}^f \circ \phi_t^f(x) = -\widetilde{\phi}_{-t}^f \frac{\partial f}{\partial x} \circ \phi_t^f(x).$$

Additionally,

$$\frac{d}{dt}(f_i \circ \phi_t^f(x)) = \frac{\partial f_i}{\partial x} f \circ \phi_t^f(x).$$

Therefore,

$$\frac{dF_i}{dt} = \widetilde{\phi}_{-t}^f [f, f_i] \circ \phi_t^f(x).$$

Since Δ is involutive, we have

$$[f, f_i] = \sum_{j=1}^d c_j^i f_j.$$

Hence,

$$\frac{dF_i}{dt} = \sum_{j=1}^d c_j^i F_j.$$

As a result, we obtain

$$[F_1(t), \ldots, F_d(t)] = [F_1(0), \ldots, F_d(0)] \Xi(t) = [f_1(x), \ldots, f_d(x)] \Xi(t),$$

where $\Xi(t)$ is a $d \times d$ matrix. By multiplying both left sides by $\widetilde{\phi}_t^f$, we obtain

$$[f_1(\phi_t^f(x)), \ldots, f_d(\phi_t^f(x))] = [\widetilde{\phi}_t^f f_1(x), \ldots, \widetilde{\phi}_t^f f_d(x)] \Xi(t).$$

2.4. Lie groups

We now replace x by $\phi^f_{-t}(x)$ and obtain

$$[f_1(x), \ldots, f_d(x)] = [\widetilde{\phi}^f_t f_1 \circ \phi^f_{-t}(x), \ldots, \widetilde{\phi}^f_t f_d \circ \phi^f_{-t}(x)]\Xi(t).$$

Since $\Xi(t)$ is nonsingular, the vectors

$$\widetilde{\phi}^f_t f_1 \circ \phi^f_{-t}(x), \ \ldots, \ \widetilde{\phi}^f_t f_d \circ \phi^f_{-t}(x)$$

generate Δ. ◄

2.4 ▪ Lie groups

A group G is a set of elements equipped with an internal binary operation $G \times G \to G$, i.e., that associates with any pair $g_1, g_2 \in G$ an element $g = g_1 g_2 \in G$. This operation satisfies the following properties:

- There exists an element $e \in G$ such that, for any $g \in G$, $eg = ge = g$.
- For any $g \in G$, there exists an element g^{-1} such that $gg^{-1} = g^{-1}g = e$.
- For any $g_1, g_2, g_3 \in G$, we have $(g_1 g_2)g_3 = g_1(g_2 g_3)$ (associativity).

Groups for which commutativity with respect to the internal operation also holds are called *Abelian* groups. Namely, for any $g_1, g_2 \in G$, we have $g_1 g_2 = g_2 g_1$.

A Lie group, G, is a group endowed with the structure of a C^∞ manifold of a given dimension n and such that the inversion map

$$G \to G : g \to g^{-1}$$

and the multiplication map

$$G \times G \to G : (g_1, g_2) \to g = g_1 g_2$$

are smooth.[11] It is also called an n-parameter Lie group, since any element of the group is identified by n parameters.

2.4.1 ▪ Examples of Lie groups

Scale

Let us consider the transformation $\mathbb{R} \to \mathbb{R}$:

$$x' = \lambda x,$$

with λ a nonzero real number. It is easy to prove that all the transformations defined as above form a Lie group, specifically a one-parameter Lie group. It is also immediate to prove that this group is Abelian.

The above transformation can also be defined in $\mathbb{R}^n$. It is still defined by one parameter, and the corresponding group is Abelian.

---
[11] Note that it suffices to require that only the map $G \times G \to G : (g_1, g_2) \to g = g_1 g_2^{-1}$ is smooth.

Scale and shift

We consider now the transformation in $\mathbb{R}$ defined by two continuous parameters (λ and μ):

$$x' = \lambda x + \mu,$$

with $\lambda \neq 0$. It is easy to prove that all the transformations defined as above form a two-parameter Lie group. In this case, however, the group is not Abelian.

The above transformation can also be defined in $\mathbb{R}^n$. In this case, λ is still a scalar while $\mu \in \mathbb{R}^n$. Hence, the group becomes an $(n+1)$-parameter Lie group.

$GL(n, \mathbb{R})$

For a given integer n, we consider the set of all the square real-valued matrices $n \times n$, with nonvanishing determinant. It is immediate to prove that they form an n^2-parameter Lie group (by defining the internal binary operation as the matrix multiplication). The element e is the identity matrix (I), and the existence of the inverse is guaranteed by the fact that each matrix has nonvanishing determinant.

More in general, $GL(n, K)$ is the set of all the square matrices $n \times n$, with nonvanishing determinant and with the entries that belong to the field K. K can be $\mathbb{Q}$, $\mathbb{R}$, and $\mathbb{C}$. In all these cases, $GL(n, K)$ is a Lie group.

$SO(n)$

A real and square matrix O is *orthogonal* if it is invertible and its inverse coincides with its transpose:

$$O^T O = O O^T = I. \tag{2.23}$$

In other words, the columns of an orthogonal matrix are unit vectors orthogonal to each other. Obviously, the inverse of an orthogonal matrix is an orthogonal matrix. Additionally, it is immediate to prove that the product of two orthonormal matrices is an orthogonal matrix:

$$((O_1 O_2)^T (O_1 O_2) = O_2^T O_1^T O_1 O_2 = I).$$

This proves that the set of all the orthogonal matrices form a group, which is the *orthogonal group*. The determinant of an orthogonal matrix can be 1 or -1. The set of all the orthogonal matrices with determinant equal to 1 is denoted by $SO(n)$. Since this set is closed with respect to the product (i.e., the product of two orthogonal matrices with determinant equal to 1 is an orthogonal matrix with determinant 1), $SO(n)$ is a group (the special orthogonal group). In particular, it is possible to show that it is an $\frac{n(n-1)}{2}$-parameter Lie group (see section 2.4.5).

$SO(2)$

When $n = 2$, we have $\frac{n(n-1)}{2} = 1$. We denote the parameter by θ. The condition in (2.23) provides two types of solution:

$$R_\theta = \begin{bmatrix} \cos\theta & -\sin\theta \\ \sin\theta & \cos\theta \end{bmatrix}, \quad S_\theta = \begin{bmatrix} \cos\theta & \sin\theta \\ \sin\theta & -\cos\theta \end{bmatrix}.$$

Only the first type belongs to $SO(2)$, since the determinant of the second one is -1. When the matrix R_θ is applied to the vectors of the two-dimensional Euclidean space, it generates a

2.4. Lie groups

rotation about the vertical axis of an angle θ. The matrix S_θ also generates a reflection with respect to the x-axis before the same rotation:

$$S_\theta = R_\theta \begin{bmatrix} 1 & 0 \\ 0 & -1 \end{bmatrix}.$$

$SU(n)$

A square matrix U with entries in $\mathcal{C}$ is *unitary* if it is invertible and its inverse coincides with its conjugate transpose:

$$U^\dagger U = UU^\dagger = I, \tag{2.24}$$

where the symbol $\dagger$ denotes the conjugate transpose. As for the orthogonal matrices, also the unitary matrices of a given dimension form a group.

The determinant of an unitary matrix has module equal to 1. The set of all the unitary matrices with determinant equal to 1 is denoted by $SU(n)$. Since this set is closed with respect to the product (i.e., the product of two unitary matrices with determinant equal to 1 is a unitary matrix with determinant 1), $SU(n)$ is a group. In particular, it is possible to show that it is an (n^2-1)-parameter Lie group (see section 2.4.5).

2.4.2 ▪ Group representation

The intuitive idea of a group representation is to "realize" the properties of a given abstract group by a set of square matrices. In particular, the group internal operation is represented by the matrix multiplication. Representations of groups are very useful because they allow us to express in linear algebra the theoretic properties that characterize the group.

We now provide the formal definition of a group representation. Let us consider a group G and a vector space $\mathcal{V}$ of dimension m over a field K (e.g., $K = \mathbb{Q}$, $K = \mathbb{R}$, $K = \mathbb{C}$). A representation of G on $\mathcal{V}$ is a homomorphism from G to the general linear group on $\mathcal{V}$ ($GL(\mathcal{V})$). In other words, a representation is a map:

$$\begin{aligned} \mathcal{R}: \quad & G \to GL(\mathcal{V}), \\ & g \to R(g). \end{aligned}$$

Since the above map is a homomorphism, it must preserve the internal operation. Namely,

$$R(g_1 g_2) = R(g_1) R(g_2).$$

We call $\mathcal{V}$ the *basis space* of the representation. Its dimension m is also called the dimension of the representation. From now on, without loss of generality, we assume that the vector space $\mathcal{V}$ is a Euclidean space. Since the space $GL(\mathcal{V})$, i.e., the space of all the linear and invertible transformations $\mathcal{V} \to \mathcal{V}$, is isomorphic to $GL(m, K)$, a representation can be directly defined as a map $G \to GL(m, K)$. In other words, $R(g) \in GL(m, K)$.

A subspace $\mathcal{V}' \subseteq \mathcal{V}$ is *invariant* for the representation $\mathcal{R}$ if

$$R(g)\mathcal{V}' \subseteq \mathcal{V}' \qquad \forall g \in G.$$

If a given representation $\mathcal{R}$ admits nontrivial invariant subspaces (i.e., subspaces that are neither the whole $\mathcal{V}$ nor $\{0\}$), then the representation $\mathcal{R}$ is called *reducible*. If a representation is not reducible, it is called *irreducible*.

Let us suppose that $\mathcal{V}'$ is an invariant subspace for $\mathcal{R}$ and the dimension of this subspace is $m_1 < m$, where m is the dimension of $\mathcal{V}$. We can find a basis of $\mathcal{V}$,

$$b_1, b_2, \ldots, b_{m_1}, b_{m_1+1}, \ldots, b_m,$$

such that $b_1, b_2, \ldots, b_{m_1}$ is a basis of $\mathcal{V}'$. For any $g \in G$, we have

$$R(g) = \begin{bmatrix} R_1(g) & S \\ 0 & R_2(g) \end{bmatrix},$$

where $R_1(g)$ is an $m_1 \times m_1$ matrix. If in addition the matrix block S is a zero $m_1 \times m_2$ matrix ($m_2 = m - m_1$), then also the complement of $\mathcal{V}'$ (from now on $\mathcal{V}''$) is invariant for the representation $\mathcal{R}$ and any matrix $R(g)$ is a block matrix. In this case, the representation $\mathcal{R}$ is called *completely reducible* and $\mathcal{R}_1$ and $\mathcal{R}_2$ are two representations of G (with basis spaces $\mathcal{V}'$ and $\mathcal{V}''$, respectively).

Schur's lemma

Let us consider a group G and an irreducible representation $\mathcal{R}$ of it on the vector space $\mathcal{V}$ over $\mathbb{C}$. If T is a linear operator $\mathcal{V} \to \mathcal{V}$ such that

$$TR(g) = R(g)T \qquad \forall g \in G,$$

then $T = \lambda I$, $\lambda \in \mathbb{R}$; i.e., T is a scalar multiple of the identity.

Note that the validity of this result requires that the basis space of the representation is over the field of the complex numbers (i.e., $K = \mathbb{C}$). The proof is immediate. Let T be a matrix that commutes with any $R(g)$, and let us denote by λ one of its eigenvalues (any matrix has at least one eigenvalue in $\mathbb{C}$). Let us denote by V_λ the subspace spanned by the eigenvectors with eigenvalue λ. For any $w \in \mathcal{V}_\lambda$ and $\forall g \in G$,

$$T(R(g)w) = R(g)T(w) = R(g)(\lambda w) = \lambda R(g)w.$$

Hence, $R(g)w \in \mathcal{V}_\lambda$, and consequently $\mathcal{V}_\lambda$ is invariant for $\mathcal{R}$. But $\mathcal{R}$ is irreducible. As a result, $\mathcal{V}_\lambda = \mathcal{V}$. Therefore, $T(v) = \lambda v \ \forall v \in \mathcal{V}$, i.e., $T = \lambda I$. This concludes the proof.

An important consequence of the above result is that, for a given Abelian group, any representation irreducible is one dimensional. To prove this is sufficient to use Schur's lemma by taking as T any matrix of the representation.

Let us consider again the group of rotations on a plane. We know that this group is Abelian. By performing two consecutive rotations of θ_1 and θ_2, we obtain a rotation of $\theta_1 + \theta_2$ independently of the order. We provided a representation of this group by using $\mathbb{R}^2$ for the basis space. Each element of the group is represented by a matrix of $SO(2)$, i.e.,

$$\begin{bmatrix} \cos\theta & -\sin\theta \\ \sin\theta & \cos\theta \end{bmatrix}.$$

The dimension of this representation is 2. On the other hand, since the group is Abelian, the result stated by Schur's lemma ensures that it is possible to find a one-dimensional representation. The reader is invited to find a one-dimensional representation of the group of rotations on the plane.

2.4.3 ▪ Lie algebra

We provide the formal definition of a real Lie algebra. A real Lie algebra $\mathcal{A}$ is a vector space over $\mathbb{R}$ equipped with an internal operation, the *Lie product* or *Lie bracket*, that we denote by $(a_1, a_2) \to [a_1, a_2]$ and that satisfies the following properties:

- Linearity: $[a_1, \alpha a_2 + \beta a_3] = \alpha[a_1, a_2] + \beta[a_1, a_3] \; \forall \alpha, \beta \in \mathbb{R}$ and $\forall a_1, a_2, a_3 \in \mathcal{A}$.
- Anticommutativity: $[a_1, a_2] = -[a_2, a_1] \; \forall a_1, a_2 \in \mathcal{A}$.
- Jacobi identity: $[a_1, [a_2, a_3]] + [a_2, [a_3, a_1]] + [a_3, [a_1, a_2]] = 0 \; \forall a_1, a_2, a_3 \in \mathcal{A}$.

The above definition states that the Lie algebra $\mathcal{A}$ is a vector space. Let us denote by n its dimension. We can introduce a basis $b_1, \ldots, b_n$. For any pair (b_i, b_j), $i, j = 1, \ldots, n$, the Lie product $[b_i, b_j]$ belongs to $\mathcal{A}$. As a result, we have

$$[b_i, b_j] = \sum_{k=1}^{n} c_{ij}^k b_k,$$

where the coefficients c_{ij}^k are called the *structure constants* of the Lie algebra $\mathcal{A}$.

Lie algebra representation

Exactly as for the groups, we introduce the concept of Lie algebra representation. As in the case of groups, the idea is to "realize" the properties of the Lie algebra by a set of square matrices. In particular, the algebra internal operation is represented by the matrix commutator.

Let us consider a Lie algebra $\mathcal{A}$ and a vector space $\mathcal{V}$. A representation of $\mathcal{A}$ on $\mathcal{V}$ is a homomorphism from $\mathcal{A}$ to the space of all the linear transformations $L(\mathcal{V}) : \mathcal{V} \to \mathcal{V}$. In other words, it is a map:

$$\mathcal{R} : \mathcal{A} \to L(\mathcal{V}),$$
$$a \to R(a).$$

We require that the Lie product is represented by the matrix commutator, i.e.,

$$R([a_1, a_2]) = R(a_1)R(a_2) - R(a_2)R(a_1).$$

2.4.4 ▪ Correspondence between a Lie group and a Lie algebra

We start by reminding the reader of a simple result in linear algebra. Given a square matrix A $m \times m$, the series

$$\sum_{k=1}^{\infty} \frac{A^k}{k!}$$

is convergent in the norm

$$||A|| = \left(\sum_{i,j=1}^{m} |A_{ij}|^2 \right)^{\frac{1}{2}}.$$

As a result, we can define the exponential of A as follows:

$$e^A = I + \sum_{k=1}^{\infty} \frac{A^k}{k!}.$$

Therefore, we can consider the exponential map:

$$\exp: \; A \to e^A.$$

Now we consider the case of algebras and groups whose elements are matrices. We have the following fundamental result.

Given a Lie algebra $\mathcal{A}$ of dimension n, the set

$$G = \{e^A, \; A \in \mathcal{A}\}$$

is a Lie group. To prove this result, we first need to prove that, given $A, B \in \mathcal{A}$, we have

$$e^A e^B = e^C,$$

where $C \in \mathcal{A}$. The existence of such a matrix C is guaranteed by the Baker–Hausdorff formula:

$$C = A + B + \frac{1}{2}[A, B] + \frac{1}{12}[A, [A, B]] + \frac{1}{12}[B, [B, A]] + \cdots.$$

Since $[\cdot, \cdot]$ is the Lie product in $\mathcal{A}$, and since C is a sum of terms proportional to A, B and their commutators, we obtain that $C \in \mathcal{A}$. Hence, $e^A e^B \in G$ and G is closed with respect to the matrix multiplication. In addition, it is immediate to see that, for any $A \in \mathcal{A}$, $(e^A)^{-1} = e^{-A}$. Finally, the identity matrix is the element 1 of G, which is obtained by taking the zero matrix (the zero matrix belongs to $\mathcal{A}$ because $\mathcal{A}$ is a vector space). Therefore, G is a group. In particular, G is an n-parameter Lie group because any element of $\mathcal{A}$ can be parametrized in terms of n parameters ($\mathcal{A}$ being a vector space of dimension n).

The reverse also holds. Namely, given a Lie group G of dimension n it is possible to build a Lie algebra $\mathcal{A}$. To prove this, we first remark that, for any $M \in G$, there exists a parametrization

$$\Gamma: \; \mathbb{R}^n \to G,$$
$$x = (x_1, \ldots, x_n) \to \Gamma(x) = M.$$

It is then possible to prove that the matrices[12]

$$A_k = \left.\frac{\partial \Gamma}{\partial x}\right|_{x=x_0}, \quad k = 1, \ldots, n,$$

make a basis for a vector space of dimension n that we denote by $\mathcal{A}$. $\mathcal{A}$ is called the *tangent space* to the group. The matrices A_k are called the *infinitesimal generators* of $\mathcal{A}$. Finally, it is possible to prove that $\mathcal{A}$ satisfies the properties that define a Lie algebra.

The above results established a correspondence between Lie algebras and Lie groups in the case of matrices. We wonder whether the correspondence is bijective (i.e., one-to-one) and what happens in the general case, i.e., for groups which are not necessarily groups of matrices. In particular, we wonder whether, given a Lie group, there exists a one-to-one correspondence between the representations of this group and the representations of its Lie algebra. The general answer to this question is negative. In particular, the following result holds.

Given a Lie group G, if G is *simply connected*, there exists a one-to-one correspondence between the representations of G and the representations of its Lie algebra.

We do not provide the proof of this result. We remind the reader that a topological space is simply connected if it is path-connected and every closed path on it can be continuously transformed (continuously contracted) into a point (path-connected means that for any pair of points that belong to this space there exists a path that connects them).

---

[12] Here, with x_0, we denote the point such that $\Gamma(x_0) = I$. Usually, the parametrization is such that x_0 is zero. However, this is not always the case.

2.4. Lie groups

2.4.5 • Examples of Lie groups and Lie algebras

$SO(n)$ and $\mathfrak{so}(n)$

Let us consider again the group $SO(n)$. We consider elements of this group around the identity, i.e.,

$$O = I + \epsilon A.$$

We want to find the properties of the matrices A. Equation (2.23) provides

$$(I + \epsilon A)(I + \epsilon A^T) = I + \epsilon(A + A^T) + O(\epsilon^2) = I.$$

Hence, A must be skew-symmetric. On the other hand, for any skew-symmetric matrix A, by using the Baker–Hausdorff formula we obtain

$$e^A (e^A)^T = e^A e^{A^T} = e^A e^{-A} = e^{A-A} = I.$$

Therefore, the Lie algebra of $SO(n)$ is the vector space of all the skew-symmetric matrices. We denote this vector space by $\mathfrak{so}(n)$. The dimension of this vector space is $\frac{n(n-1)}{2}$. This is also the dimension of $SO(n)$.

$SO(3)$ and $\mathfrak{so}(3)$

The dimension of the vector space $\mathfrak{so}(3)$ is $\frac{3(3-1)}{2} = 3$. We introduce the following basis:

$$A_1 = \begin{bmatrix} 0 & 0 & 0 \\ 0 & 0 & -1 \\ 0 & 1 & 0 \end{bmatrix}, \quad A_2 = \begin{bmatrix} 0 & 0 & 1 \\ 0 & 0 & 0 \\ -1 & 0 & 0 \end{bmatrix}, \quad A_3 = \begin{bmatrix} 0 & -1 & 0 \\ 1 & 0 & 0 \\ 0 & 0 & 0 \end{bmatrix}.$$

Let us derive the structure constants of $\mathfrak{so}(3)$. We have

$$A_1 A_2 - A_2 A_1 = A_3, \quad A_3 A_1 - A_1 A_3 = A_2, \quad A_2 A_3 - A_3 A_2 = A_1,$$

that is,

$$A_i A_j - A_j A_i = \epsilon_{ijk} A_k,$$

where ϵ_{ijk} is the Levi-Civita symbol in three dimensions that is 1 if (i, j, k) is an even permutation of $(1, 2, 3)$ and -1 if it is an odd permutation. Additionally, it is zero whenever two of its indices take the same value.

Let us consider again elements of $SO(3)$ around the identity. We know that we have three independent elements:

$$I + \epsilon A_1, \quad I + \epsilon A_2, \quad I + \epsilon A_3.$$

We consider the following limits:

$$\lim_{n\to\infty} \left(I + \frac{\theta}{n} A_1\right)^n, \quad \lim_{n\to\infty} \left(I + \frac{\phi}{n} A_2\right)^n, \quad \lim_{n\to\infty} \left(I + \frac{\psi}{n} A_3\right)^n,$$

where θ, ϕ, ψ are three parameters. By remarking that if

$$A = \begin{bmatrix} 0 & -1 \\ 1 & 0 \end{bmatrix}$$

it holds that

$$(A)^2 = \begin{bmatrix} -1 & 0 \\ 0 & -1 \end{bmatrix}, \quad (A)^3 = \begin{bmatrix} 0 & 1 \\ -1 & 0 \end{bmatrix}, \quad (A)^4 = \begin{bmatrix} 1 & 0 \\ 0 & 1 \end{bmatrix},$$

we obtain

$$e^{\theta A_1} = \begin{bmatrix} 1 & 0 & 0 \\ 0 & \cos\theta & -\sin\theta \\ 0 & \sin\theta & \cos\theta \end{bmatrix}, \quad e^{\phi A_2} = \begin{bmatrix} \cos\phi & 0 & \sin\phi \\ 0 & 1 & 0 \\ -\sin\phi & 0 & \cos\phi \end{bmatrix},$$

$$e^{\psi A_3} = \begin{bmatrix} \cos\psi & -\sin\psi & 0 \\ \sin\psi & \cos\psi & 0 \\ 0 & 0 & 1 \end{bmatrix}.$$

These matrices correspond to a rotation of θ, ϕ, and ψ around the x-axis, the y-axis, and the z-axis, respectively.

$SU(n)$ and $\mathfrak{su}(n)$

We proceed as in the previous section. We consider elements of the group $SU(n)$ around the identity, i.e.,

$$U = I + i\epsilon H.$$

We want to find the properties of the matrices H. Equation (2.24) provides

$$(I + i\epsilon H)(I - i\epsilon H^\dagger) = I + \epsilon i(H - H^\dagger) + O(\epsilon^2) = I.$$

Hence, H must be Hermitian. On the other hand, for any Hermitian matrix H, we obtain

$$e^{iH}(e^{iH})^\dagger = e^{iH}e^{-iH^\dagger} = e^{i(H-H^\dagger)} = I.$$

We remark that a Hermitian matrix is defined by $2\frac{n(n-1)}{2} + n = n^2$ parameters. In particular, the elements on the diagonal must have vanishing imaginary part. Additionally,

$$1 = \det\left(e^{iH}(e^{iH})^\dagger\right) = \det\left(e^{iH}\right)\det\left(e^{-iH^*}\right) = \left|\det\left(e^{iH}\right)\right|^2.$$

Equation (2.24) alone does not suffice to define $SU(n)$: we must also require that the determinant is equal to 1. We remind the reader of the following equation, which holds for any complex matrix A:

$$\det\left(e^A\right) = e^{tr(A)}.$$

Therefore, the Lie algebra of $SU(n)$ is the vector space of all the Hermitian matrices with zero trace. We denote this vector space by $\mathfrak{su}(n)$. The dimension of this vector space is $n^2 - 1$. This is also the dimension of $SU(n)$.

$SU(2)$ and $\mathfrak{su}(2)$

The dimension of the vector space $\mathfrak{su}(2)$ is $2^2 - 1 = 3$. We introduce the following basis:

$$\sigma_1 = \begin{bmatrix} 0 & 1 \\ 1 & 0 \end{bmatrix}, \quad \sigma_2 = \begin{bmatrix} 0 & -i \\ i & 0 \end{bmatrix}, \quad \sigma_3 = \begin{bmatrix} 1 & 0 \\ 0 & -1 \end{bmatrix}.$$

These matrices are known in quantum physics by the name of *Pauli matrices*. Let us derive the structure constants of $\mathfrak{su}(2)$. We have

$$\sigma_1\sigma_2 - \sigma_2\sigma_1 = i\sigma_3, \qquad \sigma_3\sigma_1 - \sigma_1\sigma_3 = i\sigma_2, \qquad \sigma_2\sigma_3 - \sigma_3\sigma_2 = i\sigma_1,$$

that is,

$$\sigma_i\sigma_j - \sigma_j\sigma_i = i\epsilon_{ijk}\sigma_k.$$

$SO(3)$ and $SU(2)$

We show that these two groups are homomorphic. In other words, there exists a correspondence between the two groups that preserves the internal operation.

For any vector $x = [x^1, x^2, x^3]^T \in \mathbb{R}^3$, we define the matrix

$$X = x^j \sigma_j = \begin{bmatrix} x^3 & x^1 - ix^2 \\ x^1 + ix^2 & -x^3 \end{bmatrix}, \qquad (2.25)$$

which belongs to $\mathfrak{su}(2)$ and, consequently, $\text{tr}(X) = 0$. Additionally, we have $\det(X) = -|x|^2$.

For any $U \in SU(2)$, we define the matrix

$$X_U = UXU^\dagger.$$

$X_U^\dagger = UX^\dagger U^\dagger = X_U$. Additionally, $\text{tr}(X_U) = \text{tr}(UXU^\dagger) = \text{tr}(XUU^\dagger) = \text{tr}(X) = 0$. Hence, $X_U \in \mathfrak{su}(2)$ (we remind the reader that $\text{tr}(AB) = \text{tr}(BA)$). As a result, there exists a vector $x_U \in \mathbb{R}^3$ such that

$$X_U = x_U^j \sigma_j.$$

Finally, $\det(X_U) = \det(X) = -|x|^2$. Hence, the transformation

$$x \to x_U = R_U x$$

preserves the distance. As a result, $R_U \in O(3)$.

We consider the following correspondence:

$$\Psi: \quad SU(2) \to O(3),$$
$$U \to R_U.$$

By a direct computation, it is possible to verify that Ψ preserves the group operation. On the other hand, it is immediate to realize that $R_{-U} = R_U$. Hence, Ψ is not injective. Additionally, if $U \in SU(2)$ is such that $R(U) = I$, then $U = I$ or $U = -I$. In other words, $\ker \Psi = \{-I, I\}$. Since $\{-I, I\}$ is isomorphic to $\mathbb{Z}_2$, we obtain that $\ker \Psi$ is isomorphic to $\mathbb{Z}_2$. As a result, $SU(2)/\mathbb{Z}_2$ is isomorphic to $O(3)$. Finally, since $\det(R_U) = 1$, we obtain that $SU(2)/\mathbb{Z}_2$ is isomorphic to the group $SO(3)$.

2.5 ▪ Tensors associated with a group of transformations

In this section, we extend the concept of tensor introduced in section 2.2. Given a manifold $\mathcal{M}$ of dimension n, a tensor was defined as a set of components that change in a given manner under a change of coordinates in an open set of $\mathcal{M}$. For instance, in the case of tensors of rank 1 (vectors and covectors) we had

$$S^j = \frac{\partial y^j}{\partial x^l} T^l, \qquad S_i = \frac{\partial x^k}{\partial y^i} T_k,$$

where $\frac{\partial y^j}{\partial x^l}$ is the Jacobian of the coordinates' change $x \to y$ in (2.3). We remark that this Jacobian is a nonsingular square matrix of dimension n. For tensors of higher rank, the components change according to (2.7).

Now, let us consider a k-parameter Lie group G and an m-dimensional representation of it. Let us denote by $\tilde{g}$ the $m \times m$ matrix that represents the element $g \in G$. We define a *contravariant* vector (or simply vector) with respect to the group G as an object that consists of m components,

$$v^1, \ v^2, \ \ldots, \ v^m,$$

which, under the action of $g \in G$, change as follows:

$$v^i \to w^i = (\tilde{g}^{-1})^i_j v^j. \tag{2.26}$$

Note that, in general, $m \neq n$. For any index that appears in a given formula, the context must clarify to which group of transformations it refers. Note that a given component of the above tensor v (i.e., v^i for a given $i = 1, \ldots, m$) can be a tensor of any rank with respect to another group of transformations. In other words, a given formula can simultaneously contain indices that refer to a distinct type of tensors (i.e., tensors that refer to distinct groups of transformations). The context must clarify this. In addition, also in this case we adopt the Einstein notation.

In a similar manner, we define a *covariant* vector (or simply covector) with respect to the group G as an object that consists of m components,

$$a_1, \ a_2, \ \ldots, \ a_m,$$

which, under the action of $g \in G$, change as follows:

$$a_i \to b_i = \tilde{g}^j_i a_j. \tag{2.27}$$

Exactly as in the case of coordinates' change for manifolds, we say that vectors are tensors of type $(0, 1)$ and covectors are tensors of type $(1, 0)$. Additionally, we define tensors with respect to the group G of rank larger than 1 as follows. A tensor of generic rank $q + p$, and in particular of type (q, p), is an object that consists of $m^{(q+p)}$ components,

$$P^{i_1,\ldots,i_p}_{j_1,\ldots,j_q},$$

that, under the action of $g \in G$, change as follows:

$$P^{i_1,\ldots,i_p}_{j_1,\ldots,j_q} \to Q^{i_1,\ldots,i_p}_{j_1,\ldots,j_q} = (\tilde{g}^{-1})^{i_1}_{k_1} \cdots (\tilde{g}^{-1})^{i_p}_{k_p} \tilde{g}^{l_1}_{j_1} \cdots \tilde{g}^{l_q}_{j_q} P^{k_1,\ldots,k_p}_{l_1,\ldots,l_q}. \tag{2.28}$$

Note that the Kronecker delta δ^i_j does not change under the action of G. This can be directly obtained by using (2.28) with $p = q = 1$:

$$\delta^i_j \to (\tilde{g}^{-1})^i_k \tilde{g}^l_j \delta^k_l = (\tilde{g}^{-1})^i_k \tilde{g}^k_j = \delta^i_j.$$

Note that a tensor of rank k consists of m^k elements. In the case of an Abelian group, as we saw in section 2.4.2, we have $m = 1$. Hence, any tensor, independently of its rank, has a single component, precisely as a scalar. This means that, for an Abelian group, we do not need to introduce tensors.

Part II
Nonlinear Observability

Chapter 3
Group of Invariance of Observability

The goal of this chapter is to characterize all the transformations under which a complete theory of observability must be invariant. In section 1.3.5 we identified the following three distinct groups of transformations:

- the state coordinates transformations' group,
- the output transformations' group, and
- the input transformations' group.

On the other hand, a complete and thorough characterization of these groups requires us to deal with the variable time in a new manner. In particular, this holds when the dynamics of the state are characterized by a nonvanishing drift (i.e., in (1.3), the term f^0 has at least one non-vanishing component, or, in the presence of unknown inputs, in (1.14), the term g^0 has at least one nonvanishing component). In section 1.4, we pointed out that time plays two distinct roles. From one side, it is an index (continuous or discrete) that synchronizes all the system inputs and outputs. This was called the chronological time (or simply time when it is not ambiguous). From another side, for systems with a nonvanishing drift, time acts as a system input (with the particularity that it is ineluctably assigned). This was called the ordinary time or, when it is not ambiguous, the time.

This chapter starts by introducing a new framework that is able to separate these two roles (section 3.1). In this new framework, called the *chronospace*, the chronological time is included in the state and it is also a system output.

In the chronospace, the chronological time will be separated from the ordinary time. For instance, we can operate changes of coordinates that only act on the chronological time and leave unaffected the ordinary time, or vice versa. In particular, we can characterize our system by a new set of outputs that are functions of the chronological time and the original system outputs. Similarly, we can characterize our system by a new set of inputs that are functions of the ordinary time and the original system inputs.

The chronospace is the appropriate framework to fully characterize the group of invariance of observability. In the remaining sections of this chapter, we characterize the output transformations' group and the input transformations' group in the chronospace. We start by considering the case without unknown inputs (section 3.2) and then the case with unknown inputs (section 3.3).

One of the main goals of this chapter is the identification of the class of input-output systems for which the theory of observability is simpler than (but still representative of) the general case. In other words, the goal is to determine special systems for which the derivation of the analytic procedure to obtain the state observability is easier than in the general case but similar to the derivation in the general case. Indeed, as for many scientific problems, it is very helpful to proceed by intermediate steps and, before the investigation of the general case, it is very advantageous to investigate simplified classes of systems. The ability is to identify the suitable simplifications, i.e., simplified systems for which the solution is easier but conceptually equivalent to the solution of the general case. This is obtained by starting from the study of the groups of invariance of the theory and by determining the conditions under which these groups become Abelian. Indeed, because of Schur's lemma (see section 2.4.2), an Abelian group can be represented by scalars (i.e., 1×1 matrices). Hence, in the presence of Abelian groups, we do not need to introduce tensors, as discussed at the end of section 2.5.

In sections 3.2.3 and 3.2.4 we carry out this investigation in the case without unknown inputs, and we provide the aforementioned class of simplified systems. In sections 3.3.3 and 3.3.4 we carry out the same investigation in the presence of unknown inputs.

On the other hand, to deal with the case with unknown inputs, instead of considering separately the output transformations' group and the input transformations' group, we found it very convenient to introduce a new group of transformations that basically merges together these groups and, concerning the inputs, it only includes the unknown inputs and the ordinary time. In other words, the known inputs will be scalars with respect to this transformations' group. Obviously, the input transformations' group and the output transformations' group are independent and the input transformations' group also includes the known inputs. Hence, by proceeding in this manner, we lose generality. However, we gain in simplicity and we feel that this is the best trade-off between generality and simplicity to deal with these systems.

We call this new group of transformations the *Simultaneous Unknown Input-Output transformations' group*, from now on $\mathcal{SUIO}$.

3.1 ▪ The chronostate and the chronospace

Let us refer to a nonlinear control system with m inputs (in this section it is unnecessary to specify whether these inputs are known or unknown). Let us denote by x the state that characterizes our system, and let us characterize its dynamics (i.e., the link between the evolution of the state and the system inputs) and its output as follows:

$$\begin{cases} \dot{x} & = g^0(x,t) + \sum_{i=1}^m f^i(x,t)u_i, \\ y & = [h_1(x,t), \ldots, h_p(x,t)]^T, \end{cases} \quad (3.1)$$

where $g^0(x,t)$, $f^1(x,t), \ldots, f^m(x,t)$, are vector fields and $h_1(x,t), \ldots, h_p(x,t)$ are scalar functions. Let us denote the dimension of the state by n (≥ 1). Equation (3.1) generalizes (1.3) (or (1.14) in the presence of unknown inputs) by accounting for a possible explicit dependence on time in the dynamics and in the output.

We can also express the dynamics in terms of differentials:

$$dx = g^0(x,t)dt + \sum_{i=1}^m f^i(x,t)dU_i, \quad (3.2)$$

where $dU_i = u_i dt$.

3.1. The chronostate and the chronospace

We start by setting $x^0 = t$, and we include x^0 in the state. We call the new extended state the *chronostate*. The chronostate belongs to a new state space that will be called the *chronospace*. Its dimension exceeds by 1 the dimension of the corresponding space of states (i.e., the dimension of the chronospace in the case of our system is $n + 1$). For the above system, the chronostate is

$$\underline{x} \triangleq \begin{bmatrix} x^0 \\ x^1 \\ \ldots \\ x^n \end{bmatrix}. \tag{3.3}$$

We want to derive the new equations that characterize the dynamics (i.e., the link between the evolution of the chronostate and the system inputs) and the output in the chronospace. Concerning the dynamics, we obtain the new equations directly from the first in (3.1) (or from (3.2)). We have

$$d\underline{x} = \underline{g}^0(\underline{x})dt + \sum_{i=1}^{m} \underline{f}^i(\underline{x})dU_i, \tag{3.4}$$

with

$$\underline{g}^0(\underline{x}) \triangleq \begin{bmatrix} 1 \\ g^0 \end{bmatrix}, \quad \underline{f}^i(\underline{x}) \triangleq \begin{bmatrix} 0 \\ f^i \end{bmatrix}. \tag{3.5}$$

Regarding the outputs, we remark that we need to include a new output that is $h_0(\underline{x}) = t$. Indeed, it is a common (and implicit) assumption that all the system inputs and the outputs are synchronized. For instance, in a real system, the inputs and outputs are measured by sensors. The sensors provide their measurements together with the time when each measurement has occurred. This means that our system is also equipped with an additional sensor that is the clock (i.e., a sensor that measures time). Therefore, a full description of our system is given by the following equations:

$$\begin{cases} d\underline{x} = \underline{g}^0(\underline{x})dt + \sum_{i=1}^{m} \underline{f}^i(\underline{x})dU_i, \\ y = [h_0(\underline{x}), h_1(\underline{x}), \ldots, h_p(\underline{x})]^T. \end{cases} \tag{3.6}$$

Since the choice of a reference frame in the chronospace is arbitrary, the first component of the chronostate is not, in general, the time. This is the case only for special reference frames (i.e., the ones for which $x^0 = t$). From now on, we say that these special frames are *synchronous*. In particular, a synchronous frame is any frame in the chronospace such that the vector fields that characterize the dynamics have the structure given in (3.5) and

$$h_0(\underline{x}) = x^0 = t.$$

In this book we often use a synchronous frame. On the other hand, we could use other reference frames, which are obtained by a generic coordinates' change in the chronospace. The change of coordinates is given by the following set of transformations:

$$x^0 \to \xi^0 = \xi^0(x^0, x^1, \ldots, x^n), \quad x^1 \to \xi^1 = \xi^1(x^0, x^1, \ldots, x^n), \quad \ldots,$$
$$x^n \to \xi^n = \xi^n(x^0, x^1, \ldots, x^n).$$

We denote the inverse transformations as follows:

$$x^0 = x^0(\xi^0, \xi^1, \ldots, \xi^n), \quad x^1 = x^1(\xi^0, \xi^1, \ldots, \xi^n), \quad \ldots, \quad x^n = x^n(\xi^0, \xi^1, \ldots, \xi^n).$$

We denote the Jacobian of this transformation by

$$\frac{\partial \underline{\xi}}{\partial \underline{x}} = \begin{bmatrix} \frac{\partial \xi^0}{\partial x^0} & \frac{\partial \xi^0}{\partial x^1} & \cdots & \frac{\partial \xi^0}{\partial x^n} \\ \frac{\partial \xi^1}{\partial x^0} & \frac{\partial \xi^1}{\partial x^1} & \cdots & \frac{\partial \xi^1}{\partial x^n} \\ \cdots & \cdots & \cdots & \cdots \\ \frac{\partial \xi^n}{\partial x^0} & \frac{\partial \xi^n}{\partial x^1} & \cdots & \frac{\partial \xi^n}{\partial x^n} \end{bmatrix},$$

which is an $(n+1) \times (n+1)$ matrix. This change transforms the chronostate as follows:

$$\underline{x} \to \underline{\xi} \triangleq \begin{bmatrix} \xi^0 \\ \xi^1 \\ \cdots \\ \xi^n \end{bmatrix}, \qquad (3.7)$$

and its dynamics and output function become

$$\begin{cases} d\underline{\xi} = \underline{\eta}^0(\underline{\xi})dt + \sum_{i=1}^m \underline{\eta}^i(\underline{\xi})dU_i, \\ y = [\theta_0(\underline{\xi}),\, \theta_1(\underline{\xi}), \ldots, \theta_p(\underline{\xi})]^T, \end{cases} \qquad (3.8)$$

where

$$\underline{\eta}^0(\underline{\xi}) \triangleq \frac{\partial \underline{\xi}}{\partial \underline{x}} \underline{g}^0(\underline{x}), \quad \underline{\eta}^i(\underline{\xi}) \triangleq \frac{\partial \underline{\xi}}{\partial \underline{x}} \underline{f}^i(\underline{x}), \quad \theta_0(\underline{\xi}) \triangleq x^0(\xi^0, \xi^1, \ldots, \xi^n), \qquad (3.9)$$

$$\theta_i(\underline{\xi}) \triangleq h_i(x^0(\xi^0, \xi^1, \ldots, \xi^n),\, x^1(\xi^0, \xi^1, \ldots, \xi^n),\, \ldots,\, x^n(\xi^0, \xi^1, \ldots, \xi^n))$$

for $i = 1, \ldots, p$.

In the chronospace, the role of time is exactly the same as the one of any other component of the chronostate with a fundamental exception. Time also drives the dynamics. By setting

$$dU_0 = dt,$$

we can write the dynamics in (3.8) as $d\underline{\xi} = \underline{\eta}^0(\underline{\xi})dU_0 + \sum_{i=1}^m \underline{\eta}^i(\underline{\xi})dU_i$ and, by adopting the Einstein notation, the equations of our system read as follows:

$$\begin{cases} d\underline{\xi} = \underline{\eta}^\alpha(\underline{\xi})dU_\alpha, \\ y = [\theta_0(\underline{\xi}),\, \theta_1(\underline{\xi}), \ldots, \theta_p(\underline{\xi})]^T, \end{cases} \qquad (3.10)$$

where α is a dummy index.

In the next sections, we will see that the index α that appears in (3.10) refers to a group of transformations under which the theory of observability is invariant. If we want to specify the component β ($\beta = 0, 1, \ldots, n$) of one of the vector fields $\underline{\eta}^\alpha$, for a given $\alpha = 0, 1, \ldots, m$, we write $(\eta^\alpha)^\beta$.

We conclude this section with the following three remarks:

1. The introduction of the chronospace is required when the dynamics of the state are characterized by a nonzero drift (i.e., the term g^0 in (3.1) has at least one nonzero component). Only in this case does the role of time as a system input arise, and it is important to account for this aspect in the input transformations' group.

2. The chronospace is a very powerful framework that allows us to obtain a complete theory of observability. In particular, in this book, it is exploited to extend the observability rank condition to two new classes of systems, namely

- nonlinear time-variant systems (section 4.7) and
- systems also driven by inputs that are unknown (Chapter 8). This also includes the case time-variant (section 8.4).

The two above extensions of the observability rank condition represent the solutions of two fundamental open problems in control theory (open for half a century).

3. We do not believe that this framework can be used for a practical estimation of the state. However, we cannot exclude the existence of other contexts where the chronospace is advantageous.

3.2 • Group of invariance in the absence of unknown inputs

We consider an input-output system characterized by the equations given in (1.3) but with a generic number of outputs and the m known inputs $u_1, \ldots, u_m$. In other words, our system is characterized by the following equations:

$$\begin{cases} \dot{x} &= f^0 + f^i u_i, \\ y &= [h_1(x), \ldots, h_p(x)]^T, \end{cases}$$

where the vector fields $f^0, f^1, \ldots, f^m$ and the scalar functions $h_1, \ldots, h_p$ can also depend explicitly on time (as in (3.1)) (note that, when we do not have unknown inputs, we denote the drift by f^0 instead of g^0). As we saw in section 3.1, we can describe the above system in the chronospace as follows:

$$\begin{cases} d\underline{\xi} &= \eta^\alpha(\underline{\xi}) dU_\alpha, \\ y &= [\theta_0(\underline{\xi}), \theta_1(\underline{\xi}), \ldots, \theta_p(\underline{\xi})]^T. \end{cases} \quad (3.11)$$

We assume that all the inputs are known. We discuss separately the output transformation's group and the input transformation's group. These groups have already been introduced in section 1.3.5. Here, we also provide these groups in the chronospace and we discuss further properties.

3.2.1 • Output transformations' group

We consider the following transformation in the space of the outputs:

$$\theta_0 \to \theta_0' = \theta_0'(\theta_0, \theta_1, \ldots, \theta_p),\ \theta_1 \to \theta_1' = \theta_1'(\theta_0, \theta_1, \ldots, \theta_p),\ \ldots,\ \theta_p \to \theta_p' = \theta_p'(\theta_0, \theta_1, \ldots, \theta_p). \quad (3.12)$$

We denote the Jacobian of this transformation by

$$\frac{\partial \theta'}{\partial \theta} = \begin{bmatrix} \frac{\partial \theta_0'}{\partial \theta_0} & \frac{\partial \theta_0'}{\partial \theta_1} & \cdots & \frac{\partial \theta_0'}{\partial \theta_p} \\ \frac{\partial \theta_1'}{\partial \theta_0} & \frac{\partial \theta_1'}{\partial \theta_1} & \cdots & \frac{\partial \theta_1'}{\partial \theta_p} \\ \cdots & \cdots & \cdots & \cdots \\ \frac{\partial \theta_p'}{\partial \theta_0} & \frac{\partial \theta_p'}{\partial \theta_1} & \cdots & \frac{\partial \theta_p'}{\partial \theta_p} \end{bmatrix}.$$

From the inverse function theorem, we know that if the above Jacobian is nonsingular at a given $\underline{\xi}_0 = [\xi_0^0, \xi_0^1, \ldots, \xi_0^n]^T$, there exists an open ball of $\underline{\xi}_0$ where the transformation in (3.12) can be inverted. This basically means that if we a priori know that $\underline{\xi}_0$ belongs to the above open ball, the knowledge of the functions $\theta_0, \theta_1, \ldots, \theta_p$ at $\underline{\xi}_0$ is equivalent to the knowledge of $\theta'_0, \theta'_1, \ldots, \theta'_p$ at $\underline{\xi}_0$. In Chapter 4, we introduce the concept of *weak* observability (Definition 4.2), which is weaker than observability. Roughly speaking, if a system is weakly observable at $\underline{\xi}_0$, it means that the knowledge of the inputs and outputs during a given interval of time allows us to obtain the initial state ($\underline{\xi}_0$), provided that we also know that the initial state belongs to an a priori known open ball. In contrast, if the system is observable at $\underline{\xi}_0$, we can reconstruct the initial state without the need of further knowledge than the inputs and the outputs. Hence, the group of transformations described by (3.12) certainly belongs to the group of invariance of the weak observability. In the case of observability, we need to restrict the class of transformations. Specifically, we only consider linear output changes:

$$\theta_\alpha \to \theta'_\alpha = A^\beta_\alpha \theta_\beta, \tag{3.13}$$

where A is a nonsingular $(p+1) \times (p+1)$ matrix. Note that, while the group of transformations in (3.12) is infinite-dimensional, the one in (3.13) is a $(p+1)^2$- Lie group. Specifically, it is $GL(p+1, \mathbb{R})$. Even if in this book we mainly deal with the concept of weak observability, we refer to the group provided in (3.13). Clearly, a complete theory of weak observability should refer to the group given in (3.12). However, as it will be seen, referring to the group in (3.13) is easier and general enough to allow us to derive the general criterion to check the weak observability.

3.2.2 ▪ Input transformations' group

We already introduced this group in section 1.3.5. Basically, we can operate a change in the inputs without altering the information on the initial state. Specifically, in section 1.3.5 we characterized this group by (1.10)–(1.11). In the new framework introduced in section 3.1 and for systems characterized by a nonzero drift, (1.10) becomes

$$dU_\alpha \to dU'_\alpha = M^\beta_\alpha dU_\beta, \tag{3.14}$$

where M is any nonsingular $(m+1) \times (m+1)$ known matrix. Equation (1.11) becomes

$$\underline{\eta}^\alpha \to \underline{\eta}'^\alpha = N^\alpha_\gamma \underline{\eta}^\gamma, \tag{3.15}$$

with

$$N^\alpha_\gamma M^\beta_\alpha = \delta^\beta_\gamma.$$

The above condition guarantees that $d\underline{\xi}$ is a scalar with respect to this new group of transformations:

$$d\underline{\xi} = \underline{\eta}^\alpha dU_\alpha \to \underline{\eta}'^\alpha dU'_\alpha = N^\alpha_\gamma \underline{\eta}^\gamma M^\beta_\alpha dU_\beta = d\underline{\xi}.$$

3.2.3 ▪ The Abelian case

We study the conditions under which the two groups of transformations defined in sections 3.2.1 and 3.2.2 become Abelian. This analysis will provide the structure of simplified systems for which the derivation of the observability properties significantly simplifies even if remaining representative of the general case.

Output transformations' group

In the chronospace, this group is defined by the transformations given in (3.13). Hence, it is the group $GL(p+1, \mathbb{R})$, which is Abelian for $p = 0$. On the other hand, from the observability perspective, the case when $p = 0$ does not have interest. Indeed, in this case, no component of the state is observable. We remark that the output transformations' group coincides with $GL(p, \mathbb{R})$ if we work in the usual space of states, and not in the chronospace. This group is Abelian for $p = 1$, i.e., when the output consists of a single scalar function. In accordance with our remarks at the end of section 3.1, this means that the simplified system is obtained by considering a single output and driftless dynamics.

Input transformations' group

In the chronospace, this group is defined by the transformations given in (3.14) and (3.15). This is the group $GL(m + 1, \mathbb{R})$. It becomes Abelian in the following two cases:

1. f^0 vanishes and $m = 1$.

2. Nonvanishing f^0 and $m = 0$.

3.2.4 ▪ Simplified systems in the absence of unknown inputs

We conclude this section by providing the class of simplified systems in the case when all the inputs are known. They are the following two:

1. Null f^0, $m = 1$ and $p = 1$.

2. Nonzero f^0, $m = 0$ and $p = 1$.

Both these systems are characterized by an Abelian input transformations' group. Additionally, also the output transformations' group is Abelian in the usual space of states. In Chapter 4, we start our study by first analyzing these systems. The general case can be dealt with by following the same basic steps with the same conceptual complexity.

3.3 ▪ Group of invariance in the presence of unknown inputs

Now we consider systems whose dynamics are also driven by unknown inputs. We distinguish the case when the dynamics are driftless from the case when the dynamics are also driven by a nonzero drift. In the former case, we can simply work in the space of the states, while in the latter case, we need to use the chronospace. In the rest of this section, we assume that the number of outputs coincides with the number of unknown inputs. In other words, we have $p = m_w$ outputs.[13]

---

[13]Behind this hypothesis there is actually a much deeper theoretical property that allows us to set $p = m_w$. This is the *system canonization*, which is the analytic procedure provided in Appendix C. At this stage, we can only say that this procedure provides the true number of unknown inputs, which cannot exceed m_w. In particular, if this number is strictly smaller than m_w, some of the original unknown inputs are spurious and we can write the equations of our system in terms of a reduced number of unknown inputs. The new unknown inputs can be expressed in terms of the original unknown inputs together with their time derivatives up to a given order. The canonization also provides a set of scalar functions, starting from the $p(\geq 1)$ outputs of the system. When the canonization is completed, the number of these scalar functions coincides with the number of the new unknown inputs. The canonization returns a new system, characterized by a given number of unknown inputs (which cannot exceed m_w) and by new outputs that include the above scalar functions. The new system has exactly the same observability properties of the original system. If the number of outputs of this new system exceeds the number of the above scalar functions, the exceeding output functions do not intervene in the $\mathcal{SUIO}$ and, in this section, they will be ignored. Hence, we are allowed to assume that the number of outputs coincides with the number of unknown inputs.

3.3.1 • The $\mathcal{SUIO}$ for driftless systems

A general characterization of the dynamics, in terms of differentials, is given by the following equation:

$$dx = \sum_{k=1}^{m_u} f^k(x)dU_k + \sum_{j=1}^{m_w} g^j(x)dW_j,$$

where m_u is the number of inputs which are known and m_w the number of inputs which are unknown.

The output transformations' group remains the same as in the case without unknown inputs, and it is defined by (3.13), where A is now a $p \times p$ nonsingular matrix and the indices that appear in (3.13) are Latin. The input transormations' group is defined by (3.14) and (3.15), with $\alpha = k = 1, \ldots, m_u, m_u + 1, \ldots, m_u + m_w$ and

$$dU_k = \begin{cases} dU_k, & 1 \leq k \leq m_u, \\ dW_{k-m_u}, & m_u + 1 \leq k \leq m_u + m_w. \end{cases}$$

On the other hand, as mentioned at the beginning of this chapter, to deal with the case with unknown inputs we found it very convenient to introduce a new group of transformations that merges together these groups, and, concerning the inputs, it only includes the unknown inputs and the ordinary time. This is the $\mathcal{SUIO}$.

Since we are setting $p = m_w$, the description of our system is given by the following equations:

$$\begin{cases} dx &= \sum_{k=1}^{m_u} f^k(x)dU_k + \sum_{j=1}^{m_w} g^j(x)dW_j, \\ y &= [h_1(x), \ldots, h_{m_w}(x)]^T. \end{cases} \quad (3.16)$$

The $\mathcal{SUIO}$ is defined by the following three equations:

$$\begin{cases} h_j \to h'_j &= S_j^k h_k, \\ dW_j \to dW'_j &= S_j^k dW_k, \\ g^j \to g'^j &= (S^{-1})_k^j g^k, \end{cases} \quad (3.17)$$

where S is a nonsingular $m_w \times m_w$ matrix and all the Latin indices can take the values $1, \ldots, m_w$. The index k is a dummy index in all three above equations. In accordance with the Einstein notation, this implies the sum $\sum_{k=1}^{m_w}$.

3.3.2 • The $\mathcal{SUIO}$ in the presence of a drift

In the presence of a drift we must work in the chronospace to account for the input role of time. Specifically, the presence of a drift makes the differential dt a new input. Note that dt behaves as a known input because it is known, but it behaves as an unknown input because it cannot be assigned. We found that it is fundamental to include dt in the $\mathcal{SUIO}$. Hence, we set $dW_0 = dt$. A general characterization of the dynamics, in terms of differentials, and in general coordinates, is given by the following equation:

$$d\underline{\xi} = \underline{\zeta}^0(\underline{\xi})dW_0 + \sum_{k=1}^{m_u} \underline{\eta}^k(\underline{\xi})dU_k + \sum_{j=1}^{m_w} \underline{\zeta}^j(\underline{\xi})dW_j,$$

where $\underline{\xi}$ is the chronostate in general coordinates (i.e., ξ^0 is not necessarily the chronological time).

3.3. Group of invariance in the presence of unknown inputs

We proceed as in section 3.3.1. We assume that we have $p = m_w$ outputs. Hence, the description of our system, in the chronospace, is given by the following equations:

$$\begin{cases} d\underline{\xi} = \underline{\zeta}^0(\underline{\xi})dW_0 + \sum_{k=1}^{m_u} \underline{\eta}^k(\underline{\xi})dU_k + \sum_{j=1}^{m_w} \underline{\zeta}^j(\underline{\xi})dW_j, \\ y = [\theta_0(\underline{\xi}),\ \theta_1(\underline{\xi}), \ldots, \theta_{m_w}(\underline{\xi})]^T. \end{cases} \quad (3.18)$$

The $\mathcal{SUIO}$ is now defined by the following three equations:

$$\begin{cases} \theta_\alpha \to \theta'_\alpha = S_\alpha^\beta \theta_\beta, \\ dW_\alpha \to dW'_\alpha = S_\alpha^\beta dW_\beta, \\ \underline{\zeta}^\alpha \to \underline{\zeta}'^\alpha = (S^{-1})_\beta^\alpha \underline{\zeta}^\beta, \end{cases} \quad (3.19)$$

where S is a nonsingular $(m_w + 1) \times (m_w + 1)$ matrix and all the Greek indices can take the values $0, 1, \ldots, m_w$. The index β is a dummy index in all three above equations. In accordance with the Einstein notation, this implies the sum $\sum_{\beta=0}^{m_w}$.

In the next section we study the conditions under which the $\mathcal{SUIO}$ becomes Abelian, both in the driftless case and in the presence of a drift.

3.3.3 ▪ The Abelian case

Driftless case

The $\mathcal{SUIO}$ is characterized by the matrix S that appears in (3.17). Hence, this group coincides with $GL(m_w, \mathbb{R})$ and it is an m_w^2-Lie group. It is Abelian when $m_w = 1$.

The case with a drift

The $\mathcal{SUIO}$ is characterized by the matrix S that appears in (3.19). Hence, this group coincides with $GL(m_w + 1, \mathbb{R})$ and it is an $(m_w + 1)^2$-Lie group. It is Abelian when $m_w = 0$. On the other hand, this case is characterized by a dynamics without unknown inputs and, consequently, has already been discussed in section 3.2.

3.3.4 ▪ Simplified systems in the presence of unknown inputs

We conclude this section by providing the class of simplified systems in the presence of unknown inputs. From the previous analysis we conclude that we have a single case that is the one characterized by a null drift, a single unknown input ($m_w = 1$), and a single output.

Chapter 4
Theory of Nonlinear Observability in the Absence of Unknown Inputs

The goal of this chapter is to provide basic results about state observability. In particular, one of the most important results is the analytic test to automatically check whether the state that characterizes a given system is observable or not. In the case when the dynamics are not driven by unknown inputs (that is, the case studied in the present chapter), this automatic test was introduced in the 1970s [19]: it is the *observability rank condition*.[14]

In this chapter we refer to input-output systems characterized by a state $x \in \mathcal{M}$, where $\mathcal{M}$ is a differentiable manifold of dimension n. The equations that characterize our systems are the following:

$$\begin{cases} \dot{x} = f^0(x) + \sum_{i=1}^m f^i(x) u_i, \\ y = [h_1(x), \ldots, h_p(x)]^T, \end{cases} \quad (4.1)$$

where $u_1, \ldots, u_m$ are the system inputs and $f^0(x), f^1(x), \ldots, f^m(x)$ are smooth vector fields.

In Chapter 3, by studying systems defined by (4.1), we obtained two simpler systems that we called simplified systems (see section 3.2.4). They were obtained by requiring that the groups of invariance of observability become Abelian groups (section 3.2). They are special cases of the system in (4.1). Specifically, the first system is obtained by setting $m = p = 1$ and a zero drift (i.e., a zero f^0), while the second system is obtained by setting $m = 0$ and $p = 1$. In this chapter, in addition to the general case defined by (4.1), we refer to the first system,[15] and we refer to it as the *simplified system*. It is characterized by the following equations:

$$\begin{cases} \dot{x} = f(x) u, \\ y = h(x). \end{cases} \quad (4.2)$$

We will see that the derivation of the observability rank condition is much simpler for systems characterized by (4.2) than for systems characterized by (4.1). On the other hand, the derivation in the general case follows exactly the same steps and it is conceptually equivalent. In section 4.2 we provide the derivation of the observability rank condition for the simplified system. Then, in section 4.3 we deal with the general case. We actually refer to the case $p = 1$. The extension to the multiple outputs is very trivial, and it will be done at the end.

Observability is a structural property of the system, defined as the possibility of inferring the state x from the knowledge of the inputs and the outputs during a given time interval. We

---

[14] The first criterion only accounted for linear systems [25, 26], and later, at the beginning of the 1970s, some criteria for observability in the nonlinear case were also discovered [14, 28, 42].

[15] We could choose also the second system, whose complexity is comparable.

emphasize the strong link between observability and a very well known problem in differential calculus, the *inverse function problem*, and, as a result, the strong link between the observability rank condition and the *inverse function theorem*. More generally, we wish to emphasize the following fundamental aspect. Independently of the manner that it is used to introduce the concept of observability, the derivation of the analytic criterion to check whether a system is observable or not (i.e., the observability rank condition) is based on two fundamental results from differential calculus:

1. the Taylor theorem and

2. the inverse function theorem.

Our approach makes manifest the above feature, in particular by dealing with the simplified system.

In section 4.5, we also introduce the concept of indistinguishability and we discuss its link with the observability.

Finally, in section 4.7, we provide the extension of the observability rank condition to nonlinear time-variant systems.

4.1 • Theory based on a constructive approach

We consider a given time interval that we denote by $\mathcal{I}$. Without loss of generality, we set $\mathcal{I} = [0, T]$. We also denote the initial state by x_0 (i.e., $x(0) = x_0$). We define observability as follows.

Definition 4.1 (State Observability). *An input-output system is observable at a given $x_0 \in \mathcal{M}$ if there exists at least one choice of inputs $u_1(t), \ldots, u_m(t)$ such that x_0 can be obtained from the knowledge of the output $y(t)$ and the inputs $u_1(t), \ldots, u_m(t)$ on the time interval $\mathcal{I}$. In addition, the system is observable on a given set $\mathcal{U} \subseteq \mathcal{M}$, if it is observable at any $x \in \mathcal{U}$.*

In accordance with this definition, if a system is observable on a given $\mathcal{U} \subseteq \mathcal{M}$, it means that there exists the possibility of reconstructing the initial state, by only knowing the system inputs and outputs on a given time interval (provided that we set the inputs in a suitable manner). We do not need further knowledge than the system inputs and output on a given time interval.

We want to define a less strong property. Basically, we want to account for possible further information, which consists of the knowledge that the initial state belongs to a given open subset of $\mathcal{M}$.

Definition 4.2 (State Weak Observability). *An input-output system is weakly observable at a given $x_0 \in \mathcal{M}$ if there exists an open set B of x_0 such that, by knowing that $x_0 \in B$, there exists at least one choice of inputs $u_1(t), \ldots, u_m(t)$ such that x_0 can be obtained from the knowledge of the output $y(t)$ and the inputs $u_1(t), \ldots, u_m(t)$ on the time interval $\mathcal{I}$. In addition, the system is weakly observable on a given set $\mathcal{U} \subseteq \mathcal{M}$ if it is weakly observable at any $x \in \mathcal{U}$.*

4.2 • Observability rank condition for the simplified system

In this section we refer to the simplified system, which is the system characterized by (4.2). We provide the derivation of the observability rank condition by showing its simplicity with respect

4.2. Observability rank condition for the simplified system

to the derivation carried out for the general case in the next section. On the other hand, it is conceptually equivalent.

The system in (4.2) is characterized by the output function $h(x)$. Starting from this function, we can compute its Lie derivative along $f(x)$. In other words, we can compute the following scalar function:

$$L^1(x) \triangleq \mathcal{L}_f h(x).$$

We simply refer to this function as the first-order Lie derivative of the system in (4.2). Starting from it, we can compute the second-order Lie derivative:

$$L^2(x) \triangleq \mathcal{L}_f L^1(x).$$

We can repeat the procedure and compute any order Lie derivative. All are scalar functions of x. Finally, the zero-order Lie derivative is the output function itself: $L^0(x) = h(x)$.

Our goal is to understand whether we can obtain the value of the initial state, i.e., x_0, by exploiting the following knowledge/degrees of freedom:

- knowledge of all the values that the two functions $u(t)$ and $y(t)$ take on the entire time interval $\mathcal{I}$ and

- the possibility of setting any time assignment for the function $u(t)$.

The problem we want to solve seems strange because we want to convert the knowledge of two functions of time into the knowledge of the initial state. First of all, we remark that the knowledge of the function $y(t)$ on $\mathcal{I}$ includes the knowledge of its initial value $y(0)$. On the other hand,

$$y(0) = h(x_0). \tag{4.3}$$

Since $y(0)$ is known, the above equality can be considered as a scalar equation where the unknowns are the components of x_0. In the special case where the dimension of the state is $n = 1$ and the function h can be inverted, we can conclude that the state is observable (in particular, $x_0 = h^{-1}(y(0))$). But this is a very special case.

The idea is to exploit the knowledge of $y(t)$ also for $t > 0$ and the knowledge of $u(t)$ to build additional equations in the same unknown x_0. Then, by using the aforementioned possibility to set the time assignment for $u(t)$, we check whether we can make the resulting equations' system invertible, at least locally.

We start by using the Taylor theorem. In accordance with this theorem, we know that the knowledge of an analytic real function on a given interval $[a, b]$ is equivalent to the knowledge of all its derivatives computed at a. This allows us to convert the knowledge of the two functions $u(t)$ and $y(t)$ on the entire time interval $\mathcal{I}$ into the knowledge of an infinite (but countable) set of real coefficients (the time derivatives of the above functions computed at $t = 0$). More quantitatively, for the Taylor theorem we have

$$y(t) = \sum_{k=0}^{\infty} \left.\frac{d^k y}{dt^k}\right|_{t=0} \frac{t^k}{k!}$$

and

$$u(t) = \sum_{k=0}^{\infty} \left.\frac{d^k u}{dt^k}\right|_{t=0} \frac{t^k}{k!}.$$

We can reformulate the observability problem as follows. Determine whether we can obtain the value of the initial state, i.e., x_0, by exploiting the following knowledge/degrees of freedom:

- Knowledge of all the values $\frac{d^k y}{dt^k}\big|_{t=0}$ and $\frac{d^k u}{dt^k}\big|_{t=0}$ for any positive integer $k = 0, 1, \ldots$ and

- the possibility of setting all the values $\frac{d^k u}{dt^k}\big|_{t=0}$.

Let us start by computing $\frac{dy}{dt}$. We have

$$\frac{dy}{dt} = \frac{\partial}{\partial x} h \cdot \dot{x} = \frac{\partial}{\partial x} h \cdot f(x) u = L^1(x) u.$$

By computing the above equality at $t = 0$ we obtain

$$\frac{dy}{dt}\bigg|_{t=0} = L^1(x_0) u(0).$$

By setting $u(0) = u_0 \neq 0$ we obtain

$$\frac{1}{u_0} \frac{dy}{dt}\bigg|_{t=0} = L^1(x_0), \tag{4.4}$$

which is a new equation in x_0, exactly as (4.3) (note that the left-hand-side member is known).

We can repeat further the same procedure:

$$\frac{d^2 y}{dt^2} = \frac{d}{dt}\left(L^1(x) u\right) = \left(\frac{\partial}{\partial x} L^1(x) \cdot f(x) u\right) u + L^1(x) \dot{u} = L^2(x) u^2 + L^1(x) \dot{u}.$$

By computing the above equality at $t = 0$ we obtain

$$\frac{d^2 y}{dt^2}\bigg|_{t=0} = L^2(x_0) u_0^2 + L^1(x) \frac{du}{dt}\bigg|_{t=0}.$$

By setting $\frac{du}{dt}\big|_{t=0} = \dot{u}_0 = 0$ we obtain

$$\frac{1}{u_0^2} \frac{d^2 y}{dt^2}\bigg|_{t=0} = L^2(x_0), \tag{4.5}$$

which is a new equation in x_0.

We set to zero all the time derivatives of u at the initial time, with the exception of the zero-order time derivative. In other words,

$$u(0) = u_0 \neq 0, \quad \frac{d^k u}{dt^k}\bigg|_{t=0} = 0, \quad k = 1, 2, \ldots. \tag{4.6}$$

By repeating the above procedure we obtain

$$\frac{1}{u_0^k} \frac{d^k y}{dt^k}\bigg|_{t=0} = L^k(x_0). \tag{4.7}$$

Therefore, we can easily obtain the value that every Lie derivative takes at the initial state. By denoting this known value with L_0^k, we can build the following system of equations in the unknown x_0:

$$\begin{bmatrix} L^0(x_0) = L_0^0, \\ L^1(x_0) = L_0^1, \\ \ldots \\ L^k(x_0) = L_0^k, \\ \ldots \end{bmatrix} \tag{4.8}$$

4.2. Observability rank condition for the simplified system

This is a system of infinite equations in the unknown x_0. Our goal is to obtain x_0 from the above system. If this is possible, it means that the system is observable, in accordance with the definition of observability given by Definition 4.1. The reverse also holds. Namely, if it is not possible to obtain x_0 from the equations in (4.8), it means that the system is not observable. This result will be directly proved for the general case, i.e., for systems defined by (4.1). In particular, in the first part of section 4.3 (Theorem 4.3), we prove that the knowledge of x_0 that we can gather from $u(t)$ and $y(t)$ is enclosed in the knowledge of the values that all the Lie derivatives of the system take at x_0. In other words, the knowledge of the values that all the Lie derivatives of the system take at x_0 is an upper bound on the knowledge of x_0. By setting the input as in (4.6), we showed that we can attain this upper bound. Hence, we do not have further information about x_0 in addition to the one provided by the system in (4.8). If we cannot obtain x_0 from this system, it means that we cannot obtain x_0.

Therefore, by only using Taylor's theorem, we have converted the problem of observability into a very well known problem in differential calculus, which has been intensely investigated by many mathematicians. It is the *inverse function problem*. It consists in finding general conditions under which a given function can be inverted. In particular, our problem is to retrieve x_0 from the equations in (4.8). The main result for this problem is the *inverse function theorem*,[16] which provides a sufficient condition for a function to be invertible in a neighborhood of a point in its domain. Based on this theorem, we provide the criterion to check the state observability, i.e., the observability rank condition. Hence, the observability rank condition is directly obtained from the inverse function theorem and provides a sufficient condition for the weak observability. On the other hand, even if there is a vast literature on the inverse function problem, obtaining necessary conditions for a function to be invertible is much more complex and there is not a simple and general result. This also holds for observability. Obtaining necessary conditions is much more difficult. However, as we will see, the observability rank condition is *almost* a necessary condition.

First of all, we do not need to consider all the (infinite) equations in (4.8). By running Algorithm 2.2 with $d = k = 1$, $f_1(x) = f(x)$, and $h_1(x) = h(x)$, we can compute the smallest codistribution that contains $\frac{\partial}{\partial x} h$ and that is invariant under $f(x)$. This codistribution, and more importantly its generators, can be easily built by running Algorithm 2.2. This algorithm, for the system defined in (4.2), is Algorithm 4.1 below (for clarity's sake, we add an index on the codistribution computed at each step).

ALGORITHM 4.1. Observable codistribution for the simplified system in the absence of unknown inputs.

Set $k = 0$
Set $\Omega_k = \text{span}\{\nabla h\}$
Set $k = k + 1$
Set $\Omega_k = \Omega_{k-1} \oplus \mathcal{L}_f \Omega_{k-1}$
while $\dim(\Omega_k) > \dim(\Omega_{k-1})$ **do**
    $k = k + 1$
    Set $\Omega_k = \Omega_{k-1} \oplus \mathcal{L}_f \Omega_{k-1}$
end while
Set $\Omega = \Omega_k$ and $s = \dim(\Omega)$

---

[16] This theorem, together with the *implicit function theorem*, can be seen as a special case of the constant rank theorem, which states that a smooth map with constant rank near a point can be put in a particular normal form near that point.

where we denoted by ∇ the operator $\frac{\partial}{\partial x}$, when applied to a scalar function of the state x. By construction, the algorithm converges when the dimension of Ω_k remains the same after one step (and this occurs in at most $n-1$ steps). Let us suppose that the dimension of the convergent codistribution Ω is s on an open neighborhood of x_0 (in other words, we assume that x_0 is a regular point for Ω). We know that a possible choice of the generators of Ω is given by the Lie derivatives of h along f up to the $s-1$ order, i.e., $L^0 = h$, L^1, L^2, ..., L^{s-1}. From Corollary 2.5 we know that the value that any Lie derivative takes at x_0 can be obtained from the values that the generators L^0, L^1, L^2, ..., L^{s-1} take at x_0. Hence, it suffices to consider the following system, which consists of s equations:

$$\begin{bmatrix} L^0(x_0) & = L_0^0, \\ L^1(x_0) & = L_0^1, \\ \cdots \\ L^{s-1}(x_0) & = L_0^{s-1}. \end{bmatrix} \tag{4.9}$$

When the inverse function theorem is applied to our specific problem, namely to the system of equations in (4.9), we obtain that if $s = n$, there exists a neighborhood of x_0 where we can invert the equations in (4.9) to express x_0 in terms of L_0^0, L_0^1, ..., L_0^{s-1} (which are known). In other words, we obtain that *the condition $s = n$ is a sufficient condition for the weak observability of the state*, where the definition of *weak observability* is Definition 4.2.

On the other hand, to determine x_0 from the system in (4.9), we need that this system consists of at least n equations. Therefore, the condition $s = n$ is also a necessary condition for the weak observability. Note that this is true provided that x_0 is a regular point for Ω. In particular, the result stated by Corollary 2.5 (used above) also uses the inverse function theorem and, for this reason, it requires that x_0 is a regular point.

The analysis of the case when x_0 is not a regular point for Ω can be carried out by using the results in the literature on the inverse function problem, specifically the ones that provide necessary conditions for a function to be invertible. By doing this, we basically obtain that if the state is weakly observable at x_0, then there exists a neighborhood B_0 of x_0 such that the subset $B \subset B_0$ of points where the dimension of Ω is smaller than n has Lebesgue measure equal to zero. We do not provide the derivation of this result. We only remark that, intuitively, it is easy to understand that, when x_0 is not regular, if the dimension of Ω at x_0 is smaller than n, we cannot conclude that the system is not weakly observable. Indeed, since x_0 is not regular, there are points which are as close as we want to x_0 where the dimension of Ω is different from its dimension at x_0.

The condition $s = n$ is precisely the observability rank condition. It is a sufficient condition for the weak observability and almost a necessary condition as explained above. When $s < n$ in a given open set of $\mathcal{M}$ we can conclude that the state is not weakly observable in this open set. In this case, the structure of the observable codistribution provides very useful information about the observability properties of our system. We discuss this aspect directly for the general case (see sections 4.4 and 4.6).

Examples

The application of the observability rank condition is simple and automatic. It consists of the computation of the observable codistribution Ω by running Algorithm 4.1 and by computing its dimension at each step. We provide two examples. Further examples will be provided in sections 4.3.3 and 4.4 and in Chapter 5.

4.2. Observability rank condition for the simplified system

First example

Let us consider a system characterized by the state $x = [x_1, x_2]^T$ whose dynamics are driven by a single input u, according to the following equations:

$$\begin{cases} \dot{x}_1 = x_2 u, \\ \dot{x}_2 = x_1 x_2 u. \end{cases}$$

Additionally, the system has the following output:

$$y = \frac{x_1^2}{2}.$$

This system is a special case of (4.2), where

$$f(x) = \begin{bmatrix} x_2 \\ x_1 x_2 \end{bmatrix}$$

and

$$h(x) = \frac{x_1^2}{2}.$$

The state is defined on $\mathcal{M} = \mathbb{R}^2$. We apply the observability rank condition to check whether and where the system is weakly observable. We must compute the observable codistribution by running Algorithm 4.1. We obtain, at the initialization,

$$\Omega_0 = \text{span}\{\nabla h\} = \text{span}\{[x_1, 0]\}.$$

Its dimension is 1, with the exception of the points for which $x_1 = 0$ (the set of all these points has zero measure in $\mathbb{R}^2$). Then, we need to compute $\mathcal{L}_f \Omega_0$. It suffices to compute the Lie derivative of its generator, i.e., $\mathcal{L}_f \nabla h = \nabla \mathcal{L}_f h$ (where the last equality follows from (2.19)). We obtain

$$\mathcal{L}_f h = \nabla h \cdot f = [x_1, 0] \cdot \begin{bmatrix} x_2 \\ x_1 x_2 \end{bmatrix} = x_1 x_2.$$

Hence, after the first step of Algorithm 4.1 we obtain

$$\Omega_1 = \text{span}\{[x_1, 0], [x_2, x_1]\}.$$

We compute its dimension by computing the determinant of the matrix:

$$\begin{bmatrix} x_1 & 0 \\ x_2 & x_1 \end{bmatrix}.$$

We obtain x_1^2. Again, the set of points for which $x_1^2 = 0$ has zero measure in $\mathbb{R}^2$. Hence, the dimension of Ω_1 is equal to the dimension of the state everywhere ($n = 2$), with the exception of a subset with zero measure. We conclude that the state is weakly observable everywhere in $\mathbb{R}^2$.

Second example

Let us consider the same system but with a different vector $f(x)$, specifically

$$f(x) = \begin{bmatrix} x_1 \\ x_1 x_2 \end{bmatrix}.$$

At the initialization we clearly obtain the same codistribution $\Omega_0 = \text{span}\{\nabla h\} = \text{span}\{[x_1,\ 0]\}$. On the other hand, this time,

$$\mathcal{L}_f h = \nabla h \cdot f = [x_1,\ 0] \cdot \begin{bmatrix} x_1 \\ x_1 x_2 \end{bmatrix} = x_1^2.$$

Hence, after the first step of Algorithm 4.1 we still have

$$\Omega_1 = \text{span}\{[x_1,\ 0]\}.$$

This means that the algorithm has converged and the dimension of the observable codistribution is $1 < 2$. This means that the state is not weakly observable everywhere in $\mathbb{R}^2$.

4.3 • Observability rank condition in the general case

In this section we refer to the system defined by (4.1). The derivation of the observability rank condition is conceptually equivalent to the one for the simplified system. We start by considering the case of a single output (section 4.3.1). The extension to multiple outputs is trivial and will be done in section 4.3.2.

4.3.1 • Analytic derivation

We start our derivation by considering a kind of inverse problem other than observability. We assume that we know the value of x_0. It is immediate to realize that we can determine the output $y(t) \ \forall t \in \mathcal{I}$ for any choice of the inputs $u_1(t), \ldots, u_m(t)$. Indeed, $y(t) = h(x(t))$ and we can determine $x(t)$ by a direct integration of (4.1), from the initial time $t = 0$ up to any $t \in \mathcal{I}$. We can express this fact by the following equation:

$$y(t) = F([u_1, \ldots, u_m], t, x_0), \tag{4.10}$$

where we adopted the square brackets for $u_1(t), \ldots, u_m(t)$ because the function F depends on them as functions of the entire time interval $[0, t]$ (i.e., F is a functional with respect to these quantities). In what follows, we will refer to this expression as the *input-output map*.

Let us go back to our observability problem, i.e., the problem of determining x_0 from the system inputs and outputs in a given time interval. We can look at (4.10) as a set of many (infinite) equations in x_0, one for each value of the time t. We can reformulate the definition of observability as follows. The considered system is observable if there exists at least one choice of inputs $u_1(t), \ldots, u_m(t)$ such that the input-output map can be inverted with respect to the last argument (x_0). Note that we do not need to explicitly invert this map but to answer whether the inversion is possible or not. This reformulation of observability highlights again its close relation with the inverse function problem, exactly as we have shown in section 4.2, where we dealt with the simplified system. Also in this case, the observability rank condition is directly obtained from the inverse function theorem and provides a sufficient condition for the weak observability. Obtaining necessary conditions is much more difficult. However, exactly as for the simplified system, the observability rank condition is *almost* a necessary condition.

Unfortunately, the analytic expression of F cannot be obtained in general. On the other hand, we can characterize more quantitatively in which manner the function F depends on x_0 and, as we will see, this will be enough to give a complete answer to our problem.

The system defined by (4.1) with $p = 1$ is characterized by the function $h(x) = h_1(x)$. Starting from this function, we can compute its Lie derivative along one of the $m + 1$ vector

4.3. Observability rank condition in the general case

fields $f^0(x), f^1(x), \ldots, f^m(x)$. In other words, we can compute the following $m+1$ scalar functions:

$$L_0^1(x) \triangleq \mathcal{L}_{f^0} h(x), \quad L_1^1(x) \triangleq \mathcal{L}_{f^1} h(x), \quad \cdots \quad L_m^1(x) \triangleq \mathcal{L}_{f^m} h(x),$$

namely

$$L_\alpha^1(x) \triangleq \mathcal{L}_{f^\alpha} h(x), \quad \alpha = 0, 1, \ldots, m.$$

These are the $m+1$ first-order Lie derivatives of the output along the vector fields $f^0(x)$, $f^1(x), \ldots, f^m(x)$. We simply refer to them as the $m+1$ first-order Lie derivatives of the system in (4.1). Starting from them, we can compute the second-order Lie derivatives. Since the Lie derivative operator is noncommutative, in general

$$L_{\alpha\beta}^2 \triangleq \mathcal{L}_{f^\beta} \mathcal{L}_{f^\alpha} h \neq L_{\beta\alpha}^2 \triangleq \mathcal{L}_{f^\alpha} \mathcal{L}_{f^\beta} h, \quad \alpha \neq \beta.$$

As a result, we have in general $(m+1)^2$ different second-order Lie derivatives.

We can repeat the procedure and compute any order Lie derivative. In particular, we have $(m+1)^k$ kth-order Lie derivatives. Finally, the zero-order Lie derivative is the output function itself, i.e., $L^0(x) = h(x)$.

We have the following result.

Theorem 4.3. *In the input-output map, the dependence on x_0 only takes place via the Lie derivatives of the system (up to any order) computed at x_0. In other words,*

$$y(t) = F([u_1, \ldots, u_m], t, L^0(x_0), L_0^1(x_0), \ldots, L_{\alpha_1 \cdots \alpha_j}^j(x_0), \ldots), \tag{4.11}$$

where $\alpha_1, \ldots, \alpha_j = 0, 1, \ldots, m$.

Proof. From the Taylor theorem we have

$$y(t) = \sum_{k=0}^{\infty} \left. \frac{d^k y}{dt^k} \right|_{t=0} \frac{t^k}{k!}, \tag{4.12}$$

$$u_i(t) = \sum_{k=0}^{\infty} u_i^{(k)}(0) \frac{t^k}{k!} \quad (i = 1, \ldots, m),$$

where $u_i^{(k)}(\tau) = \left. \frac{d^k u_i}{dt^k} \right|_{t=\tau}$.

We prove that the kth-order time derivative of the output (i.e., $\frac{d^k y}{dt^k}$) only depends on the time derivatives of the inputs up to the $(k-1)$th order (i.e., $u_1^{(0)}, \ldots, u_m^{(k-1)}$) and the Lie derivatives of the system up to the kth order. In other words, we prove the following equality:

$$\frac{d^k y}{dt^k} = F_k(u_1^{(0)}, \ldots, u_m^{(k-1)}, L^0, \ldots, L_{\alpha_1 \cdots \alpha_k}^k). \tag{4.13}$$

Once we have proved this, by computing the previous equality at $t=0$ and by substituting in (4.12) we obtain the validity of (4.11).

To prove (4.13) we proceed by induction. Obviously, the equality holds for $k=0$, since $y(t) = h(x(t))$, which is the zero-order Lie derivative of the system (i.e., L^0). Let us assume that the equality holds for $k \geq 0$. We compute the $(k+1)$th-order time derivative. We obtain

$$\frac{d^{k+1} y}{dt^{k+1}} = \frac{d}{dt} \frac{d^k y}{dt^k} = \sum_{i=1}^{m} \sum_{j=0}^{k-1} \frac{\partial F_k}{\partial u_i^{(j)}} u_i^{(j+1)} + \sum_{j=0}^{k} \sum_{\alpha_1, \ldots, \alpha_j = 0}^{m} \frac{\partial F_k}{\partial L_{\alpha_1 \cdots \alpha_j}^j} \frac{d}{dt} L_{\alpha_1 \cdots \alpha_j}^j.$$

On the other hand,

$$\frac{d}{dt}L^j_{\alpha_1\cdots\alpha_j} = \frac{\partial L^j_{\alpha_1\cdots\alpha_j}}{\partial x}\dot{x} = \frac{\partial L^j_{\alpha_1\cdots\alpha_j}}{\partial x}\left(f^0(x) + \sum_{i=1}^m f^i(x)u_i\right),$$

which provides the $(j+1)$th-order Lie derivatives of the system multiplied by the inputs (i.e., multiplied by the zero-order time derivatives of the inputs). As a result, we obtain

$$\frac{d^{k+1}y}{dt^k} = F_{k+1}(u_1^{(0)},\ldots,u_m^{(k)},L^0,\ldots,L^{k+1}_{\alpha_1\cdots\alpha_{k+1}})$$

and the theorem is proved. ◂

The result stated by Theorem 4.3 provides an upper bound on the knowledge that we can gather on the initial state x_0. In particular, if there exists a state $x \neq x_0$ such that all the Lie derivatives take at x the same values that they take at x_0, we cannot establish whether the true initial state was x or x_0.

In general, we do not need to consider all the Lie derivatives of our system. In particular, we can easily obtain a more important result than the one stated by Theorem 4.3.

By running Algorithm 2.2 we can compute the smallest codistribution that contains ∇h and that is invariant under $f^0(x), f^1(x),\ldots,f^m(x)$. Let us denote by Ω this codistribution.

The more important result holds for the very common case when x_0 is a regular point for Ω, namely when the dimension of Ω is equal to s in a given neighborhood of x_0. Let us denote by $l_1(x),\ldots,l_s(x)$ the generators of Ω. From Theorem 4.3 we immediately obtain the following fundamental result.

Corollary 4.4. *If x_0 is a regular point for Ω, in the input-output map, the dependence on x_0 only takes place via the scalars $l_1,\ldots,l_s$ computed at x_0. In other words,*

$$y(t) = F([u_1,\ldots,u_m],t,l_1(x_0),\ldots,l_s(x_0)). \tag{4.14}$$

Proof. From Theorem 4.3 we know that in the input-output map, the dependence on x_0 only takes place via the Lie derivatives of the system (up to any order) computed at x_0. On the other hand, if x_0 is a regular point for Ω, we know, from Corollary 2.5, that the value that any Lie derivative takes at x_0 can be obtained from the values that the generators $l_1,\ldots,l_s$ take at x_0. Hence, (4.14) holds. ◂

The result stated by Corollary 4.4 provides an upper bound on the knowledge that we can gather on the initial state x_0. In particular, if there exists a state $x \neq x_0$ such that $l_1,\ldots,l_s$ take at x the same values that they take at x_0, we cannot establish whether the true initial state was x or x_0.

Our next question is to understand whether, by a suitable choice of the inputs, we can achieve this upper bound. In other words, we want to understand whether we can acquire the values that the functions $l_1(x),\ldots,l_s(x)$ take at $x = x_0$. The answer is positive. Proving this fact is immediate for the simplified system (see section 4.2). Proving the same in the general case is more complex, even if equivalent from a conceptual point of view. The difficulty is to select suitable system inputs. For the simplified system, this selection is trivial (one possibility is the one given in (4.6)). For the general case, the selection is very complex and it is based on some

4.3. Observability rank condition in the general case

results from number theory (i.e., the results stated by Lemmas A.1 and A.2, which are used in the proof of Theorem 4.5).

We have the following result.

Theorem 4.5. *There exists a suitable choice of the inputs such that we can obtain the value that all the generators take at x_0 (i.e., $l_1(x_0), \ldots, l_s(x_0)$) from the knowledge of the output in $\mathcal{I}$.*

Proof. The proof is given in Appendix A. ◂

The results stated by Corollary 4.4 and Theorem 4.5 tell us that the information on x_0 that we gather from the output during any time interval is equivalent to the knowledge of the values that the scalar functions $l_1, \ldots, l_s$ take at x_0. These functions are the generators of the smallest codistribution that contains ∇h, and it is invariant under the vector fields $f^0, f^1, \ldots, f^m$. Note that we require that x_0 is a regular point for this codistribution, i.e., that it exists a neighborhood of x_0 where the codistribution has dimension s. This codistribution, and more importantly its generators, can be easily built by running Algorithm 2.2. This algorithm, for the system defined in (4.1), is Algorithm 4.2 below.

ALGORITHM 4.2. Observable codistribution in the absence of unknown inputs.

Set $k = 0$
Set $\Omega_k = \text{span}\{\nabla h\}$
Set $k = k + 1$
Set $\Omega_k = \Omega_{k-1} \oplus \mathcal{L}_{f^0}\Omega_{k-1} \oplus \bigoplus_{i=1}^{m} \mathcal{L}_{f^i}\Omega_{k-1}$
while $\dim(\Omega_k) > \dim(\Omega_{k-1})$ **do**
    Set $k = k + 1$
    Set $\Omega_k = \Omega_{k-1} \oplus \mathcal{L}_{f^0}\Omega_{k-1} \oplus \bigoplus_{i=1}^{m} \mathcal{L}_{f^i}\Omega_{k-1}$
end while
Set $\Omega = \Omega_k$ and $s = \dim(\Omega)$

We remind the reader that $\bigoplus_{i=1}^{m} \mathcal{L}_{f^i}\Omega = \mathcal{L}_{f^1}\Omega \oplus \mathcal{L}_{f^2}\Omega \oplus \cdots \oplus \mathcal{L}_{f^m}\Omega$.

On the basis of the results stated by Corollary 4.4 and Theorem 4.5, we call the codistribution returned by the above algorithm, after its convergence, the *observable codistribution*. A set of generators is any set of s scalar functions computed by the algorithm, whose gradients are independent. Note that the algorithm converges in at most $n - 1$ steps.

We can summarize the results stated by Corollary 4.4 and theorem 4.5 by saying that all the information that we can gather on x_0 is enclosed in the values that the functions $l_1(x), \ldots, l_s(x)$ take at x_0. Let us denote these values by $l_1^0, \ldots, l_s^0$. As for the simplified system, x_0 satisfy a set of s equations. For the simplified system, the s equations were the ones in (4.9). The new equations' system is

$$\begin{bmatrix} l_1(x_0) &= l_1^0, \\ l_2(x_0) &= l_2^0, \\ \ldots \\ l_s(x_0) &= l_s^0, \end{bmatrix} \quad (4.15)$$

provided that x_0 is a regular point for Ω. By using the inverse function theorem, we obtain that *if $s = n$, there exists a neighborhood of x_0 where we can invert the equations in (4.15) to express x_0 in terms of $[l_1^0, \ldots, l_s^0]^T$*, which are known. Hence, also in this case, the condition $s = n$

is a sufficient condition for the weak observability. On the other hand, to determine x_0 from the system in (4.15), we need that this system consists at least of n equations. Therefore, the condition $s = n$ is also a necessary condition for the weak observability. Note that this is true provided that x_0 is a regular point for Ω. In particular, the result stated by Corollary 2.5 also uses the inverse function theorem and, for this reason, it requires that x_0 is a regular point. This result is used to obtain the results stated by Corollary 4.4 and Theorem 4.5.

Exactly as for the simplified system dealt with in section 4.2, the analysis of the case when x_0 is not a regular point for Ω can be carried out by using the results in the literature on the inverse function problem, specifically the ones that provide necessary conditions for a function to be invertible. By doing this, we obtain that if the state is weakly observable at x_0, then there exists a neighborhood B_0 of x_0 such that the subset $B \subset B_0$ of points where the dimension of Ω is smaller than n has Lebesgue measure equal to zero.

4.3.2 ▪ Extension to the case of multiple outputs

We discuss the case of multiple outputs, namely the case when the system is defined by (4.1) with $p > 1$. We can repeat all the same steps of section 4.3.1 to obtain the observability rank condition when $p > 1$. The only difference is that the input-output map in (4.10) consists now of p components instead of 1. As a result, we must compute the Lie derivatives of all the outputs along all the vector fields that define the dynamics of the state. The observability rank condition remains the same. The observable codistribution is provided by Algorithm 4.2, the only difference being that the initialization, i.e., Ω_0, must include the gradients of all the outputs. In other words, the second line of Algorithm 4.2 (i.e., Set $\Omega_0 = \text{span}\{\nabla h\}$) must be replaced by

$$\text{Set } \Omega_0 = \text{span}\{\nabla h_1, \nabla h_2, \ldots, \nabla h_p\}.$$

4.3.3 ▪ Examples of applications of the observability rank condition

The application of the observability rank condition consists in calculating the observable codistribution and its dimension. This can be done automatically by following the steps of Algorithm 4.2. Obviously, the preliminary step to be done to implement this algorithm consists of the determination of the analytic expression of the vector fields $f^0(x), f^1(x), \ldots, f^m(x)$ and the scalar functions $h_1(x), \ldots, h_p(x)$. We provide this implementation for several systems described by the equations in (4.1). Specifically, we refer to the elementary input-output system studied in the first part of Chapter 1 and for which we obtained the observability properties by applying heuristic procedures. We will show that the application of the observability rank condition provides the same results.

Our system is defined in section 1.1. It is a vehicle that moves on a plane by satisfying the unicycle constraint. Its state $[x_v, y_v, \theta_v]^T$ includes the vehicle position and orientation. The dynamics of this state are given by (1.1). We consider the following three cases of outputs (see also Figure 1.1 for an illustration):

$$y = \rho = \sqrt{x_v^2 + y_v^2}, \tag{4.16}$$

$$y = \beta = \pi - \theta_v + \text{atan2}(y_v, x_v), \tag{4.17}$$

$$y = \phi = \text{atan2}(y_v, x_v). \tag{4.18}$$

We remark that all these outputs take a very simple analytic expression in polar coordinates. For this reason, it is much more convenient to adopt these coordinates in order to apply the observability rank condition. Note that, by working in Cartesian coordinates, we would obtain exactly

4.3. Observability rank condition in the general case

the same results (with some more laborious computation). Our system will be characterized by the state (see Figure 1.1):

$$x = [\rho,\ \phi,\ \theta_v]^T.$$

Note that, in these coordinates, we compute the gradient by differentiating with respect to ρ, ϕ, and θ_v. By using the dynamics in (1.1) we obtain the following dynamics:

$$\begin{bmatrix} \dot{\rho} &= v\cos(\theta_v - \phi), \\ \dot{\phi} &= \frac{v}{\rho}\sin(\theta_v - \phi), \\ \dot{\theta}_v &= \omega. \end{bmatrix} \tag{4.19}$$

Let us apply the observability rank condition in four distinct cases. The first three cases are defined by the three above outputs, i.e., (4.16), (4.17), and (4.18), respectively. The fourth case is defined by the first two outputs, i.e., (4.16) and (4.17) simultaneously.

First case

The preliminary step is obtained by comparing (4.19) and (4.16) with (4.1). We obtain $m = 2$, $p = 1$, $u_1 = v$, $u_2 = \omega$,

$$f_1(x) = \begin{bmatrix} \cos(\theta_v - \phi) \\ \frac{\sin(\theta_v - \phi)}{\rho} \\ 0 \end{bmatrix}, \quad f_2(x) = \begin{bmatrix} 0 \\ 0 \\ 1 \end{bmatrix}, \quad h(x) = \rho.$$

We compute the observable codistribution by running Algorithm 4.2. We obtain, at the initialization,

$$\Omega_0 = \text{span}\{\nabla h\} = \text{span}\{[1,\ 0,\ 0]\}.$$

Its dimension is 1. Then we need to compute $\mathcal{L}_{f_1}\Omega_0$ and $\mathcal{L}_{f_2}\Omega_0$. It suffices to compute the Lie derivatives of the generator of Ω_0, i.e., $\mathcal{L}_{f_1}\nabla h$ and $\mathcal{L}_{f_2}\nabla h$. Because of the equality (2.19), we compute $\nabla \mathcal{L}_{f_1} h$ and $\nabla \mathcal{L}_{f_2} h$. We have

$$\mathcal{L}_{f_1} h = \nabla h \cdot f_1 = [1,\ 0,\ 0] \cdot \begin{bmatrix} \cos(\theta_v - \phi) \\ \frac{\sin(\theta_v - \phi)}{\rho} \\ 0 \end{bmatrix} = \cos(\theta_v - \phi),$$

$$\mathcal{L}_{f_2} h = \nabla h \cdot f_1 = [1,\ 0,\ 0] \cdot \begin{bmatrix} 0 \\ 0 \\ 1 \end{bmatrix} = 0.$$

Hence, after the first step of Algorithm 4.2 we obtain

$$\Omega_1 = \text{span}\{[1,\ 0,\ 0],\ [0,\ \sin(\theta_v - \phi),\ -\sin(\theta_v - \phi)]\}, \tag{4.20}$$

whose dimension is 2. We need to repeat the recursive step of Algorithm 4.2 (since $2 > 1$, which is the dimension at the previous step, and $2 < 3$, which is the state dimension). We need to compute $\mathcal{L}^2_{f_1 f_1} h$ and $\mathcal{L}^2_{f_1 f_2} h$ (where we are adopting the notation defined by (2.12), (2.13)).

We obtain

$$\mathcal{L}^2_{f_1 f_1} h = (\nabla \mathcal{L}_{f_1} h) \cdot f_1 = [0,\ \sin(\theta_v - \phi),\ -\sin(\theta_v - \phi)] \cdot \begin{bmatrix} \cos(\theta_v - \phi) \\ \frac{\sin(\theta_v - \phi)}{\rho} \\ 0 \end{bmatrix} = \frac{\sin^2(\theta_v - \phi)}{\rho},$$

$$\mathcal{L}^2_{f_1 f_2} h = (\nabla \mathcal{L}_{f_1} h) \cdot f_2 = [0,\ \sin(\theta_v - \phi),\ -\sin(\theta_v - \phi)] \cdot \begin{bmatrix} 0 \\ 0 \\ 1 \end{bmatrix} = -\sin(\theta_v - \phi)$$

and

$$\nabla \mathcal{L}^2_{f_1 f_1} h = \left[-\frac{\sin^2(\theta_v - \phi)}{\rho^2},\ -2\frac{\sin(\theta_v - \phi)\cos(\theta_v - \phi)}{\rho},\ 2\frac{\sin(\theta_v - \phi)\cos(\theta_v - \phi)}{\rho} \right],$$

$$\nabla \mathcal{L}^2_{f_1 f_2} h = [0,\ \cos(\theta_v - \phi),\ -\cos(\theta_v - \phi)].$$

It is immediate to check that both the two above covectors belong to the codistribution Ω_1 at the previous step (i.e., the one given in (4.20)). Hence, the dimension of Ω_2 is 2 and this means that Algorithm 4.2 has converged. We conclude that the state is not weakly observable. This result agrees with the result provided in section 1.2 for the same system, obtained by following a heuristic geometric reasoning.

Second case

The preliminary step is obtained by comparing (4.19) and (4.17) with (4.1). We obtain the same expression for all the quantities that define the previous system with the exception of the output, which is now

$$h(x) = \pi - \theta_v + \phi.$$

We compute the observable codistribution by running Algorithm 4.2. We obtain, at the initialization,

$$\Omega_0 = \text{span}\{\nabla h\} = \text{span}\{[0,\ 1,\ -1]\}.$$

Its dimension is 1. Additionally, by proceeding as in the previous case, we obtain

$$\mathcal{L}_{f_1} h = \frac{\sin(\theta_v - \phi)}{\rho},$$

$$\mathcal{L}_{f_2} h = -1.$$

Hence, after the first step of Algorithm 4.2 we obtain

$$\Omega_1 = \text{span}\left\{ [0,\ 1,\ -1],\ \left[-\frac{\sin(\theta_v - \phi)}{\rho^2},\ -\frac{\cos(\theta_v - \phi)}{\rho},\ \frac{\cos(\theta_v - \phi)}{\rho} \right] \right\},$$

whose dimension is 2. By repeating the recursive step, it is possible to check that the dimension of Ω_2 is 2, and this means that Algorithm 4.2 has converged. We conclude that the state is not weakly observable. This result agrees with the result provided in section 1.2 for the same system, obtained by following a heuristic geometric reasoning.

4.3. Observability rank condition in the general case

Third case

The preliminary step is obtained by comparing (4.19) and (4.18) with (4.1). We obtain the same expression for all the quantities that define the previous system with the exception of the output, which is now

$$h(x) = \phi.$$

We compute the observable codistribution by running Algorithm 4.2. We obtain, at the initialization,

$$\Omega_0 = \text{span}\{\nabla h\} = \text{span}\{[0,\ 1,\ 0]\}.$$

Its dimension is 1. Additionally, by proceeding as in the previous case, we obtain

$$\mathcal{L}_{f_1}h = \frac{\sin(\theta_v - \phi)}{\rho},$$

$$\mathcal{L}_{f_2}h = 0.$$

Hence, after the first step of Algorithm 4.2 we obtain

$$\Omega_1 = \text{span}\left\{[0,\ 1,\ 0],\ \left[-\frac{\sin(\theta_v - \phi)}{\rho^2},\ -\frac{\cos(\theta_v - \phi)}{\rho},\ \frac{\cos(\theta_v - \phi)}{\rho}\right]\right\},$$

whose dimension is 2. We need to repeat the recursive step. We will show that the dimension of Ω_2 is 3. To prove this, it is sufficient to compute one of the two Lie derivatives of $\mathcal{L}_{f_1}h$ and show that its gradient does not belong to Ω_1. This can be done by computing $\mathcal{L}^2_{f_1 f_2}h$. We obtain

$$\mathcal{L}^2_{f_1 f_2}h = \frac{\cos(\theta_v - \phi)}{\rho},$$

whose gradient does not belong to Ω_1. We conclude that the state is weakly observable. This result agrees with the result provided in section 1.2 for the same system, obtained by following a heuristic geometric reasoning.

Fourth case

The preliminary step is obtained by comparing (4.19), (4.16), and (4.17) with (4.1). We obtain the same expression for all the quantities that define the previous system with the exception of the output. We have $p = 2$ and

$$h_1(x) = \rho, \quad h_2(x) = \pi - \theta_v + \phi.$$

We compute the observable codistribution by running Algorithm 4.2. We obtain, at the initialization,

$$\Omega_0 = \text{span}\{\nabla h_1,\ \nabla h_2\} = \text{span}\{[1,\ 0,\ 0],\ [0,\ 1,\ -1]\}.$$

Its dimension is 2. Additionally, by proceeding as in the previous cases (i.e., by computing the Lie derivatives of the generators of Ω_0), we obtain that the dimension of Ω_1 is 2, and this means that algorithm 4.2 has converged. We conclude that the state is not weakly observable. This result agrees with the result provided in section 1.2 for the same system, obtained by following a heuristic geometric reasoning.

4.4 ▪ Observable function

When a system is neither observable nor weakly observable we can ask whether we can obtain the value that a given scalar function takes at the initial state. Let us consider a given scalar function $\theta(x)$. Obviously, if the system is observable at x_0, we can obtain the value $\theta(x_0)$. We are interested in understanding what happens when the system is not weakly observable. In particular, we introduce the following two definitions.

Definition 4.6 (Observable Function). *The scalar function $\theta(x)$ is observable at x_0 if there exists at least one choice of inputs $u_1(t), \ldots, u_m(t)$ such that $\theta(x_0)$ can be obtained from the knowledge of the output $y(t)$ and the inputs $u_1(t), \ldots, u_m(t)$ on the time interval $\mathcal{I}$. In addition, $\theta(x)$ is observable on a given set $\mathcal{U} \subseteq \mathcal{M}$ if it is observable at any $x \in \mathcal{U}$.*

Definition 4.7 (Weak Observable Function). *The scalar function $\theta(x)$ is weakly observable at x_0 if there exists an open neighborhood B of x_0 such that, by knowing that $x_0 \in B$, there exists at least one choice of inputs $u_1(t), \ldots, u_m(t)$ such that $\theta(x_0)$ can be obtained from the knowledge of the output $y(t)$ and the inputs $u_1(t), \ldots, u_m(t)$ on the time interval $\mathcal{I}$. In addition, $\theta(x)$ is weakly observable on a given set $\mathcal{U} \subseteq \mathcal{M}$ if it is weakly observable at any $x \in \mathcal{U}$.*

The result stated by Theorem 4.5 ensures that any order Lie derivative of the outputs along the vector fields that define the dynamics is an observable function.

In the following, we provide a simple criterion that allows us to check whether a given function is weakly observable. Let us denote by Ω the observable codistribution (i.e., the codistribution computed by Algorithm 4.2). We have the following result.

Theorem 4.8. *Let us assume that the codistribution Ω returned by Algorithm 4.2 is nonsingular on an open set $B \subset \mathcal{M}$. Then, $\nabla \theta \in \Omega$ if and only if $\theta(x)$ is weakly observable at any $x \in B$.*

Proof. Let us denote by $l_1(x), \ldots, l_s(x)$ the generators of Ω. We know that the gradients of $l_1(x), \ldots, l_s(x)$ are independent. Hence, at any $x \in B$, there exists a local coordinate change such that in the new coordinates, the first s are $l_1(x), \ldots, l_s(x)$ and the last $n - s$ coordinates are $n - s$ coordinates among $x_1, \ldots, x_n$. Without loss of generality, we assume that they are the last $n - s$ (this is obtained by reordering the variables $x_1, \ldots, x_n$). Hence, our coordinate change is

$$(x_1, \ldots, x_n) \to (l_1, \ldots, l_s, x_{s+1}, \ldots, x_n).$$

We know that all the knowledge we can gather on the initial state is enclosed in the vector $[l_1^0, \ldots, l_s^0]^T$. If we know that $x \in B$, we are allowed to work in the new coordinates defined above. If $\nabla \theta \in \Omega$, in the new coordinates this means that θ only depends on the first s coordinates (i.e., $\theta(x) = F(l_1(x), \ldots, l_s(x))$). Hence, its initial value can be obtained and θ is weakly observable (weakly because we knew that $x \in B$). Vice versa, let us suppose that θ is weakly observable, and let us prove that $\nabla \theta \in \Omega$. We proceed by contradiction, and we assume that $\nabla \theta \notin \Omega$. In the new coordinates, this would mean that θ cannot be expressed only in terms of the first s coordinates: it would also depend on at least one of the last $n - s$ coordinates. On the other hand, the values that these last $n - s$ coordinates take at the initial time cannot be determined. As a result, also the value that θ takes at the initial time could not be determined and θ would not be weakly observable. ◀

Note that the above definition generalizes the definition of observability. In particular, by considering the n scalar functions $\theta_1(x), \ldots, \theta_n(x)$ that are the components of the state (i.e.,

4.4. Observable function

$\theta_1(x) = x_1, \ldots, \theta_n(x) = x_n$), we obtain that Definition 4.1 is a special case of Definition 4.6 and that Definition 4.2 is a special case of Definition 4.7.

We conclude this section with the following remark, which can be fundamental to reduce the computational burden when implementing Algorithm 4.2. When we run the kth step of Algorithm 4.2, we compute the k-order Lie derivatives of the output (this was precisely what we have done in our examples). In many cases, the analytic expression of these Lie derivatives can be very complicated (especially when k becomes large). If we are able to find simple scalar functions whose gradient belongs to Ω_k, we are allowed to set these functions among the generators of Ω_k; i.e., we can substitute the most complicated Lie derivatives with these scalar functions (provided that we maintain a set of generators of the entire Ω_k). Then, at the kth step, we simply compute the Lie derivative of these (simpler) generators. In practice, at each step of the algorithm, we can try to select the simplest generators for Ω. This can be done by selecting simple functions and by checking that their gradients belong to Ω. In Chapter 5, and in particular in sections 5.6, 5.7, and 5.8, we will use this astuteness.

Examples

First example

Let us go back to the system analyzed in section 4.3.3. It is immediate to realize that the two functions

$$\theta(x) = \rho,$$

$$\theta(x) = \phi - \theta_v$$

are weakly observable functions in all four cases. This can be proved by checking that their gradient belongs to the observable codistributions in all four cases.

Second example

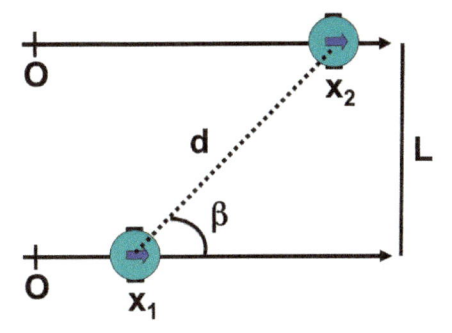

Figure 4.1. *Two vehicles constrained to move along two parallel lines.*

Two vehicles are constrained to move along two parallel lines whose distance, denoted by L, is unknown (see Figure 4.1). We know the speed of both the vehicles. In other words, both the vehicles are equipped with odometry sensors. Additionally, we assume that they are also equipped with range sensors able to return their reciprocal distance (denoted by d). We wish to know whether, by using the measurements provided by these sensors during a given interval of time, we can obtain the distance L between the lines. Note that the exercise is deliberately very simple and could be solved by applying heuristic procedures. We want to show that it can be solved by simply applying the automatic procedure that consists of the following steps:

1. characterize the system with the equations given in (4.1),

2. compute its observable codistribution, and

3. check whether the gradient of the function $\theta = L$ belongs to the above codistribution.

We can characterize our system with the following state:

$$x = [x_1, \, x_2, \, L]^T.$$

Its dynamics are

$$\begin{bmatrix} \dot{x}_1 &= v_1, \\ \dot{x}_2 &= v_2, \\ \dot{L} &= 0, \end{bmatrix}$$

where v_1 and v_2 are the two known speeds. Additionally, for the sake of simplicity, we set the following output:

$$y = \frac{d^2}{2} = \frac{(x_2 - x_1)^2 + L^2}{2}.$$

This expression avoids more laborious computation because we do not have the square root. It is allowed because if we know the distance, we also know its square value divided by 2. Note that this expression only simplifies the computation. The result about the observability of L (and more broadly the state x) is unaffected.

We compute the observable codistribution of our system by running Algorithm 4.2. The preliminary step is obtained by comparing the previous expression for the dynamics and for the output with (4.1). We obtain $m = 2$, $p = 1$, $u_1 = v_1$, $u_2 = v_2$,

$$f_1(x) = \begin{bmatrix} 1 \\ 0 \\ 0 \end{bmatrix}, \quad f_2(x) = \begin{bmatrix} 0 \\ 1 \\ 0 \end{bmatrix}, \quad h(x) = \frac{(x_2 - x_1)^2 + L^2}{2}.$$

We compute the observable codistribution by running Algorithm 4.2. We obtain, at the initialization,

$$\Omega_0 = \text{span}\{\nabla h\} = \text{span}\{[x_1 - x_2, \, x_2 - x_1, \, L]\}.$$

At the next step we obtain

$$\Omega_1 = \text{span}\{\nabla h, \nabla \mathcal{L}_{f_1} h\} = \text{span}\{[x_1 - x_2, \, x_2 - x_1, \, L], \, [1, \, -1, \, 0]\}.$$

By running the next step it is immediate to realize that the dimension of Ω_2 is the same and, as a result, the above codistribution is the observable codistribution. Since its dimension is 2 (i.e., smaller than the dimension of the state), we conclude that the system is not weakly observable. On the other hand, it is immediate to check that the gradient of the function

$$\theta(x) = L$$

(that is, $\nabla \theta = [0, \, 0, \, 1]^T$) belongs to Ω. This means that the parameter L is weakly observable. Note that we cannot conclude whether it is observable since the result stated by Theorem 4.8 only

4.5. Theory based on the standard approach

provides a criterion for the weak observability. In the specific case, L is only weakly observable since we cannot decide whether the second line is above or below the first line, by analyzing all the measurements.

The same results also hold for the system characterized by the output $y = \beta$ (in this case L is observable since we can also distinguish whether the second line is above or below the first line).

4.5 • Theory based on the standard approach

In this section we still refer to the system defined on the manifold $\mathcal{M}$ by the equations in (4.1) during the time interval $\mathcal{I} = [0, T]$ and we introduce the concept of indistinguishability. We actually refer to the case of a single output ($p = 1$) for the sake of simplicity. The extension to the multiple outputs case is trivial. Based on the concept of indistinguishability we provide the standard definition of observability and we show that it is equivalent to our previous definition.

4.5.1 • Indistinguishability and observability

Let us denote by $x(t; u; x_0)$ the state of the system at time $t \in \mathcal{I}$, when the initial state (i.e., the state at time $t = 0$) was x_0 and the system was driven by the inputs $u = [u_1(\tau), \ldots, u_m(\tau)]^T$, $\tau \in \mathcal{I}$. We introduce the following definition.

Definition 4.9 (State Indistinguishability). x_a and x_b are indistinguishable if $h(x(t; u; x_a)) = h(x(t; u; x_b)) \; \forall t \in \mathcal{I}$ and for any u.

We denote by $I(x)$ the subset of $\mathcal{M}$ of all the points which are indistinguishable from x. In what follows, we often refer to $I(x)$ as the indistinguishable set. We can introduce the standard definition of observability and weak observability as follows.

Definition 4.10. *An input-output system is observable at a given $x_0 \in \mathcal{M}$ if $I(x_0) = \{x_0\}$. In addition, it is observable on a given set $\mathcal{U} \subseteq \mathcal{M}$ if it is observable at any $x \in \mathcal{U}$.*

Definition 4.11. *An input-output system is weakly observable at a given $x_0 \in \mathcal{M}$ if there exists an open neighborhood B of x_0 such that $I(x_0) \cap B = \{x_0\}$. In addition, it is weakly observable on a given set $\mathcal{U} \subseteq \mathcal{M}$ if it is weakly observable at any $x \in \mathcal{U}$.*

Similarly, we introduce the concept of observable function and weakly observable function as follows.

Definition 4.12. *The scalar function $\theta(x)$ is observable at x_0 if it is constant on $I(x_0)$. In addition, $\theta(x)$ is observable on a given set $\mathcal{U} \subseteq \mathcal{M}$ if it is observable at any $x \in \mathcal{U}$.*

Definition 4.13. *The scalar function $\theta(x)$ is weakly observable at x_0 if there exists an open neighborhood B of x_0 such that $\theta(x)$ is constant on $I(x_0) \cap B$. In addition, $\theta(x)$ is weakly observable on a given set $\mathcal{U} \subseteq \mathcal{M}$ if it is weakly observable at any $x \in \mathcal{U}$.*

The idea behind the last two definitions is that if two or more states are indistinguishable, we cannot determine which one was the initial state. However, for a scalar function that takes the same value on these states, we can determine its initial value since we do not need to distinguish between the indistinguishable states to know its initial value. Note that if the indistinguishable sets consist of single points (i.e, the system is observable), all the state components, regarded as scalar functions of the state, are observable functions.

As we remarked at the end of section 4.4, also in this standard approach to observability the two above definitions generalize the definitions of observability and weak observability. In particular, by considering the n scalar functions $\theta_1(x), \ldots, \theta_n(x)$ that are the components of the state (i.e., $\theta_1(x) = x_1, \ldots, \theta_n(x) = x_n$), we obtain that Definition 4.10 is a special case of Definition 4.12 and Definition 4.11 a special case of Definition 4.13.

4.5.2 ▪ Equivalence of the approaches

In the following we will always assume that the codistribution returned by Algorithm 4.2 is nonsingular on a given open set and we restrict our analysis to this open set. This means that its dimension is constant on this open set. Under this assumption, we show the equivalence of the definitions of observability provided in the constructive approach (i.e., the definitions introduced in section 4.1 and in section 4.4) and the definitions provided in section 4.5.

We start by proving the following result.

Theorem 4.14. x_a and x_b are indistinguishable if and only if all the Lie derivatives of the system take at x_a the same values that they take at x_b.

Proof. From Theorem 4.3 and the definition of indistinguishability it is immediate to realize that if all the Lie derivatives of the system take at x_a the same value that they take at x_b, then x_a and x_b are indistinguishable. Let us prove the reverse. If x_a and x_b are indistinguishable, $h(x(t; x_a)) = h(x(t; x_b))$ $\forall t \in \mathcal{I}$ and for any $u(t)$. As in the proof of Theorem 4.5, by suitable choices of u, we can retrieve the Lie derivatives of the system at the initial state only in terms of the system inputs and the output. Hence, their values at x_a coincide with the ones at x_b. ◂

This result tells us that all the Lie derivatives of the system are observable functions according to Definition 4.12. On the other hand, from Theorem 4.5, we also know that they are observable according to Definition 4.6. More in general, we prove the following result.

Theorem 4.15. Let us assume that the codistribution Ω returned by Algorithm 4.2 is nonsingular on a given open set $\mathcal{U} \subseteq \mathcal{M}$. Then the following hold:

- Definition 4.1 is equivalent to Definition 4.10.
- Definition 4.2 is equivalent to Definition 4.11.
- Definition 4.6 is equivalent to Definition 4.12.
- Definition 4.7 is equivalent to Definition 4.13.

Proof. As we remarked at the end of section 4.4, Definitions 4.1 and 4.2 are special cases of Definitions 4.6 and 4.7. Additionally, as we remarked at the end of section 4.5.1, Definitions 4.10 and 4.11 are special cases of Definitions 4.12 and 4.13. Therefore, it suffices to prove the last two above statements. We start from the last one.

From Theorem 4.8 we know that all the functions θ that are weakly observable according to Definition 4.7 are the functions such that $\nabla \theta \in \Omega$.

Let us consider the same coordinates' change adopted in the proof of Theorem 4.8. Because of Theorem 4.14, in the new coordinates, the indistinguishable sets are the ones where the first

4.6 • Unobservability and continuous symmetries

s coordinates are constant. If $\nabla\theta \in \Omega$, in the new coordinates θ only depends on the first s coordinates, and consequently it is constant on the indistinguishable sets. On the other hand, if $\nabla\theta \notin \Omega$, θ also depends on some of the last $n - s$ coordinates and is not constant on the indistinguishable sets.

Finally, regarding the third statement, it suffices to refer to the case when $\mathcal{U}$ coincides with $\mathcal{M}$. ◀

4.6 • Unobservability and continuous symmetries

Let us suppose that, for a given system characterized by (4.1), the observable codistribution has dimension equal to $s < n$ on a given open set $\mathcal{U} \subseteq \mathcal{M}$. We know that, on $\mathcal{U}$, the state is not weakly observable. Let us denote the generators of the observable codistribution by $l_1(x), \ldots, l_s(x)$. These scalar functions are observable (and consequently also weakly observable). We have

$$\Omega = \text{span}\left\{\nabla l_1, \ldots, \nabla l_s\right\}.$$

In this section we want to discuss the following two issues:

- We explore the possibility of redefining the state in such a way that, at least locally, the new state is observable.

- On the basis of the discussion carried out in sections 1.2 and 1.3.3, we know that there is a continuous transformation such that, when the initial state is changed according to this transformation, the analysis of the system inputs and outputs cannot reveal this change. We want to better formalize this concept.

We discuss these two issues in the next two subsections. In Chapter 5 we also illustrate these concepts through several examples.

4.6.1 • Local observable subsystem

By construction, the generators $l_1, \ldots, l_s$ automatically returned by Algorithm 4.2 are among the Lie derivatives of the system (of order smaller than s). These include the zero-order Lie derivatives, i.e., the output functions $h_1, \ldots, h_p$. Additionally, we know that Ω is invariant with respect to all the vector fields $f^0, f^1, \ldots, f^m$ simultaneously. As a result, given a generator l_k, we have

$$\nabla \mathcal{L}_{f^0} l_k \in \Omega, \quad \nabla \mathcal{L}_{f^i} l_k \in \Omega \;\; \forall i = 1, \ldots, m, \;\; \forall k = 1, \ldots, s. \tag{4.21}$$

Since Ω is nonsingular in the open set $\mathcal{U}$, we are allowed to perform a local coordinate change such that, in the new coordinates, the first s are $l_1(x), \ldots, l_s(x)$ and the last $n - s$ coordinates are $n - s$ coordinates among $x_1, \ldots, x_n$. Without loss of generality, we assume that they are the last $n - s$ (this is obtained by reordering the variables $x_1, \ldots, x_n$). Hence, our coordinate change is

$$(x_1, \ldots, x_n) \to (l_1, \ldots, l_s, x_{s+1}, \ldots, x_n). \tag{4.22}$$

In the new coordinates, because of (4.21), every Lie derivative of l_k along any vector field among $f^0, f^1, \ldots, f^m$ only depends on the first s new coordinates. In addition, since we have

$$\dot{l}^k = \frac{\partial l_k}{\partial x} \cdot f^0 + \sum_{i=1}^{m} \frac{\partial l_k}{\partial x} \cdot f^i u_i = \mathcal{L}_{f^0} l_k + \sum_{i=1}^{m} \mathcal{L}_{f^i} l_k u_i,$$

the quantity l^k can be expressed only in terms of the first s new coordinates (and the system inputs). Hence, by defining the new state,

$$l = [l_1, \ldots, l_s]^T,$$

we have

$$\begin{cases} \dot{l} = \tilde{g}^0(l) + \sum_{i=1}^m \tilde{f}^i(l) u_i, \\ y = [\tilde{h}_1(l), \ldots, \tilde{h}_p(l)]^T, \end{cases} \quad (4.23)$$

where

$$\tilde{g}^0 = \begin{bmatrix} \mathcal{L}_{f^0} l_1 \\ \mathcal{L}_{f^0} l_2 \\ \cdots \\ \mathcal{L}_{f^0} l_s \end{bmatrix}, \quad \tilde{f}^i = \begin{bmatrix} \mathcal{L}_{f^i} l_1 \\ \mathcal{L}_{f^i} l_2 \\ \cdots \\ \mathcal{L}_{f^i} l_s \end{bmatrix}, \quad i = 1, \ldots, m,$$

and the functions $\tilde{h}_1(l), \ldots, \tilde{h}_p(l)$ are the expressions of $h_1, \ldots, h_p$ in terms of the state l. In most cases, if $h_1, \ldots, h_p$ are independent, we set the first p generators of Ω equal to $h_1, \ldots, h_p$. In this case, we have

$$\tilde{h}_1(l) = l_1, \ldots, \tilde{h}_p(l) = l_p.$$

The equations given in (4.23) provides the *local observable subsystem* of our original system. We conclude with the following three remarks:

1. The observable subsystem given by (4.23) is *local* because it only holds on the subset of the original space of the states where the local change of coordinates in (4.22) holds.

2. The observable subsystem given by (4.23) has a fundamental practical importance. If the original state that characterizes our system is not observable, estimating directly the state (e.g., by using an extended Kalman filter) brings us to divergence. On the other hand, in order to estimate the observable state, we need to refer to the equations in (4.23), i.e., the equations that provide the analytic expression of the link between the time evolution of the state and the system inputs only in terms of the observable state and the analytic expression of the outputs only in terms of the observable state (e.g., these analytic expressions are required to implement the prediction phase and the estimation phase of an extended Kalman filter).

3. An automatic procedure able to provide the observable subsystem in (4.23) from the original system (4.1) does not exist. In particular, finding the expression of the Lie derivatives $\mathcal{L}_{f^i} l_k$ (for $i = 1, \ldots, m$, $k = 1, \ldots, s$ and also $\mathcal{L}_{f^0} l_k$) in terms of $l_1, \ldots, l_s$, can be a very demanding and ambitious task. The key is to select the generators of Ω whose expression in terms of the original state is as simple as possible.[17] In Chapter 5 we provide several examples by showing the complexity of this task.

4.6.2 ▪ Continuous symmetries

The dimension of Ω is $s < n$. We denote by Δ its orthogonal distribution. The dimension of Δ is $n - s \geq 1$. We introduce the following definition.

Definition 4.16 (Continuous Symmetry). *A vector field ξ that belongs to the orthogonal distribution Δ is a continuous symmetry of the system.*

---

[17] According to this, we can choose functions which are not necessarily Lie derivatives of the original system. This will be shown in some examples provided in Chapter 5.

4.6. Unobservability and continuous symmetries

Let us denote by x_0 the initial state. Let us consider the following infinitesimal transformation:

$$x_0 \to x' = x_0 + \epsilon \xi, \tag{4.24}$$

with $\xi \in \Delta$ and ϵ an infinitesimal parameter. Let us denote by $\mathcal{L}(x)$ a Lie derivative (of any order) of the system. We have

$$\mathcal{L}(x') = \mathcal{L}(x_0 + \epsilon \xi) \simeq \mathcal{L}(x_0) + \epsilon \nabla \mathcal{L} \cdot \xi = \mathcal{L}(x_0),$$

where the last equality holds because $\nabla \mathcal{L}(x) \in \Omega$ and ξ is a continuous symmetry. Since the above equality holds for any Lie derivative of the system, because of Theorem 4.14, we proved that x' and x_0 are indistinguishable. We conclude that, given a continuous symmetry, we can build an infinitesimal transformation that transforms the initial state into a state which is indistinguishable. We wonder whether we can also build a finite transformation, starting from the one given in (4.24). The answer is positive. It suffices to apply repetitively the above infinitesimal transform. Each time we have $dx = \epsilon \xi$. In other words, we need to integrate the following differential equations:

$$\begin{cases} \frac{dx}{d\tau} = \xi(x), \\ x(0) = x_0. \end{cases}$$

We know that $x(\tau)$ is indistinguishable from x_0. In general, the dimension of the distribution Δ is $d = n - s > 1$. We account for this by generalizing the above differential equations as follows:

$$\begin{cases} \frac{dx}{d\tau} = \sum_{i=1}^{d} c_i(\tau) \xi^i(x), \\ x(0) = x_0, \end{cases} \tag{4.25}$$

where $c_1(\tau), \ldots, c_d(\tau)$ are smooth functions of the parameter τ. The finite transformation is

$$x_0 \to x' = x(\tau). \tag{4.26}$$

We introduce the following definition.

Definition 4.17 (Indistinguishable Set). *Given $x_0 \in \mathcal{M}$ the indistinguishable set of x_0 is the set of all the points $x(t)$, which are the solutions of (4.25) for a given choice of the functions $c_1(\tau), \ldots, c_d(\tau)$ and a given $t \in \mathbb{R}$.*

Since the dimension of Δ is d, the dimension of the indistinguishable set of x_0 is also equal to d.

We conclude this section with the following remark. By construction, the observable codistribution Ω is integrable (see section 2.3.2). As a result, because of Theorem 2.6, its orthogonal distribution Δ is involutive. For this reason, the vector space Δ is a Lie algebra with respect to the Lie bracket. Indeed, the Lie bracket satisfies the three properties required to define a Lie algebra: linearity, anticommutativity, and the Jacobi identity. In addition, the fact that Δ is involutive makes Δ closed with respect to the Lie bracket and allows us to define the Lie bracket as the internal binary operation that defines the Lie algebra.

On the basis of the discussion carried out in section 2.4, this suggests to regard the infinitesimal transformation in (4.24) as the action of an element of a d-parameter Lie group. In particular, since the transformation is infinitesimal, this element of the group must be close to the identity element of the group. Finally, the finite transformation in (4.26) can be considered the action of an element of this d-parameter Lie group.

4.7 • Extension of the observability rank condition to time-variant systems

In this section we extend the observability rank condition to a more general class of systems than the ones characterized by (4.1). In particular, the new systems explicitly depend on time; namely, all the functions that characterize their dynamics and/or their outputs explicitly depend on time. The new systems are characterized by the following equations:

$$\begin{cases} \dot{x} = f^0(x,\, t) + \sum_{i=1}^m f^i(x,\, t) u_i, \\ y = [h_1(x,\, t), \ldots, h_p(x,\, t)]^T. \end{cases} \quad (4.27)$$

The observability properties are obtained by computing the observable codistribution. Once computed, if its dimension is smaller than the state dimension, the rest of the analysis follows the same steps as in the case characterized by (4.1), which are (i) determination of all the independent system symmetries; (ii) determination of an observable state; and (iii) determination of a local observable subsystem. Therefore, the goal of our study is to find the new algorithm that computes the observable codistribution and that extends Algorithm 4.2.

We need to work in the chronospace (see section 3.1). In particular, the derivation can be carried out by using a synchronous frame, i.e., characterized by the following equations:

$$\begin{cases} \underline{\dot{x}} = \underline{f}^0(\underline{x}) + \sum_{i=1}^m \underline{f}^i(\underline{x}) u_i, \\ y = [h_0(\underline{x}),\, h_1(\underline{x}), \ldots, h_p(\underline{x})]^T, \end{cases} \quad (4.28)$$

where

$$\underline{x} \triangleq \begin{bmatrix} t \\ x^1 \\ \ldots \\ x^n \end{bmatrix}, \quad \underline{f}^0(\underline{x}) = \begin{bmatrix} 1 \\ f^0 \end{bmatrix}, \quad \underline{f}^i(\underline{x}) = \begin{bmatrix} 0 \\ f^i \end{bmatrix} \quad (4.29)$$

and $h_0(\underline{x}) = t$.

The above system now has the same structure of the system characterized by (4.1). We only need to account for the fact that the state also includes t. This means that if we have a scalar function

$$\theta = \theta(x,\, t)$$

and we want to compute its Lie derivative along one of the vector fields that characterize the dynamics in (4.28) (i.e., one of the vector fields $\underline{f}^\alpha$, $\alpha = 0, 1, \ldots, m$), we need to compute the gradient also with respect to t, i.e.,

$$\mathcal{L}_{\underline{f}^\alpha} \theta = \underline{\nabla}\theta \cdot \underline{f}^\alpha = \left[\frac{\partial}{\partial t},\, \nabla\right] \theta \cdot \underline{f}^\alpha = \left[\frac{\partial \theta}{\partial t},\, \nabla\theta\right] \cdot \underline{f}^\alpha, \quad (4.30)$$

where we denoted by $\underline{\nabla}$ the gradient in the chronospace, i.e.,

$$\underline{\nabla} \triangleq \left[\frac{\partial}{\partial t},\, \nabla\right].$$

By repeating all the steps carried out in section 4.3 to obtain Algorithm 4.2, we immediately obtain the following algorithm, which computes the observable codistribution in the chronospace for the system defined by (4.28).

4.7. Extension of the observability rank condition to time-variant systems

ALGORITHM 4.3. Observable codistribution in the chronospace (in the absence of unknown inputs).

Set $k = 0$
Set $\underline{\Omega}_k = \text{span}\{\underline{\nabla}h_0, \underline{\nabla}h_1, \ldots, \underline{\nabla}h_p\}$
Set $k = k + 1$
Set $\underline{\Omega}_k = \underline{\Omega}_{k-1} \oplus \mathcal{L}_{\underline{f}^0}\underline{\Omega}_{k-1} \oplus \bigoplus_{i=1}^m \mathcal{L}_{\underline{f}^i}\underline{\Omega}_{k-1}$
while $\dim(\underline{\Omega}_k) > \dim(\underline{\Omega}_{k-1})$ **do**
  Set $k = k + 1$
  Set $\underline{\Omega}_k = \underline{\Omega}_{k-1} \oplus \mathcal{L}_{\underline{f}^0}\underline{\Omega}_{k-1} \oplus \bigoplus_{i=1}^m \mathcal{L}_{\underline{f}^i}\underline{\Omega}_{k-1}$
end while

Our final objective is to obtain the observable codistribution in the space of the states, i.e., the codistribution that includes all the observability properties of the state (and not the chronostate). On the other hand, we know that the time is observable. In particular, the function h_0 only depends on t ($h_0 = t$) and, as a result, $\underline{\nabla}h_0 = [1, 0_n] \in \underline{\Omega}_0$. Hence, for each k, we can split $\underline{\Omega}_k$ into two codistributions as follows:

$$\underline{\Omega}_k = [1, 0_n] \oplus [0, \Omega_k],$$

where $[0, \Omega_k]$ includes covectors whose first component is zero. As a result, Ω_k is precisely the observable codistribution we want to compute, i.e., the one that only includes the observability properties of the state. Our objective is to derive the algorithm that directly provides the above Ω_k. We proceed as follows. From (4.30) and the structure of $\underline{f}^\alpha$ given in (4.29), we obtain

$$\mathcal{L}_{\underline{f}^\alpha}\theta = \begin{bmatrix} \frac{\partial \theta}{\partial t} + \mathcal{L}_{f^0}\theta, & \alpha = 0, \\ \mathcal{L}_{f^i}\theta, & \alpha = i = 1, \ldots, m. \end{bmatrix} \quad (4.31)$$

As a result, for any θ such that $\underline{\nabla}\theta \in \underline{\Omega}_k$, the following $m+1$ functions belong to $\underline{\Omega}_{k+1}$:

$$\frac{\partial \theta}{\partial t} + \mathcal{L}_{f^0}\theta, \qquad \mathcal{L}_{f^i}\theta, \ i = 1, \ldots, m.$$

Hence, the observable codistribution in the space of the states is given by the following algorithm.

ALGORITHM 4.4. Observable codistribution for time-variant systems (in the absence of unknown inputs).

Set $k = 0$
Set $\Omega_k = \text{span}\{\nabla h_1, \ldots, \nabla h_p\}$
Set $k = k + 1$
Set $\Omega_k = \Omega_{k-1} \oplus \widetilde{\mathcal{L}}_{f^0}\Omega_{k-1} \oplus \bigoplus_{i=1}^m \mathcal{L}_{f^i}\Omega_{k-1}$
while $\dim(\Omega_k) > \dim(\Omega_{k-1})$ **do**
  Set $k = k + 1$
  Set $\Omega_k = \Omega_{k-1} \oplus \widetilde{\mathcal{L}}_{f^0}\Omega_{k-1} \oplus \bigoplus_{i=1}^m \mathcal{L}_{f^i}\Omega_{k-1}$
end while
Set $\Omega = \Omega_k$ and $s = \dim(\Omega)$

where we introduced the following operator:

$$\widetilde{\mathcal{L}}_{f^0} \triangleq \frac{\partial}{\partial t} + \mathcal{L}_{f^0}. \tag{4.32}$$

We conclude this section by providing a property that plays a key role for the implementation of Algorithm 4.4. Specifically, as for Algorithm 4.2, at each step k, it suffices to compute the generators of Ω_k by performing simple operations (Lie derivatives and time derivatives) on the generators of Ω_{k-1}. Additionally, as for Algorithm 4.2, these generators are gradients of scalar fields.

We need to extend the result of Proposition 2.3 to the case when, instead of the operator $\mathcal{L}_f$, we consider the operator defined in (4.32).

We have the following new property.

Proposition 4.18. *Let us consider a nonsingular codistribution Ω spanned by covectors $\omega_1, \ldots, \omega_s$ and a smooth vector field f. We have*

$$\Omega \oplus \widetilde{\mathcal{L}}_f \Omega = \mathrm{span}\{\omega_1, \ldots, \omega_s, \widetilde{\mathcal{L}}_f \omega_1, \ldots, \widetilde{\mathcal{L}}_f \omega_s\}.$$

Proof. Obviously, $\mathrm{span}\{\omega_1, \ldots, \omega_s, \widetilde{\mathcal{L}}_f \omega_1, \ldots, \widetilde{\mathcal{L}}_f \omega_s\}$ is included in $\Omega \oplus \widetilde{\mathcal{L}}_f \Omega$.

Let us prove the reverse. Let us consider a generic covector $\mu \in \Omega$. We have

$$\mu = \sum_{i=1}^{s} c_i \omega_i.$$

We have

$$\widetilde{\mathcal{L}}_f \mu = \mathcal{L}_f \left(\sum_{i=1}^{s} c_i \omega_i \right) + \frac{\partial}{\partial t} \left(\sum_{i=1}^{s} c_i \omega_i \right)$$

$$= \sum_{i=1}^{s} \left(\mathcal{L}_f c_i + \frac{\partial c_i}{\partial t} \right) \omega_i + \sum_{i=1}^{s} c_i \widetilde{\mathcal{L}}_f \omega_i,$$

which belongs to $\mathrm{span}\{\omega_1, \ldots, \omega_s, \widetilde{\mathcal{L}}_f \omega_1, \ldots, \widetilde{\mathcal{L}}_f \omega_s\}$. ◂

Simple example

We consider the system given in (4.27) with $m = p = 1$,

$$f \triangleq f^1 = \begin{bmatrix} x^1 \\ x^2 \\ \vdots \\ x^n \end{bmatrix}, \quad f^0 = \begin{bmatrix} 0 \\ 0 \\ \vdots \\ 0 \end{bmatrix}, \quad h_1 \triangleq h = \sum_{i=1}^{n} x^i t^i,$$

where, in the function h, x^i is the ith component of the state and t^i is t to the power of i. We use Algorithm 4.4 to compute the observable codistribution. We obtain the following result. $\Omega_0 = \mathrm{span}\{\nabla h\}$, with

$$\nabla h = [t, \ t^2, \ t^3, \ \ldots, \ t^n].$$

4.7. Extension of the observability rank condition to time-variant systems

We compute Ω_1. We need to compute $\mathcal{L}_f h$ and $\widetilde{\mathcal{L}}_{f^0} h$. We have

$$\mathcal{L}_f h = h, \quad \widetilde{\mathcal{L}}_{f^0} h = \frac{\partial h}{\partial t} = \sum_{i=1}^{n} i x^i t^{i-1}.$$

Hence, $\Omega_1 = \text{span}\{\nabla h, \nabla \widetilde{\mathcal{L}}_{f^0} h\}$, with

$$\nabla \widetilde{\mathcal{L}}_{f^0} h = [1,\ 2t,\ 3t^2,\ \ldots,\ nt^{n-1}].$$

Since $\dim(\Omega_1) = 2 > 1 = \dim(\Omega_0)$, we need to repeat the recursive step and compute Ω_2. We need to compute the two scalar fields:

$$\mathcal{L}_f \widetilde{\mathcal{L}}_{f^0} h = \widetilde{\mathcal{L}}_{f^0} h,$$

$$\widetilde{\mathcal{L}}_{f^0} \widetilde{\mathcal{L}}_{f^0} h = \frac{\partial \widetilde{\mathcal{L}}_{f^0} h}{\partial t} = \sum_{i=2}^{n} i(i-1) x^i t^{i-2}.$$

Hence, $\Omega_2 = \text{span}\{\nabla h, \nabla \widetilde{\mathcal{L}}_{f^0} h, \nabla \widetilde{\mathcal{L}}_{f^0} \widetilde{\mathcal{L}}_{f^0} h\}$, with

$$\nabla \widetilde{\mathcal{L}}_{f^0} \widetilde{\mathcal{L}}_{f^0} h = [0,\ 2,\ 6t,\ \ldots,\ n(n-1)t^{n-2}]$$

and $\dim(\Omega_2) = 3 > 2 = \dim(\Omega_1)$. By proceeding in this manner we finally obtain $\Omega_{n-1} = \Omega_{n-2} \oplus \text{span}\{\nabla \widetilde{\mathcal{L}}_{f^0}^{n-1} h\}$ and $\dim(\Omega_{n-1}) = n$. We conclude that the state is weakly observable.

We remark that, in this driftless case, we have

$$\widetilde{\mathcal{L}}_{f^0} = \frac{\partial}{\partial t}.$$

Therefore, the observable codistribution is obtained by only considering the output and its time derivatives up to the $n - 1$ order. In other words, the result is independent of the system input. We would obtain the weak observability by setting $m = 0$. In addition, we would also obtain the weak observability for the system with $m = 0$ and characterized by the output

$$h = \sum_{i=1}^{n} h^i(x^i) t^i,$$

where $h^1, \ldots, h^n$ are n scalar functions $\mathbb{R} \to \mathbb{R}$ with nonzero derivative (the case considered above corresponds to the case $h^i(x^i) = x^i\ \forall i$).

This result is not surprising. The output $h = \sum_{i=1}^{n} x^i t^i$ weights the components of the state in a different manner. For instance, for $n = 2$, it suffices to take the output at two distinct nonvanishing times, t_1, t_2, to obtain two independent equations in the two components of the state. The same holds with the output $h = \sum_{i=1}^{n} h^i(x^i) t^i$. In this case we first obtain $h^i(x^i)\ \forall i$ and then, since the functions h^i have nonvanishing derivative, they can be inverted to give the components of the state.

Finally, note that the case characterized by $m = 0$ and $h = \sum_{i=1}^{n} x^i t^i$ can be investigated by using the method that holds in the linear case (e.g., see [41]). The result that we obtain is the same.

The reader can find more realistic applications in [39].

Chapter 5
Applications: Observability Analysis for Systems in the Absence of Unknown Inputs

In this chapter we provide several examples where we apply all the concepts introduced in the previous chapter. For each example, the analysis consists of the following four steps:

1. determination of the observable codistribution Ω,

2. determination of all the independent system symmetries,

3. determination of an observable state, and

4. determination of a local observable subsystem.

The first step is obtained by running Algorithm 4.2, as we illustrated in section 4.3.3. Note that the last three steps are unnecessary when the dimension of Ω is n, which is the dimension of the state. In this case, the system does not have symmetries and the original state is weakly observable. When the dimension of Ω is $s < n$, the system has $n - s > 0$ independent symmetries. These are obtained by computing the distribution Δ orthogonal to Ω. An observable state is then obtained by computing s independent observable functions. The key is to obtain the observable functions whose analytic expression in terms of the components of the original state is as simple as possible. This concern is fundamental to simplify as much as possible the last step. As we explained in section 4.6.1, this step consists in obtaining a complete description of our system only in terms of the observable state. This means that both the state evolution (i.e., the link between the time derivative of the observable state and the system inputs) and the system outputs must be expressed only in terms of the components of the observable state. The equations in (4.23) realize precisely this.

While the first three steps can be performed automatically, for the last step there does not exist an automatic procedure and its execution can be very challenging (even if it is always possible).

The examples studied in sections 5.1 and 5.2 are elementary, and they are chosen for educational purposes. The remaining examples play a fundamental practical role in the framework of robotics. For these examples, the determination of a local observable subsystem has a fundamental practical importance since it allows us to obtain the basic equations necessary for the implementation of any estimation scheme (e.g., filter based, optimization based, etc.) or even to obtain a closed-form solution of the problem (when it exists). Note that, in order to provide as much material as possible, we use different characterizations to describe these systems (e.g., for the rotation in 3D, we sometimes use quaternions, sometimes orthogonal matrices).

5.1 • The unicycle

We discuss again the system introduced in section 1.1 and for which we determined the observable codistribution by considering four cases depending on the outputs (section 4.3.3) and we also provided two independent weakly observable functions (section 4.4).

With the exception of the case when the output consists of the bearing angle ϕ of the vehicle in the global frame (third case in section 4.3.3), the dimension of the observable codistribution is always 2 and the orthogonal distribution is always the following (in polar coordinates):

$$\Delta = \text{span}\left\{\begin{bmatrix} 0 \\ 1 \\ 1 \end{bmatrix}\right\}.$$

The infinitesimal transformation given in (4.24) becomes, for the specific case,

$$\begin{bmatrix} \rho_0 \\ \phi_0 \\ \theta_{v0} \end{bmatrix} \to \begin{bmatrix} \rho' \\ \phi' \\ \theta'_v \end{bmatrix} = \begin{bmatrix} \rho_0 \\ \phi_0 \\ \theta_{v0} \end{bmatrix} + \epsilon \begin{bmatrix} 0 \\ 1 \\ 1 \end{bmatrix},$$

which is an infinitesimal rotation about the vertical axis. The indistinguishable set of $[\rho_0, \phi_0, \theta_0]^T$ is obtained by solving the equations in (4.25) for the specific case. We obtain that it is the set of points

$$\begin{bmatrix} \rho_0 \\ \phi_0 + t \\ \theta_{v0} + t \end{bmatrix} \quad \forall t \in \mathbb{R}.$$

We remark that these points are obtained by performing a rotation of t around the vertical axis.

Note that the same result could also be obtained by working in Cartesian coordinates, with some more laborious computation. In that case,

$$\Delta = \text{span}\left\{\begin{bmatrix} -y_v \\ x_v \\ 1 \end{bmatrix}\right\}$$

and the infinitesimal transformation becomes

$$\begin{bmatrix} x_{v0} \\ y_{v0} \\ \theta_{v0} \end{bmatrix} \to \begin{bmatrix} x'_v \\ y'_v \\ \theta'_v \end{bmatrix} = \begin{bmatrix} x_{v0} \\ y_{v0} \\ \theta_{v0} \end{bmatrix} + \epsilon \begin{bmatrix} -y_{v0} \\ x_{v0} \\ 1 \end{bmatrix},$$

which is an infinitesimal rotation about the vertical axis because

$$\begin{bmatrix} \cos\epsilon & -\sin\epsilon \\ \sin\epsilon & \cos\epsilon \end{bmatrix} \simeq \begin{bmatrix} 1 & -\epsilon \\ \epsilon & 1 \end{bmatrix}.$$

The indistinguishable set of $[x_{v0}, y_{v0}, \theta_{v0}]^T$ is obtained by solving the equations in (4.25) for the specific case. We obtain that it is the set of points

$$\begin{bmatrix} \begin{bmatrix} \cos t & -\sin t \\ \sin t & \cos t \end{bmatrix} \begin{bmatrix} x_{v0} \\ y_{v0} \end{bmatrix} \\ \theta_{v0} + t \end{bmatrix} \quad \forall t \in \mathbb{R},$$

which is the same equation that we provided in section 1.1 (see (1.5)).

5.2. Vehicles moving on parallel lines

With the exception of the case when the output consists of the bearing angle ϕ of the vehicle in the global frame (third case in section 4.3.3), the system is not weakly observable. We want to obtain a local observable subsystem. Let us refer to the first case. A possible choice of generators of Ω is given by $h(x) = \rho$ and $\mathcal{L}_{f^1} h = \cos(\theta_v - \phi)$. On the other hand, as we mentioned in section 4.6.1, it is advantageous to detect the generators whose analytic expression is as simple as possible. We remark that the function $\theta_v - \phi$ is also observable. This can be easily checked by using Theorem 4.8, which can be immediately verified by computing the gradient $\nabla(\theta_v - \phi) = [0, -1, 1]$ and by verifying that it is orthogonal to Δ. Therefore, we set

$$l_1(x) = \rho, \quad l_2(x) = \theta_v - \phi, \quad l = [l_1, l_2]^T.$$

From (4.19) we immediately obtain

$$\left[\begin{array}{l} \dot{l}_1 = v \cos l_2, \\ \dot{l}_2 = \omega - \frac{v}{l_1} \sin l_2. \end{array}\right.$$

This is a local observable subsystem for our system. The output can be also expressed only in terms of the state l. We have

$$h(l) = l_1.$$

In the other two cases discussed in section 4.3.3 for which the system is not weakly observable (second and fourth cases) we obtain the same local observable subsystem with the outputs

$$h(l) = \pi - l_2$$

in the second case and

$$h(l) = \left[\begin{array}{c} l_1 \\ \pi - l_2 \end{array}\right]$$

in the fourth case.

5.2 ▪ Vehicles moving on parallel lines

We consider again the second example from section 4.4. We know that the state $[x_1, x_2, L]$ is not observable. Additionally, we know that the observable codistribution has dimension 2. Hence, the system has one continuous symmetry that is,

$$\xi = \left[\begin{array}{c} 1 \\ 1 \\ 0 \end{array}\right].$$

The infinitesimal transformation given in (4.24) becomes

$$\left[\begin{array}{c} x_{10} \\ x_{20} \\ L_0 \end{array}\right] \rightarrow \left[\begin{array}{c} x_1' \\ x_2' \\ L' \end{array}\right] = \left[\begin{array}{c} x_{10} \\ x_{20} \\ L_0 \end{array}\right] + \epsilon \left[\begin{array}{c} 1 \\ 1 \\ 0 \end{array}\right].$$

This means that the system is invariant with respect to a shift along the direction of the lines. This result is trivial and could have been obtained by following an intuitive reasoning. We emphasize that it is automatically returned by our methodology.

We want to obtain a local observable subsystem for this system. Also in this case the observable state will have two components. In section 4.4, we checked that L can be one of them. We need to find another one. A possible choice could be the output, i.e., $\frac{d^2}{2} = \frac{(x_2-x_1)^2+L^2}{2}$. On the other hand, as we mentioned in section 4.6.1, it is advantageous to detect a function whose analytic expression is as simple as possible. We remark that the function $x_2 - x_1$ has the gradient orthogonal to ξ. Hence, we set

$$l_1(x) = x_2 - x_1, \quad l_2(x) = L, \quad l = [l_1, l_2]^T$$

and the local observable subsystem is

$$\begin{bmatrix} \dot{l}_1 &= v_2 - v_1, \\ \dot{l}_2 &= 0, \\ y &= \frac{l_1^2 + l_2^2}{2}. \end{bmatrix}$$

5.3 ▪ Simultaneous odometry and camera calibration

Again, we consider a vehicle that moves on a plane by satisfying the unicycle constraint. In other words, the dynamics of its position $[x_v, y_v]^T$ and orientation θ_v are given by the equations in (1.1) or, in polar coordinates, by the equations in (4.19). We assume that the odometry sensors do not provide directly v and ω. In particular, we assume that the vehicle is equipped by a differential drive, and the vehicle movement is based on two wheels, driven separately, placed on either side of the vehicle body. The odometry sensors (e.g., wheel encoders) provide the angular speed of each wheel. We denote them by ω_R and ω_L, for the right and the left wheels, respectively. The linear and angular speeds of the vehicle (i.e., the above v and ω) are related to ω_R and ω_L as follows:

$$v = \frac{r_R \omega_R + r_L \omega_L}{2}, \qquad \omega = \frac{r_R \omega_R - r_L \omega_L}{B}, \qquad (5.1)$$

where r_R and r_L are the radius of the right and left wheels, respectively, and B is the distance between the wheels. The above odometry model is characterized by the three parameters B, r_R, and r_L, and calibrating the odometry consists of the determination of these three parameters.

A bearing sensor (e.g. a camera) is mounted on the vehicle. We assume that its vertical axis is aligned with the z-axis of the vehicle reference frame, and therefore the transformation between the frame attached to this sensor and the one of the odometry system is characterized by the three parameters ψ_1, R, and ψ_2 (see Figure 5.1).

The camera calibration consists of the estimation of a set of parameters, which are divided into the *intrinsic* and the *extrinsic* parameters. The former characterize the optical center and focal length of the camera. The latter characterize the camera position and orientation in the 3D scene. Here, we assume that the intrinsic parameters are known and the extrinsic parameters represent the relative configuration of the camera with respect to the odometry frame. Hence, in this specific case, the camera calibration only consists of the determination of the three above parameters ψ_1, R, and ψ_2.

We assume that the camera perceives a point feature at the origin (see Figure 5.1) and provides its bearing angle β in the camera local frame.

The question we want to answer is whether we can simultaneously calibrate the odometry and the camera by only using the measurements provided by the wheel encoders and the camera during a given interval of time. In other words, we assume to know the three functions

$$\omega_R(t), \quad \omega_L(t), \quad \beta(t)$$

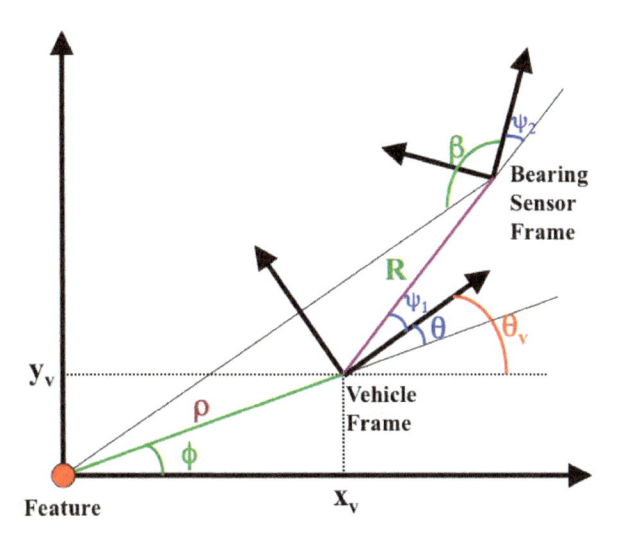

5.3. Simultaneous odometry and camera calibration

Figure 5.1. *The two reference frames respectively attached to the vehicle and to the bearing sensor.*

for $t \in \mathcal{I}$ and we want to know whether we can obtain from them the values of the six parameters

$$B, \quad r_R, \quad r_L, \quad \psi_1, \quad R, \quad \psi_2.$$

We include these parameters in a state that characterizes our system, and we study its observability properties. From the analysis carried out in section 5.1, we know that, even when the above calibration parameters are known, the state $[\rho, \phi, \theta_v]$ is not observable and the system is characterized by one continuous symmetry that is a rotation about the vertical axis. When the calibration parameters are unknown, this symmetry remains. Hence, we characterize our system by the following state:

$$x = [\rho, \theta, B, r_R, r_L, \psi_1, R, \psi_2]^T,$$

where $\theta = \theta_v - \phi$ (see Figure 5.1). This state satisfies the following dynamics:

$$\begin{bmatrix} \dot{\rho} &= v\cos\theta, \\ \dot{\theta} &= \omega - \frac{v}{\rho}\sin\theta, \\ \dot{B} &= \dot{r}_R = \dot{r}_L = \dot{\psi}_1 = \dot{R} = \dot{\psi}_2 = 0. \end{bmatrix}$$

Regarding the output, without altering the observability properties, we set

$$y = h(x) = \tan\beta.$$

By using some trigonometry algebra and up to a sign (which is inconsequential for our analysis), we obtain (see also Figure 5.1)

$$h(x) = \frac{\rho\sin(\theta + \psi_1) + (R + \rho\cos(\theta + \psi_1))\tan\psi_2}{R + \rho\cos(\theta + \psi_1) - \rho\sin(\theta + \psi_1)\tan\psi_2}.$$

To simplify the computation, we directly set the last entry of the state as the tangent of ψ_2, i.e.,

$$t_2 = \tan\psi_2.$$

We start by computing the observable codistribution. By comparing the above equations with (4.1) we obtain $m = 2$, $p = 1$, $u_1 = \omega_R$, $u_2 = \omega_L$,

$$f^1(x) = \begin{bmatrix} \frac{r_R}{2}\cos\theta \\ \frac{r_R}{B} - \frac{r_R}{2\rho}\sin\theta \\ 0 \\ 0 \\ 0 \\ 0 \\ 0 \\ 0 \end{bmatrix}, \quad f^2(x) = \begin{bmatrix} \frac{r_L}{2}\cos\theta \\ -\frac{r_L}{B} - \frac{r_L}{2\rho}\sin\theta \\ 0 \\ 0 \\ 0 \\ 0 \\ 0 \\ 0 \end{bmatrix},$$

$$h(x) = \frac{\rho\sin(\theta + \psi_1) + (R + \rho\cos(\theta + \psi_1))t_2}{R + \rho\cos(\theta + \psi_1) - \rho\sin(\theta + \psi_1)t_2}.$$

We compute the observable codistribution by running Algorithm 4.2. We do not provide the analytic expression of the observable codistribution at each step. At the initialization we have

$$\Omega_0 = \text{span}\{\nabla h\}$$

and its dimension is 1. At the first iterative step we obtain

$$\Omega_1 = \text{span}\{\nabla h, \nabla \mathcal{L}_{f^1} h, \nabla \mathcal{L}_{f^2} h\};$$

namely, all the above covectors are independent and the dimension of Ω_1 is 3. At the second iterative step we obtain

$$\Omega_2 = \text{span}\{\nabla h, \nabla \mathcal{L}_{f^1} h, \nabla \mathcal{L}_{f^2} h, \nabla \mathcal{L}^2_{f^1 f^1} h, \nabla \mathcal{L}^2_{f^1 f^2} h, \nabla \mathcal{L}^2_{f^2 f^1} h, \nabla \mathcal{L}^2_{f^2 f^2} h\};$$

namely, all the above covectors are independent and the dimension of Ω_2 is 7. At the third iterative step the dimension of Ω_3 does not change (i.e., the gradients of all the eight 3-order Lie derivatives belong to the above codistribution). This means that the algorithm has converged and the above codistribution is the observable codistribution. Since its dimension is $7 = n - 1$, the system has a single symmetry. We obtain this symmetry by computing the vector orthogonal to the above Ω. We obtain

$$\xi = \begin{bmatrix} \rho \\ 0 \\ B \\ r_R \\ r_L \\ 0 \\ R \\ 0 \end{bmatrix}.$$

The infinitesimal transformation given in (4.24) becomes

$$\begin{bmatrix} \rho_0 \\ \theta_0 \\ B \\ r_R \\ r_L \\ \psi_1 \\ R \\ t_2 \end{bmatrix} \rightarrow \begin{bmatrix} \rho' \\ \theta' \\ B' \\ r'_R \\ r'_L \\ \psi'_1 \\ R' \\ t'_2 \end{bmatrix} = \begin{bmatrix} \rho_0 \\ \theta_0 \\ B \\ r_R \\ r_L \\ \psi_1 \\ R \\ t_2 \end{bmatrix} + \epsilon \begin{bmatrix} \rho_0 \\ 0 \\ B \\ r_R \\ r_L \\ 0 \\ R \\ 0 \end{bmatrix}.$$

5.4. Simultaneous odometry and camera calibration in the case of circular trajectories

This means that the system is invariant with respect to the scale (e.g., $B \to (1+\epsilon)B$). This result could have been obtained by pursuing the following intuitive reasoning: the inputs and the output are all angular quantities, and no source of metric information is available.

We want to obtain a local observable subsystem for this system. The observable state will have seven components. A possible choice is as follows:

$$l_1(x) = \frac{R}{\rho}, \quad l_2(x) = \theta, \quad l_3(x) = \frac{r_R}{2R}, \quad l_4(x) = \frac{r_L}{r_R}, \quad l_5(x) = \frac{r_R}{B}, \quad l_6(x) = \psi_1, \quad l_7(x) = \psi_2,$$

$$l = [l_1, \; l_2, \; l_3, \; l_4, \; l_5, \; l_6, \; l_7]^T.$$

To prove the correctness of the above choice (i.e., that they are generators of Ω) it suffices to check the independence of their gradients and that their gradients are all orthogonal to ξ.

The local observable subsystem is

$$\begin{cases} \dot{l}_1 = -l_1^2 l_3(\omega_R + l_4\omega_L)\cos l_2, \\ \dot{l}_2 = l_5(\omega_R - l_4\omega_L) - l_1 l_3(\omega_R + l_4\omega_L)\sin l_2, \\ \dot{l}_3 = \dot{l}_4 = \dot{l}_5 = \dot{l}_6 = \dot{l}_7 = 0, \\ y = \beta = \begin{bmatrix} -\arctan\left(\frac{\sin(l_2+l_6)}{l_1+\cos(l_2+l_6)}\right) - l_7 + \pi, & |\theta+\phi| \leq \arccos l_1, \\ -\arctan\left(\frac{\sin(l_2+l_6)}{l_1+\cos(l_2+l_6)}\right) - l_7 & \text{otherwise.} \end{bmatrix} \end{cases} \quad (5.2)$$

Note that, to achieve the observability of the state $[l_1, l_2, l_3, l_4, l_5, l_6, l_7]^T$, the vehicle must move along all the allowed degrees of freedom (i.e., both the inputs ω_R and ω_L must be excited). In the next section we consider circular trajectories (i.e., trajectories characterized by a constant ratio $\frac{\omega_R}{\omega_L}$). For them, the state becomes unobservable.

5.4 ▪ Simultaneous odometry and camera calibration in the case of circular trajectories

We study the same system analyzed in the previous section. However, we consider the special case when the vehicle is constrained to move along circular trajectories. Since we remove one degree of freedom, the observability properties cannot increase (i.e., the observable codistribution cannot exceed the one obtained in the previous section). As we will see, the observable codistribution loses three dimensions (i.e., its dimension will be equal to 4). The interest of this analysis is not only academic but can also have a practical importance. In general, estimating a state is more precise for states with small dimension. Intuitively, setting a special trajectory allows us to focus the information provided by the measurements on a smaller number of physical quantities. In this scenario, the calibration takes place in multiple phases: each phase, which is characterized by a special trajectory, is designed to estimate a subset of all the calibration parameters. The entire estimation process becomes in general more precise and more robust.

From the academic point of view, this analysis clearly illustrates how complex the determination of a local observable subsystem can be, especially because it does not follow the steps of an automatic procedure but it relies upon our inventiveness.

For the sake of clarity, we report all the variables adopted in the considered calibration problem in Tables 5.1 and 5.2.

We consider the motion obtained by setting

$$\omega_R = \nu, \quad \omega_L = q\nu, \quad (5.3)$$

Table 5.1. *Variables adopted in our calibration problem*

Original State
$[\rho,\ \theta, B,\ r_R,\ r_L,\ \psi_1,\ R,\ \psi_2]^T$
Observable States
$l_1 \triangleq \frac{R}{\rho},\ l_2 \triangleq \theta,\ l_3 \triangleq \frac{r_R}{2R},\ l_4 \triangleq \frac{r_L}{r_R},\ l_5 \triangleq \frac{r_R}{B},\ l_6 \triangleq \psi_1,\ l_7 \triangleq \psi_2$
Observable States in a single circular trajectory (characterized by $q \triangleq \frac{\omega_L}{\omega_R}$)
$A^q \triangleq \frac{\Psi_1^q - \Psi_3}{1+\Psi_1^q \Psi_3},\quad V^q \triangleq \Psi_2^q \frac{1+\Psi_1^q \Psi_3}{1+\Psi_3^2},\quad L^q \triangleq \psi_2 - \arctan \Psi_1^q,\quad l_5^q \triangleq \frac{r_R}{B}(1 - q\frac{r_L}{r_R})$
where:
$\Psi_1^q \triangleq \frac{l_5^q - l_3^q \sin\psi_1}{l_3^q \cos\psi_1},\quad \Psi_2^q \triangleq \frac{R l_3^q \cos\psi_1}{\rho \sin(\theta+\phi_1)},\quad \Psi_3 \triangleq \frac{R+\rho\cos(\theta+\phi_1)}{\rho\sin(\theta+\phi_1)},\quad l_3^q \triangleq \frac{r_R}{2R}(1+q\frac{r_L}{r_R})$

Table 5.2. *Observable parameters*

Original Calibration Parameters
Camera: $\psi_1,\ R,\ \psi_2$ Odometry: $r_R,\ r_L,\ B$
Observable Parameters
$l_3 \triangleq \frac{r_R}{2R},\ l_4 \triangleq \frac{r_L}{r_R},\ l_5 \triangleq \frac{r_R}{B},\ l_6 \triangleq \psi_1,\ l_7 \triangleq \psi_2$
Observable Parameters in a single circular trajectory (characterized by $q \triangleq \frac{\omega_L}{\omega_R}$)
$L^q \triangleq \psi_2 - \arctan\left(\frac{2R l_5^q - r_R(1+q\frac{r_L}{r_R})\sin\psi_1}{r_R(1+q\frac{r_L}{r_R})\cos\psi_1}\right),\quad l_5^q \triangleq \frac{r_R}{B}(1 - q\frac{r_L}{r_R})$

where q is a known time-independent parameter. The corresponding trajectory is a circumference with radius $\frac{B}{2}\frac{r_R+qr_L}{r_R-qr_L}$. In this section we focus our attention on a single value of q.

We analyze the observability properties of the system characterized by (5.2) under the constraint given by (5.3). We substitute (5.3) in (5.2), and we obtain a new system, characterized by the same state as in (5.2), but with a single input (ν) instead of two (ω_R and ω_L). We obtain

$$\begin{cases} \dot{l}_1 = -l_1^2 l_3 (1+l_4 q)\cos l_2 \nu, \\ \dot{l}_2 = l_5(1-l_4 q)\nu - l_1 l_3(1+l_4 q)\sin l_2 \nu, \\ \dot{l}_3 = \dot{l}_4 = \dot{l}_5 = \dot{l}_6 = \dot{l}_7 = 0, \\ y = \beta = \begin{bmatrix} -\arctan\left(\frac{\sin(l_2+l_6)}{l_1+\cos(l_2+l_6)}\right) - l_7 + \pi, & |\theta+\phi| \leq \arccos l_1, \\ -\arctan\left(\frac{\sin(l_2+l_6)}{l_1+\cos(l_2+l_6)}\right) - l_7 & \text{otherwise.} \end{bmatrix} \end{cases} \quad (5.4)$$

We compute the new observable codistribution. By comparing the above equations with (4.1), and by taking the tangent of the output, we obtain $m = p = 1, u_1 = \nu$,

$$f^1(l) = \begin{bmatrix} -l_1^2 l_3(1+l_4 q)\cos l_2 \\ l_5(1-l_4 q) - l_1 l_3(1+l_4 q)\sin l_2 \\ 0 \\ 0 \\ 0 \\ 0 \\ 0 \end{bmatrix},\quad h(l) = \frac{\sin(l_2+l_6) + (l_1+\cos(l_2+l_6))\tan l_7}{l_1+\cos(l_2+l_6) - \sin(l_2+l_6))\tan l_7}.$$

5.4. Simultaneous odometry and camera calibration in the case of circular trajectories

We compute the observable codistribution by running Algorithm 4.2. It converges at the third iterative step. We obtain that the observable codistribution is

$$\Omega = \text{span}\left\{\nabla h, \nabla \mathcal{L}_{f^1} h, \nabla \mathcal{L}_{f^1}^2 h, \nabla \mathcal{L}_{f^1}^3 h\right\}.$$

The system has three independent symmetries ($3 = 7 - 4$). We do not provide their analytic expression because is too complex, with the exception of one, which is

$$\begin{bmatrix} 0 \\ 0 \\ -l_3 q(1 - l_4 q) \\ (1 - l_4 q)(1 + l_4 q) \\ l_5 q(1 + l_4 q) \\ 0 \\ 0 \end{bmatrix}.$$

This symmetry trivially expresses that, instead of considering the three components l_3, l_4, and l_5, we can adopt the following two:

$$l_3^q \triangleq l_3(1 + l_4 q), \quad l_5^q \triangleq l_5(1 - l_4 q).$$

We introduce the following new state (with six components):

$$l^q = [l_1,\ l_2,\ l_3^q,\ l_5^q,\ l_6,\ l_7]^T.$$

Its dynamics and the output can be easily expressed in terms of the above new state. Specifically, our system can be described by the following equations:

$$\begin{cases} \dot{l}_1 &= -l_1^2 l_3^q \cos l_2 \nu, \\ \dot{l}_2 &= l_5^q \nu - l_1 l_3^q \sin l_2 \nu, \\ \dot{l}_3^q &= \dot{l}_5^q = \dot{l}_6 = \dot{l}_7 = 0, \\ y &= \tan\beta = \frac{\sin(l_2+l_6)+(1+\cos(l_2+l_6))\tan l_7}{l_1+\cos(l_2+l_6)-\sin(l_2+l_6))\tan l_7}. \end{cases} \quad (5.5)$$

So far, we have only exploited one of the three symmetries that characterize the system in (5.4). This allowed us to reduce the state by one component. On the other hand, we know that we still need to remove two components if we want to achieve a local observable subsystem. The system in (5.5) is characterized by two independent symmetries. We know that four generators of the observable codistribution are the output together with its first three-order Lie derivatives along f^1. On the other hand, their expression is definitely intractable to achieve a local observable subsystem (although possible).

As we mentioned at the beginning of this chapter, there does not exist an automatic procedure to achieve a local observable subsystem. We need to define a new four-dimensional state whose components generate the entire observable codistribution. In addition, we must express the dynamics of this state and the output in terms of this state. We proceed by using our inventiveness/instinct. We start by analyzing the following simpler system:

$$\begin{cases} \dot{l}_1 &= -l_1^2 l_3^q \cos l_2 \nu, \\ \dot{l}_2 &= l_5^q \nu - l_1 l_3^q \sin l_2 \nu, \\ \dot{l}_3^q &= \dot{l}_5^q = \dot{l}_6 = 0, \\ y &= \frac{\sin(l_2+l_6)}{l_1+\cos(l_2+l_6)}, \end{cases} \quad (5.6)$$

Chapter 5. Observability Analysis for Systems in the Absence of Unknown Inputs

where we removed the parameter $l_7 = \psi_2$. In practice, this simpler system is obtained by setting the angle $\psi_2 = 0$. The state $[l_1, l_2, l_3^q, l_5^q, l_6]^T$ is five-dimensional. We wonder whether it is observable. We compute the observable codistribution by running Algorithm 4.2. It converges at its third iterative step. Hence, the observable codistribution has dimension equal to 4 and the system has a single continuous symmetry, which is

$$\left[l_1 \cos(l_2 + l_6) + 1, \; \sin(l_2 + l_6), \; \frac{l_5^q \sin l_6 - l_3^q}{l_1}, \; 0, \; \frac{l_5^q \cos l_6}{l_3^q l_1} \right]^T.$$

In order to obtain four generators of the observable codistribution we search for four functions $\Psi(l_1, l_2, l_3^q, l_5^q, l_6)$ that satisfy the above symmetry (and whose gradients are independent). Hence, we have

$$(l_1 \cos(l_2 + l_6) + 1) \frac{\partial \Psi}{\partial l_1} + \sin(l_2 + l_6) \frac{\partial \Psi}{\partial l_2} + \frac{l_5^q \sin l_6 - l_3^q}{l_1} \frac{\partial \Psi}{\partial l_3^q} + \frac{l_5^q \cos l_6}{l_3^q l_1} \frac{\partial \Psi}{\partial l_6} = 0.$$

Finding four independent solutions is not difficult since we know that all the Lie derivatives are solutions. However, their expression is very troublesome (with the exception of the zero order, which coincides with the output $\frac{\sin(l_2+l_6)}{l_1+\cos(l_2+l_6)}$). On the other hand, a very simple solution for the previous partial differential equation is provided by l_5^q. By using this solution and the output $\frac{\sin(l_2+l_6)}{l_1+\cos(l_2+l_6)}$, and starting from the expressions of the first- and second-order Lie derivatives, we were able to detect two other solutions: $\frac{l_5^q - l_3^q \sin l_6}{l_3^q \cos l_6}$ and $\frac{l_1 l_3^q \cos l_6}{\sin(l_2+l_6)}$. We therefore find the following four independent generators:

$$\Psi_1^q \triangleq \frac{l_5^q - l_3^q \sin l_6}{l_3^q \cos l_6}, \quad \Psi_2^q \triangleq \frac{l_1 l_3^q \cos l_6}{\sin(l_2+l_6)}, \quad \Psi_3 \triangleq \frac{l_1 + \cos(l_2+l_6)}{\sin(l_2+l_6)}, \quad l_5^q, \tag{5.7}$$

and a local observable subsystem of (5.6) is

$$\begin{cases} \dot{\Psi}_2^q = \nu \Psi_2^q (\Psi_1^q \Psi_2^q - l_5^q \Psi_3), \\ \dot{\Psi}_3 = \nu (\Psi_2^q + \Psi_1^q \Psi_2^q \Psi_3 - l_5^q - l_5^q \Psi_3^2), \\ \dot{l}_5^q = \dot{\Psi}_1^q = 0, \\ y = \frac{1}{\Psi_3}. \end{cases} \tag{5.8}$$

We now add the parameter $l_7 = \psi_2$ (with $\dot{l}_7 = 0$) to the system in (5.8), and we consider the output $y = \beta = \arctan\left(\frac{1}{\Psi_3}\right) - l_7$ instead of $y = \frac{1}{\Psi_3}$. In other words, we consider our original system described by the new five-dimensional state $[\Psi_1^q, \Psi_2^q, \Psi_3, l_5^q, l_7]^T$. We know that we still have a continuous symmetry (we know that the observable codistribution of the system in (5.4) has dimension equal to 4, and we must describe our system with a four-dimensional state). We compute the observable codistribution, and we detect the following symmetry:

$$\left[\Psi_1^{q^2} + 1, \; \Psi_2^q(\Psi_3 - \Psi_1^q), \; \Psi_3^2 + 1, \; 0, \; 1 \right]^T.$$

Again, in order to obtain four generators of the observable codistribution we search for four functions $G(\Psi_1^q, \Psi_2^q, \Psi_3, l_5^q, l_7)$ that satisfy the above symmetry (and whose gradients are independent). Hence, we have

$$(\Psi_1^{q^2} + 1) \frac{\partial G}{\partial \Psi_1^q} + (\Psi_2^q(\Psi_3 - \Psi_1^q)) \frac{\partial G}{\partial \Psi_2^q} + (\Psi_3^2 + 1) \frac{\partial G}{\partial \Psi_3} + \frac{\partial G}{\partial l_7} = 0.$$

Again, a very simple solution is provided by l_5^q. Then, by proceeding as before, we found the following four independent solutions:

$$A^q \triangleq \frac{\Psi_1^q - \Psi_3}{1 + \Psi_1^q \Psi_3}, \ V^q \triangleq \Psi_2^q \frac{1 + \Psi_1^q \Psi_3}{1 + \Psi_3^2}, \ L^q \triangleq l_7 - \arctan \Psi_1^q, \ l_5^q, \tag{5.9}$$

and, finally, a local observable subsystem is

$$\begin{cases} \dot{A}^q &= \nu(1 + A^{q^2})(l_5^q - V^q), \\ \dot{V}^q &= \nu A^q V^q (2V^q - l_5^q), \\ \dot{L}^q &= \dot{l}_5^q = 0, \\ \beta &= -\arctan(A^q) - L^q + S_p \frac{\pi}{2}, \end{cases} \tag{5.10}$$

where S_p can be ± 1 depending on the values of the system parameters and the initial vehicle configuration.

The above subsystem has a very practical importance. It tells us that, when the vehicle accomplishes circular trajectories, the information contained in the sensor measurements (i.e., the information contained in the function $\nu(t)$ and $\beta(t)$) allows us to estimate only the state $[A^q, V^q, L^q, l_5^q]^T$ and not the original state $[l_1, l_2, l_3, l_4, l_5, l_6, l_7]^T$. In addition, it provides the link between the observable state $[A^q, V^q, L^q, l_5^q]^T$ and the sensor data ν and β. In [32], we used the local subsystem in (5.10) to introduce a strategy able to simultaneously calibrate the odometry and the camera of a wheeled robot.

5.5 ▪ Visual inertial sensor fusion with calibrated sensors

The problem of fusing visual and inertial data has been extensively investigated in recent years (e.g., [13, 21, 23, 30, 31, 40]). Vision and inertial sensing have received great attention from the mobile robotics community. These sensors require no external infrastructure, and this is a key advantage for robots operating in unknown environments where GPS signals are shadowed. Additionally, these sensors have very interesting complementarities, and together provide rich information to build a system capable of vision-aided inertial navigation and mapping. It is worth noting that most mammals rely precisely on these sensing capabilities to autonomously move in the environment (and this explains the great interest in this sensor fusion problem also from the neuroscience community (e.g., [4, 10, 11])).

We investigate the observability properties of this problem. In this section we consider the easiest scenario, i.e., when the sensors are fully calibrated. Specifically, we base our observability analysis on the following assumptions:

- The inertial sensors provide bias-free measurements.

- The sensors are extrinsically calibrated; i.e., their relative configuration is known (in particular, we assume that the camera and the IMU[18] frames coincide).

- The magnitude of the gravity is known.

In section 5.6 we relax all the above assumptions. Finally, in Chapter 8 we return to this problem by investigating the very challenging case where some of the degrees of freedom of the system, e.g., some components of the acceleration and/or of the angular speed, cannot be measured by the inertial sensors and act as unknown inputs on the dynamics of the system.

---

[18]IMU is the abbreviation for inertial measurement unit. It consists of three orthogonal accelerometers and three orthogonal gyroscopes.

5.5.1 ▪ The system

We consider a rigid body that moves in the 3D environment. We denote it by $\mathcal{B}$, and we call it the *body*. We assume that it is equipped with a monocular camera and an IMU. We introduce a global frame to characterize the motion of the body. Its z-axis points vertically upwards. We will adopt lower-case letters to denote vectors in this frame (e.g., the gravity is $g = [0, 0, -|g|]^T$, where $|g|$ is the magnitude of the gravitational acceleration). We define the body local frame as the IMU frame, and we assume that it coincides with the camera frame. We will adopt upper-case letters to denote vectors in the local frame.

The IMU provides the body angular speed and acceleration. Actually, regarding the acceleration, the one perceived by the accelerometer (A) is not simply the body inertial acceleration (A_B). It also contains the gravitational acceleration (G). In particular, we have $A = A_B - G$ since, when $\mathcal{B}$ does not accelerate (i.e., A_B vanishes), the accelerometer perceives an acceleration which is the same as that of an object accelerated upward in the absence of gravity.

We assume that the camera is observing a point feature during a given time interval. We fix the global frame attached to this feature. The body and the feature are displayed in Figure 5.2.

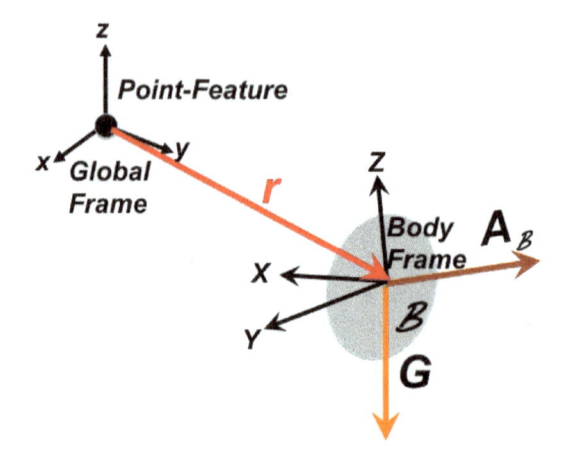

Figure 5.2. *The body moves in the 3D environment and observes a point feature at the origin by its on-board monocular camera.*

We adopt a quaternion to represent the body orientation. Indeed, even if this representation is redundant, it is very powerful since the dynamics can be expressed in a very easy and compact notation (see Appendix B).

Our system is characterized by the state

$$x = [r_x, r_y, r_z, v_x, v_y, v_z, q_t, q_x, q_y, q_z]^T, \quad (5.11)$$

where the following hold:

- $r = [r_x, r_y, r_z]^T$ is the position of the body in the global frame.

- $v = [v_x, v_y, v_z]^T$ is the speed of the body in the global frame.

5.5. Visual inertial sensor fusion with calibrated sensors

- $q = q_t + q_x i + q_y j + q_z k$ is the unit quaternion that describes the rotation between the global and the local frames.[19]

Additionally, we have the following:

- $A = [A_x, A_y, A_z]^T$ is the body acceleration perceived by the IMU (this includes both the inertial acceleration and gravity).

- $\Omega = [\Omega_x \ \Omega_y \ \Omega_z]^T$ is the angular speed expressed in the local frame.

In the following, for each vector defined in the 3D space, the subscript q will be adopted to denote the corresponding imaginary quaternion. For instance, $\Omega_q = 0 + \Omega_x \, i + \Omega_y \, j + \Omega_z \, k$. By using the properties of the unit quaternions, we can easily obtain vectors in the global frame starting from the local frame and vice versa, as it is explained in Appendix B. For instance, given $\Omega = [\Omega_x \ \Omega_y \ \Omega_z]$ in the local frame, we build $\Omega_q = 0 + \Omega_x \, i + \Omega_y \, j + \Omega_z \, k$, and then we compute the quaternion product $\omega_q = q\Omega_q q^*$. The result will be an imaginary quaternion,[20] i.e., $\omega_q = 0 + \omega_x \, i + \omega_y \, j + \omega_z \, k$. The vector $\omega = [\omega_x \ \omega_y \ \omega_z]^T$ is the body angular speed in the global frame. Conversely, to obtain this vector in the local frame starting from ω, it suffices to compute the quaternion product $\Omega_q = q^* \omega_q q$.

The dynamics of the body are

$$\begin{bmatrix} \dot{r}_q &= v_q, \\ \dot{v}_q &= q A_q q^* - |g|k, \\ \dot{q} &= \tfrac{1}{2} q \Omega_q, \end{bmatrix} \tag{5.12}$$

where k is the fourth fundamental quaternion unit ($k = 0 + 0\,i + 0\,j + 1\,k$).

The monocular camera provides the position of the point feature in the local frame up to a scale. Let us denote this position by F. We have

$$F_q = 0 + F_x \, i + F_y \, j + F_z \, k = q^*(-r_q) q. \tag{5.13}$$

Since the camera provides F up to a scale, it provides the two ratios of its components: $\frac{F_x}{F_z}, \frac{F_y}{F_z}$. These are the outputs of our system. Finally, we must account for the constraint that expresses the unity of q. For the observability analysis, we can account for this constraint by adding the further output $(q_t)^2 + (q_x)^2 + (q_y)^2 + (q_z)^2$. Therefore, our system is characterized by the following three outputs:

$$y = h(x) = \begin{bmatrix} h_1(x) \\ h_2(x) \\ h_3(x) \end{bmatrix} = \begin{bmatrix} F_x/F_z \\ F_y/F_z \\ (q_t)^2 + (q_x)^2 + (q_y)^2 + (q_z)^2 \end{bmatrix}, \tag{5.14}$$

where the components F_x, F_y, and F_z depend on the state x through (5.13).

5.5.2 ▪ Observability analysis

We study the observability properties of the system characterized by the state in (5.11), the dynamics in (5.12), and the three outputs in (5.13) and (5.14).

---

[19] A quaternion $q = q_t + q_x i + q_y j + q_z k$ is a unit quaternion if the product with its conjugate is 1, i.e., $qq^* = q^*q = (q_t + q_x i + q_y j + q_z k)(q_t - q_x i - q_y j - q_z k) = (q_t)^2 + (q_x)^2 + (q_y)^2 + (q_z)^2 = 1$ (see Appendix B).
[20] The product of a unit quaternion times an imaginary quaternion times the conjugate of the unit quaternion is always an imaginary quaternion.

Chapter 5. Observability Analysis for Systems in the Absence of Unknown Inputs

We start by computing the observable codistribution. By comparing the above equations with (4.1) we obtain $m = 6$, $p = 3$, $u_1 = A_x$, $u_2 = A_y$, $u_3 = A_z$, $u_4 = \Omega_x$, $u_5 = \Omega_y$, $u_6 = \Omega_z$,

$$f^0(x) = \begin{bmatrix} v_x \\ v_y \\ v_z \\ 0 \\ 0 \\ -|g| \\ 0 \\ 0 \\ 0 \\ 0 \end{bmatrix}, \quad f^1(x) = \begin{bmatrix} 0 \\ 0 \\ 0 \\ q_t^2 + q_x^2 - q_y^2 - q_z^2 \\ 2(q_t q_z + q_x q_y) \\ 2(q_x q_z - q_t q_y) \\ 0 \\ 0 \\ 0 \\ 0 \end{bmatrix}, \quad f^2(x) = \begin{bmatrix} 0 \\ 0 \\ 0 \\ 2(q_x q_y - q_t q_z) \\ q_t^2 - q_x^2 + q_y^2 - q_z^2 \\ 2(q_t q_x + q_y q_z) \\ 0 \\ 0 \\ 0 \\ 0 \end{bmatrix},$$

$$f^3(x) = \begin{bmatrix} 0 \\ 0 \\ 0 \\ 2(q_t q_y + q_x q_z) \\ 2(q_y q_z - q_x q_t) \\ q_t^2 - q_x^2 - q_y^2 + q_z^2 \\ 0 \\ 0 \\ 0 \\ 0 \end{bmatrix}, \quad f^4(x) = \frac{1}{2}\begin{bmatrix} 0 \\ 0 \\ 0 \\ 0 \\ 0 \\ 0 \\ -q_x \\ q_t \\ q_z \\ -q_y \end{bmatrix}, \quad f^5(x) = \frac{1}{2}\begin{bmatrix} 0 \\ 0 \\ 0 \\ 0 \\ 0 \\ 0 \\ -q_y \\ -q_z \\ q_t \\ q_x \end{bmatrix},$$

$$f^6(x) = \frac{1}{2}\begin{bmatrix} 0 \\ 0 \\ 0 \\ 0 \\ 0 \\ 0 \\ -q_z \\ q_y \\ -q_x \\ q_t \end{bmatrix}, \quad y = \begin{bmatrix} h_1 \\ h_2 \\ h_3 \end{bmatrix} = \begin{bmatrix} r_x q_t^2 - 2r_z q_t q_y + 2r_y q_t q_z + r_x q_x^2 + 2r_y q_x q_y + 2r_z q_x q_z - r_x q_y^2 - r_x q_z^2 \\ r_z q_t^2 - 2r_y q_t q_x + 2r_x q_t q_y - r_z q_x^2 + 2r_x q_x q_z - r_z q_y^2 + 2r_y q_y q_z + r_z q_z^2 \\ r_y q_t^2 + 2r_z q_t q_x - 2r_x q_t q_z - r_y q_x^2 + 2r_x q_x q_y + r_y q_y^2 + 2r_z q_y q_z - r_y q_z^2 \\ r_z q_t^2 - 2r_y q_t q_x + 2r_x q_t q_y - r_z q_x^2 + 2r_x q_x q_z - r_z q_y^2 + 2r_y q_y q_z + r_z q_z^2 \\ q_t^2 + q_x^2 + q_y^2 + q_z^2 \end{bmatrix}.$$

We compute the observable codistribution by running Algorithm 4.2. We do not provide the analytic expression of the observable codistribution at each step. At the initialization we have

$$\Omega_0 = \text{span}\{\nabla h_1, \nabla h_2, \nabla h_3\}$$

and its dimension is 3. At the first iterative step we obtain

$$\Omega_1 = \text{span}\{\nabla h_1, \nabla h_2, \nabla h_3, \nabla \mathcal{L}_{f^0} h_1, \nabla \mathcal{L}_{f^0} h_2\};$$

namely, all the above covectors are independent and the dimension of Ω_1 is 5. At the second iterative step we obtain

$$\Omega_2 = \text{span}\{\nabla h_1, \nabla h_2, \nabla h_3, \nabla \mathcal{L}_{f^0} h_1, \nabla \mathcal{L}_{f^0} h_2, \nabla \mathcal{L}_{f^0}^2 h_1, \nabla \mathcal{L}_{f^0 f^1}^2 h_1, \nabla \mathcal{L}_{f^0}^2 h_2\};$$

namely, all the above covectors are independent and the dimension of Ω_2 is 8. At the third iterative step we obtain

$$\Omega_3 = \text{span}\{\nabla h_1, \nabla h_2, \nabla h_3, \nabla \mathcal{L}_{f^0} h_1, \nabla \mathcal{L}_{f^0} h_2, \nabla \mathcal{L}_{f^0}^2 h_1, \nabla \mathcal{L}_{f^0 f^1}^2 h_1, \nabla \mathcal{L}_{f^0}^2 h_2, \nabla \mathcal{L}_{f^0}^3 h_1\};$$

5.5. Visual inertial sensor fusion with calibrated sensors

namely, all the above covectors are independent and the dimension of Ω_3 is 9. At the fourth iterative step Ω_4 remains the same. Hence, we conclude that the observable codistribution is $\Omega = \Omega_3$. Since its dimension is 9, the system has $1(= 10 - 9)$ symmetry (10 is the state dimension). We obtain this symmetry by computing the vector orthogonal to Ω. We obtain

$$\begin{bmatrix} -r_y \\ r_x \\ 0 \\ -v_y \\ v_x \\ 0 \\ -q_z/2 \\ -q_y/2 \\ q_x/2 \\ q_t/2 \end{bmatrix}.$$

The infinitesimal transformation given in (4.24) becomes, for the specific case,

$$\begin{bmatrix} r_x \\ r_y \\ r_z \end{bmatrix} \rightarrow \begin{bmatrix} r_x \\ r_y \\ r_z \end{bmatrix} + \epsilon \begin{bmatrix} -r_y \\ r_x \\ 0 \end{bmatrix} \simeq \begin{bmatrix} \cos\epsilon & -\sin\epsilon & 0 \\ \sin\epsilon & \cos\epsilon & 0 \\ 0 & 0 & 1 \end{bmatrix} \begin{bmatrix} r_x \\ r_y \\ r_z \end{bmatrix},$$

$$\begin{bmatrix} v_x \\ v_y \\ v_z \end{bmatrix} \rightarrow \begin{bmatrix} v_x \\ v_y \\ v_z \end{bmatrix} + \epsilon \begin{bmatrix} -v_y \\ v_x \\ 0 \end{bmatrix} \simeq \begin{bmatrix} \cos\epsilon & -\sin\epsilon & 0 \\ \sin\epsilon & \cos\epsilon & 0 \\ 0 & 0 & 1 \end{bmatrix} \begin{bmatrix} v_x \\ v_y \\ v_z \end{bmatrix},$$

$$\begin{bmatrix} q_t \\ q_x \\ q_y \\ q_z \end{bmatrix} \rightarrow \begin{bmatrix} q_t \\ q_x \\ q_y \\ q_z \end{bmatrix} + \frac{\epsilon}{2} \begin{bmatrix} -q_z \\ -q_y \\ q_x \\ q_t \end{bmatrix},$$

which is an infinitesimal rotation about the vertical axis (regarding the quaternion, this can be verified starting from the last equation in (5.12) that provides $\dot{q} = \frac{1}{2}\omega q$ (where ω is the angular speed in the global frame) and by setting $\omega_x = \omega_y = 0$ and $\omega_z dt = \epsilon$).

We want to describe our system with an observable state. We have many choices. Since we want to achieve a local observable subsystem, we are interested in introducing very simple observable functions that generate the observable codistribution. First of all, we remark that the distance of the point feature is an observable function. This can be verified by checking that the gradient of the function $r_x^2 + r_y^2 + r_z^2$ belongs to Ω (actually, we do not even need to check this: it suffices to remark that the scale is rotation invariant and, consequently, it satisfies the above symmetry and it is observable). Now, if the scale is observable, by combining this knowledge with the two outputs $h_1 = \frac{F_x}{F_z}$ and $h_2 = \frac{F_y}{F_z}$ (and by knowing that $F_x^2 + F_y^2 + F_z^2 = r_x^2 + r_y^2 + r_z^2$), we can select the following three observable functions:

$$F_x, \; F_y, \; F_z.$$

In a similar manner, we can also select the following three observable functions:

$$V_x, \; V_y, \; V_z,$$

which are the components of the body speed expressed in the local frame. Finally, regarding the orientation, we know that the yaw, which characterizes a rotation about the vertical axis, is unobservable. Conversely, the roll and the pitch angles are observable. We denote the roll and pitch with ψ_R and ψ_P, respectively. Hence, we introduce the following observable state:

$$X = [F_x,\ F_y,\ F_z,\ V_x,\ V_y,\ V_z,\ \psi_R,\ \psi_P]^T. \tag{5.15}$$

These components can be expressed in terms of the components of the original state in (5.11) as follows. The first three components are given by (5.13). The last five are

$$V_q = 0 + V_x\,i + V_y\,j + V_z\,k = q^* v_q q, \qquad \psi_R = \arctan\left(2\frac{q_t q_x + q_y q_z}{1 - 2(q_x^2 + q_y^2)}\right),$$

$$\psi_P = \arcsin\left(2(q_t q_y - q_x q_z)\right).$$

From the above equations, we obtain the following description of the local observable subsystem:

$$\begin{bmatrix} \dot{F} &= -\Omega \wedge F - V, \\ \dot{V} &= -\Omega \wedge V + A + G, \\ \dot{\psi}_R &= \Omega_x + \Omega_y \tan\psi_P \sin\psi_R + \Omega_z \tan\psi_P \cos\psi_R, \\ \dot{\psi}_P &= \Omega_y \cos\psi_R - \Omega_z \sin\psi_R, \\ y &= [F_x/F_z,\ F_y/F_z]^T, \end{bmatrix} \tag{5.16}$$

where the symbol "$\wedge$" denotes the cross product and G is the gravity in the local frame and it only depends on the roll and pitch angles:

$$G = |g| \begin{bmatrix} \sin\psi_P \\ -\cos\psi_P \sin\psi_R \\ -\cos\psi_P \cos\psi_R \end{bmatrix}. \tag{5.17}$$

We conclude this section with the following remarks:

- In the case of multiple features, the observability properties remain the same. The yaw angle remains unobservable. To prove this it is unnecessary to repeat the computation of the observable codistribution. Since the gravity is invariant to the yaw, the system maintains the same continuous symmetry that describes a rotation about the vertical axis. In the presence of M point features, a local observable subsystem is given by (5.16), where the first equation $\dot{F} = -\Omega \wedge F - V$ must be replaced by the M equations $\dot{F}^j = -\Omega \wedge F^j - V$, $j = 1, \ldots, M$, and the last equation $y = [F_x/F_z,\ F_y/F_z]^T$ by

$$y = [F_x^1/F_z^1,\ F_y^1/F_z^1,\ \ldots,\ F_x^M/F_z^M,\ F_y^M/F_z^M]^T.$$

- We obtain for our system a single symmetry which expresses the unobservability of the absolute yaw angle. On the other hand, if we do not a priori know the position of the point feature in the global frame, we have three further symmetries which express the unobservability of the absolute position of the point feature (i.e., its position in this global frame). In our analysis, by introducing a global frame whose origin coincides with the point feature, we are implicitly assuming that the position of the point feature is a priori known.

- The equations given in (5.16) could be used for a practical implementation, (e.g., to implement an extended Kalman filter). In this case, we recommend the use of a different output. The camera provides F up to a scale. Instead of the two ratios, $\frac{F_x}{F_z}$ and $\frac{F_y}{F_z}$, which are singular when $F_z = 0$, it is better to introduce two angles (e.g., by setting $[F_x,\ F_y,\ F_z]^T = |F|[\cos\alpha_1\cos\alpha_2,\ \cos\alpha_1\sin\alpha_2,\ \sin\alpha_1]^T$).

- We could perform additional observability analyses to explore further scenarios. For instance, we can repeat the computation of the observable codistribution by only considering the output h_1 or a single angle (e.g., the above α_1). By doing this, it is possible to prove that the observability properties remain the same, meaning that a linear camera (i.e., a camera that only provides the azimuth or the zenith of a point feature) provides the same observable state.

- We can also explore other scenarios by avoiding, in the computation of the observable codistribution, the computation of the Lie derivatives along some of the vector fields that define the dynamics. By doing this, if we obtain the same observable codistribution, we have proved that not all the degrees of freedom must be excited to obtain the same observability properties (for instance, in our computation, the generators of Ω only include the Lie derivative along f^0 and f^1).[21] Conversely, if we found that by avoiding a given vector field we do not reach the same observable codistribution, we have proved that this degree of freedom must be excited. This aspect is very important. In a practical implementation, we can have a divergence (with consequences which can be dramatic) simply because we are not exciting the necessary degrees of freedom in order to achieve the state observability. Unfortunately, the observability analysis does not automatically provide all these singular excitations. We can only a priori set special settings for the system inputs and then compute the corresponding observable codistribution.

- On the basis of the previous remark, we cannot obtain automatically an exhaustive analysis of all the singularities of a given problem by performing an observability analysis. Regarding the visual inertial sensor fusion problem studied in this section, this exhaustive analysis was obtained in [33]. As in the case of other computer vision problems, the key of the analysis was the establishment of an equivalence between the visual-inertial sensor fusion problem and a polynomial equation system (PES). Specifically, [33] established that this sensor fusion problem is equivalent to a very simple PES. In particular, this PES consists of a single polynomial equation of second degree and several linear equations. This PES can be easily solved in closed form, and this solution has the advantage to provide the state without initialization. Even more importantly, this PES contains all the structural properties of the problem. In particular, by studying this PES, we obtain a detailed analysis of the problem by providing all the system singularities and minimal cases depending on the trajectory, the number of camera images, and the features layout. The problem can have up to two solutions in its minimal cases.

Exercise: Prove that, when the body moves at constant speed, the absolute scales becomes unobservable.

5.6 • Visual inertial sensor fusion with uncalibrated sensors

We consider the same problem studied in section 5.5 but when all the sensors are not calibrated and the magnitude of the gravity is unknown. In other words, we assume that all the measurements provided by the inertial sensors (i.e., acceleration and angular speed) are biased. In addition, the monocular camera and the IMU do not share the same frame and the transformation between the two frames is unknown (camera extrinsic calibration).

---

[21] We could even avoid f^1. This means that we have full observability when the body is in free fall.

5.6.1 • The system

We need to introduce several new parameters to characterize our problem. Specifically, the following hold:

1. $A^b = [A_x^b, A_y^b, A_z^b]^T$ is the bias vector that affects the measurements from the three-axial accelerometer. In other words, the true acceleration perceived by an ideal sensor would be $A - A^b$. A^b is assumed constant on the considered time interval.[22]

2. $\Omega^b = [\Omega_x^b, \Omega_y^b, \Omega_z^b]^T$ is the bias vector that affects the measurements from the three-axial gyroscope (we adopt the same conventions that hold for the acceleration).

3. $P^c = [P_x^c, P_y^c, P_z^c]^T$ is the position of the camera optical center in the IMU frame. This consists of three independent (constant) parameters.

4. R^c is the orthogonal matrix that characterizes the different orientation between the camera and the IMU frame. This is a 3×3 matrix that is fully characterized by three independent parameters. However, we adopt a quaternion representation, which includes four nonindependent parameters. Specifically,

$$R^c = \begin{bmatrix} a^2 + b^2 - c^2 - d^2 & 2(bc - ad) & 2(ac + bd) \\ 2(ad + bc) & a^2 - b^2 + c^2 - d^2 & 2(cd - ab) \\ 2(bd - ac) & 2(ab + cd) & a^2 - b^2 - c^2 + d^2 \end{bmatrix},$$

with $a^2 + b^2 + c^2 + d^2 = 1$.

With respect to the problem studied in section 5.5, we have 14 additional parameters for its characterization: 6 for the two bias vectors (A^b, Ω^b), 7 for the camera extrinsic calibration (P^c and a, b, c, d), and 1 for the magnitude of the gravity. Due to the large state dimension, carrying out an observability analysis for this system could be prohibitive from a computational point of view. In order to simplify the computation of the observable codistribution, we need to characterize our system by a state such that the analytic expression of the dynamics and of the output function is as simple as possible. Certainly, the expression of the output function is very simple by introducing in the state the three coordinates of the point feature in the camera frame (i.e., the vector F). Note that, since the camera and the IMU frame do not coincide, F is not the position of the point feature in the IMU frame (see also Figure 5.3). By exploiting the results of the study carried out in section 5.5, and in particular the equations that characterize the observable subsystem given in (5.16), we introduce the following state to characterize our system:

$$x = [F_x, F_y, F_z, V_x, V_y, V_z, \psi_R, \psi_P, A_x^b, A_y^b, A_z^b, \Omega_x^b, \Omega_y^b, \Omega_z^b, P_x^c, P_y^c, P_z^c, a, b, c, d, |g|]^T. \tag{5.18}$$

The dynamics of the vector F depend on the rotation of the body and the speed of the body, as described by the first equation in (5.16). However, we need to change the rotational speed to account for the gyroscope bias and we need also to account for the different orientation between

---
[22] Note that, in practical applications, we performed the estimation of the bias in a few seconds.

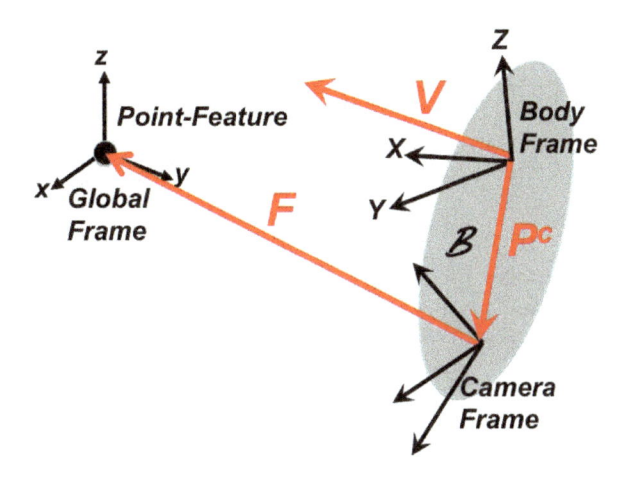

5.6. Visual inertial sensor fusion with uncalibrated sensors

Figure 5.3. *The camera frame does not coincide with the body frame, defined as the IMU frame. F is the position of the point feature in the camera frame. V is the speed of the body in the local (body) frame.*

the IMU and the camera frame. Hence, the dynamics of F due to the rotation of the body is

$$-[R^c(\Omega - \Omega^b)] \wedge F.$$

For the second component, we need to compute the speed of the camera center, expressed in the camera frame. The camera center speed expressed in the IMU frame is $V + (\Omega - \Omega^b) \wedge P^c$. To express this speed in the camera frame, we must premultiply it by R^c.

Hence, the dynamics of F due to the speed of the body is

$$-R^c[V + (\Omega - \Omega^b) \wedge P^c].$$

Therefore, the equations that characterize our system (dynamics and output functions) are

$$\begin{bmatrix} \dot{F} &= -[R^c(\Omega - \Omega^b)] \wedge F - R^c[V + (\Omega - \Omega^b) \wedge P^c], \\ \dot{V} &= -(\Omega - \Omega^b) \wedge V + A - A^b + G, \\ \dot{\psi}_R &= \Omega_x - \Omega_x^b + (\Omega_y - \Omega_y^b)\tan\psi_P \sin\psi_R + (\Omega_z - \Omega_z^b)\tan\psi_P \cos\psi_R, \\ \dot{\psi}_P &= (\Omega_y - \Omega_y^b)\cos\psi_R - (\Omega_z - \Omega_z^b)\sin\psi_R, \\ \dot{A}^b &= \dot{\Omega}^c = \dot{P}^c = [0,0,0]^T, \\ \dot{a} &= \dot{b} = \dot{c} = \dot{d} = \dot{|g|} = 0, \\ y &= [F_x/F_z,\ F_y/F_z,\ a^2 + b^2 + c^2 + d^2]^T, \end{bmatrix} \quad (5.19)$$

where the vector G that appears at the second equation depends on the state through the components ψ_R, ψ_P, and $|g|$, as described by (5.17).

5.6.2 • Observability analysis

We study the observability properties of the system characterized by the state in (5.18) and the equations in (5.19).

We start by computing the observable codistribution. By comparing the above equations with (4.1) we obtain $m = 6$, $p = 3$, $u_1 = A_x$, $u_2 = A_y$, $u_3 = A_z$, $u_4 = \Omega_x$, $u_5 = \Omega_y$, $u_6 = \Omega_z$,

$$f^0(x) = \begin{bmatrix} F_z(\Omega_x^b(2ad+2bc) - \Omega_z^b(2ab-2cd) + \Omega_y^b(a^2-b^2+c^2-d^2)) \\ -F_y(\Omega_y^b(2ab+2cd) - \Omega_x^b(2ac-2bd) + \Omega_z^b(a^2-b^2-c^2+d^2)) \\ -(V_x + \Omega_z^b P_y^c - \Omega_y^b P_z^c)(a^2+b^2-c^2-d^2) - (2ac+2bd)(V_z + \Omega_y^b P_x^c - \Omega_x^b P_y^c) \\ +(2ad-2bc)(V_y - \Omega_z^b P_x^c + \Omega_x^b P_z^c) \\[4pt]
F_x(\Omega_y^b(2ab+2cd) - \Omega_z^b(2ac-2bd) + \Omega_z^b(a^2-b^2-c^2+d^2)) \\ -(V_y - \Omega_z^b P_x^c + \Omega_x^b P_z^c)(a^2-b^2+c^2-d^2) \\ -F_z(\Omega_z^b(2ac+2bd) - \Omega_y^b(2ad-2bc) + \Omega_x^b(a^2+b^2-c^2-d^2)) \\ +(2ab-2cd)(V_z + \Omega_y^b P_x^c - \Omega_x^b P_y^c) - (2ad+2bc)(V_x + \Omega_z^b P_y^c - \Omega_y^b P_z^c) \\[4pt]
F_y(\Omega_z^b(2ac+2bd) - \Omega_y^b(2ad-2bc) + \Omega_x^b(a^2+b^2-c^2-d^2)) \\ -F_x(\Omega_x^b(2ad+2bc) - \Omega_z^b(2ab-2cd) + \Omega_y^b(a^2-b^2+c^2-d^2)) \\ -(V_z + \Omega_y^b P_x^c - \Omega_x^b P_y^c)(a^2-b^2-c^2+d^2) - (2ab+2cd)(V_y - \Omega_z^b P_x^c + \Omega_x^b P_z^c) \\ +(2ac-2bd)(V_x + \Omega_z^b P_y^c - \Omega_y^b P_z^c) \\[4pt]
\Omega_y^b V_z - \Omega_z^b V_y - A_x^b + |g|\sin\psi_P \\[4pt]
\Omega_z^b V_x - A_y^b - \Omega_x^b V_z - |g|\cos\psi_P \sin\psi_R \\[4pt]
\Omega_x^b V_y - \Omega_y^b V_x - A_z^b - |g|\cos\psi_R \cos\psi_P \\[4pt]
-\Omega_x^b - \Omega_z^b \cos\psi_R \tan\psi_P - \Omega_y^b \sin\psi_R \tan\psi_P \\[4pt]
\Omega_z^b \sin\psi_R - \Omega_y^b \cos\psi_R \\[4pt]
0_{14} \end{bmatrix},$$

$$f^1(x) = \begin{bmatrix} 0 \\ 0 \\ 0 \\ 1 \\ 0 \\ 0 \\ 0 \\ 0 \\ 0_{14} \end{bmatrix}, \quad f^2(x) = \begin{bmatrix} 0 \\ 0 \\ 0 \\ 0 \\ 1 \\ 0 \\ 0 \\ 0 \\ 0_{14} \end{bmatrix}, \quad f^3(x) = \begin{bmatrix} 0 \\ 0 \\ 0 \\ 0 \\ 0 \\ 1 \\ 0 \\ 0 \\ 0_{14} \end{bmatrix},$$

$$f^4(x) = \begin{bmatrix} 2F_y bd - 2F_z ad - 2F_z bc - 2F_y ac - 2P_y^c ac - 2P_z^c ad + 2P_z^c bc - 2P_y^c bd \\ F_z a^2 + F_z b^2 - F_z c^2 - F_z d^2 + P_z^c a^2 - P_z^c b^2 + P_z^c c^2 - P_z^c d^2 \\ +2F_x ac - 2F_x bd + 2P_y^c ab - 2P_y^c cd \\ F_y c^2 - F_y b^2 - F_y a^2 + F_y d^2 - P_y^c a^2 + P_y^c b^2 + P_y^c c^2 - P_y^c d^2 \\ +2F_x ad + 2F_x bc + 2P_z^c ab + 2P_z^c cd \\ 0 \\ V_z \\ -V_y \\ 1 \\ 0 \\ 0_{14} \end{bmatrix},$$

5.6. Visual inertial sensor fusion with uncalibrated sensors

$$f^5(x) = \begin{bmatrix} F_z b^2 - F_z a^2 - F_z c^2 + F_z d^2 - P_z^c a^2 - P_z^c b^2 + P_z^c c^2 + P_z^c d^2 \\ +2F_y ab + 2F_y cd + 2P_x^c ac + 2P_x^c bd \\ 2F_z bc - 2F_z ad - 2F_x ab - 2F_x cd - 2P_x^c ab - 2P_z^c ad - 2P_z^c bc + 2P_x^c cd \\ F_x a^2 - F_x b^2 + F_x c^2 - F_x d^2 + P_x^c a^2 - P_x^c b^2 - P_x^c c^2 + P_x^c d^2 \\ +2F_y ad - 2F_y bc + 2P_z^c ac - 2P_z^c bd \\ -V_z \\ 0 \\ V_x \\ \sin\psi_R \tan\psi_P \\ \cos\psi_R \\ 0_{14} \end{bmatrix},$$

$$f^6(x) = \begin{bmatrix} F_y a^2 - F_y b^2 - F_y c^2 + F_y d^2 + P_y^c a^2 + P_y^c b^2 - P_y^c c^2 \\ -P_y^c d^2 + 2F_z ab - 2F_z cd + 2P_x^c ad - 2P_x^c bc \\ F_x b^2 - F_x a^2 + F_x c^2 - F_x d^2 - P_x^c a^2 + P_x^c b^2 - P_x^c c^2 + P_x^c d^2 \\ +2F_z ac + 2F_z bd + 2P_y^c ad + 2P_y^c bc \\ 2F_x cd - 2F_y ac - 2F_y bd - 2F_x ab - 2P_x^c ab - 2P_y^c ac + 2P_y^c bd - 2P_x^c cd \\ V_y \\ -V_x \\ 0 \\ \cos\psi_R \tan\psi_P \\ -\sin\psi_R \\ 0_{14} \end{bmatrix},$$

$$h_1(x) = \frac{F_x}{F_z}, \quad h_2(x) = \frac{F_y}{F_z}, \quad h_3(x) = a^2 + b^2 + c^2 + d^2,$$

where 0_{14} is the zero 14-dimensional column vector. We compute the observable codistribution by running Algorithm 4.2. We do not provide the analytic expression of the observable codistribution at each step. At the initialization we have

$$\Omega_0 = \text{span}\{\nabla h_1, \nabla h_2, \nabla h_3\}$$

and its dimension is 3. At the first iterative step we obtain

$$\Omega_1 = \text{span}\{\nabla h_1, \nabla h_2, \nabla h_3, \nabla\mathcal{L}_0 h_1, \nabla\mathcal{L}_4 h_1, \nabla\mathcal{L}_5 h_1, \nabla\mathcal{L}_6 h_1, \nabla\mathcal{L}_0 h_2, \nabla\mathcal{L}_4 h_2, \nabla\mathcal{L}_5 h_2, \nabla\mathcal{L}_6 h_2\}$$

where, for simplicity's sake, we set $\mathcal{L}_i = \mathcal{L}_{f^i}$. The dimension of Ω_1 is 11. At the second iterative step we obtain

$$\Omega_2 = \text{span}\{\nabla h_1, \nabla h_2, \nabla h_3, \nabla\mathcal{L}_0 h_1, \nabla\mathcal{L}_4 h_1, \nabla\mathcal{L}_5 h_1, \nabla\mathcal{L}_6 h_1, \nabla\mathcal{L}_0 h_2, \nabla\mathcal{L}_4 h_2, \nabla\mathcal{L}_5 h_2, \nabla\mathcal{L}_6 h_2,$$
$$\nabla\mathcal{L}_{00}^2 h_1, \nabla\mathcal{L}_{01}^2 h_1, \nabla\mathcal{L}_{04}^2 h_1, \nabla\mathcal{L}_{05}^2 h_1, \nabla\mathcal{L}_{06}^2 h_1, \nabla\mathcal{L}_{00}^2 h_2, \nabla\mathcal{L}_{04}^2 h_2\};$$

namely, all the above covectors are independent and the dimension of Ω_2 is 18.

We need to compute Ω_3. As in all the other cases, this is obtained by computing the Lie derivatives of the generators of Ω_2 along all the vector fields $f^0, f^1, \ldots, f^6$. On the other hand, the computation becomes prohibitive. It is convenient to find new generators of Ω_2, in particular observable functions whose analytic expression in terms of the state is as simple as possible.

We can verify that
$$\nabla F_x \in \Omega_2.$$

From the expression of h_1 and h_2, we obtain that
$$\nabla F_y \in \Omega_2, \quad \nabla F_z \in \Omega_2.$$

In addition, we also obtain that
$$\nabla V_x \in \Omega_2, \quad \nabla V_y \in \Omega_2, \quad \nabla V_z \in \Omega_2.$$

This provides an enormous simplification. We know that the above six covectors belong to Ω_2. Hence, since their expression is very easy, we could use them as generators of Ω_2 (and eliminate six among the old generators, whose expression is burdensome). On the other hand, we can do even better. We can consider a new system, characterized by the same state in (5.18), and the same dynamics given in (5.19). The new system differs from the original one for the output. In particular, the new output is
$$y = [F_x, F_y, F_z, V_x, V_y, V_z, a^2 + b^2 + c^2 + d^2]^T.$$

The new system has the same observability properties (the same observable codistribution) of the original system. It is characterized by $p = 7$ outputs. We set
$$h_1 = F_x, \ h_2 = F_y, \ h_3 = F_z, \ h_4 = V_x, \ h_5 = V_y, \ h_6 = V_z, \ h_7 = a^2 + b^2 + c^2 + d^2.$$

We compute the observable codistribution by running Algorithm 4.2. At the initialization we have
$$\Omega_0 = \text{span}\{\nabla h_1, \nabla h_2, \nabla h_3, \nabla h_4, \nabla h_5, \nabla h_6, \nabla h_7\}$$
and its dimension is 7. At the first iterative step we obtain
$$\Omega_1 = \text{span}\{\nabla h_1, \nabla h_2, \nabla h_3, \nabla h_4, \nabla h_5, \nabla h_6, \nabla h_7, \nabla \mathcal{L}_0 h_1, \nabla \mathcal{L}_4 h_1, \nabla \mathcal{L}_5 h_1, \nabla \mathcal{L}_6 h_1,$$
$$\nabla \mathcal{L}_0 h_2, \nabla \mathcal{L}_4 h_2, \nabla \mathcal{L}_5 h_2, \nabla \mathcal{L}_6 h_2, \nabla \mathcal{L}_0 h_4, \nabla \mathcal{L}_0 h_5, \nabla \mathcal{L}_0 h_6\}.$$

The dimension of Ω_1 is 18. We need to compute Ω_2. We obtain
$$\Omega_2 = \text{span}\{\nabla h_1, \nabla h_2, \nabla h_3, \nabla h_4, \nabla h_5, \nabla h_6, \nabla h_7, \nabla \mathcal{L}_0 h_1, \nabla \mathcal{L}_4 h_1, \nabla \mathcal{L}_5 h_1, \nabla \mathcal{L}_6 h_1,$$
$$\nabla \mathcal{L}_0 h_2, \nabla \mathcal{L}_4 h_2, \nabla \mathcal{L}_5 h_2, \nabla \mathcal{L}_6 h_2, \nabla \mathcal{L}_0 h_4, \nabla \mathcal{L}_0 h_5, \nabla \mathcal{L}_0 h_6, \nabla \mathcal{L}_{00}^2 h_1, \nabla \mathcal{L}_{00}^2 h_4, \nabla \mathcal{L}_{05}^2 h_4, \nabla \mathcal{L}_{00}^2 h_5\},$$
whose dimension is 22, which coincides with the dimension of the state. Therefore, the entire state is weakly observable. This concludes the observability analysis since the system has no symmetry and we do not need to find an observable subsystem, the original system already being observable.

Note that, also in this case with uncalibrated sensors, the yaw angle is unobservable. In other words, if instead of using the state in (5.18) to characterize the system we had used a state in which the orientation of the body also included the yaw, we would have found that the state was not observable. In particular, the distribution orthogonal to the observable codistribution would have included the same symmetry found in section 5.5, which expresses the system invariance under a rotation around the gravity axis.

5.7 ▪ Visual inertial sensor fusion in the cooperative case

In this section we study the observability properties of the visual and inertial sensor fusion problem in the cooperative case. We investigate the extreme case where no point features are available. Additionally, we consider the critical case of only two agents (from now on bodies). In other words, we are interested in investigating the minimal case.

The first questions we wish to answer are the following: *Is it possible to retrieve the absolute scale in these conditions? And the absolute roll and pitch angles?* More generally, we want to determine the entire observable state, i.e., all the physical quantities that it is possible to determine by only using the information contained in the sensor data (from the two cameras and the two IMUs) during a short time interval. We need to perform an observability analysis.

Note that, by following intuitive reasoning, we could immediately understand that the absolute position and the absolute speed of each body are not observable because no sensor is able to perceive features attached to the global frame (the observability of the two absolute orientations is less intuitive because the sensors perceive the gravity through their accelerometers). On the other hand, for educational purposes, we do not exploit these intuitive hints. We wish to obtain all the observability properties by simply applying our systematic procedure. This clearly shows the power of the procedure, since it is able to provide the answer also when an intuitive reasoning is not possible.

5.7.1 ▪ The system

We consider two rigid bodies that move in the 3D environment. Each body is equipped with an IMU and a monocular camera. We assume that, for each body, all the sensors share the same frame. Without loss of generality, we define the body local frame as this common frame. The accelerometer sensors perceive both the gravity and the inertial acceleration in the local frame. The gyroscopes provide the angular speed in the local frame. Finally, the monocular camera of each body provides the bearing of the other body in its local frame[23] (see Figure 5.4 for an illustration). Additionally, we assume that the z-axis of the global frame is aligned with the direction of the gravity.

We adopt the following notations:

- $r^1 = [r_x^1, r_y^1, r_z^1]$ and $r^2 = [r_x^2, r_y^2, r_z^2]$ are the positions of the two bodies in the global frame;

- $v^1 = [v_x^1, v_y^1, v_z^1]$ and $v^2 = [v_x^2, v_y^2, v_z^2]$ are the velocities of the two bodies in the global frame;

- $q^1 = q_t^1 + q_x^1 i + q_y^1 j + q_z^1 k$ and $q^2 = q_t^2 + q_x^2 i + q_y^2 j + q_z^2 k$ are the two unit quaternions that describe the rotations between the global and the two local frames, respectively.

As in section 5.5.1, for each vector defined in the 3D space, the subscript q will be adopted to denote the corresponding imaginary quaternion. Additionally, we denote by A^1, Ω^1, A^2, and Ω^2 the acceleration and the angular speed perceived by the IMU mounted on the first and the second bodies (regarding the accelerations, these include both the inertial accelerations and the gravity).

---

[23]Note that we do not assume that the camera observations contain metric information (due, for instance, to the known size of the observed body). The two bodies can operate far from each other, and a single camera observation only consists of the bearing of the observed body in the frame of the observer. In other words, each body acts as a moving point feature with respect to the other body.

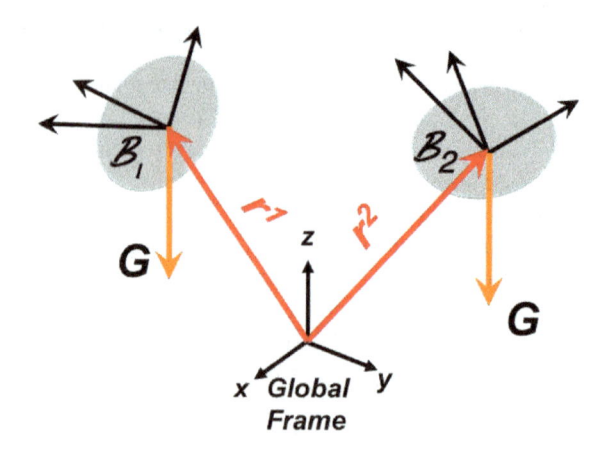

118   Chapter 5. Observability Analysis for Systems in the Absence of Unknown Inputs

Figure 5.4. *The two bodies with their local frames and the global frame.*

The dynamics of the first/second body are ($rob = 1, 2$)

$$\begin{bmatrix} \dot{r}_q^{rob} = v_q^{rob}, \\ \dot{v}_q^{rob} = q^{rob} A_q^{rob} (q^{rob})^* - |g|k, \\ \dot{q}^{rob} = \frac{1}{2} q^{rob} \Omega_q^{rob}, \end{bmatrix} \quad (5.20)$$

where k is the fourth fundamental quaternion unit ($k = 0 + 0\,i + 0\,j + 1\,k$).

Our system can be characterized by the following state:

$$X = [r_x^1\ r_y^1\ r_z^1\ v_x^1\ v_y^1\ v_z^1\ q_t^1\ q_x^1\ q_y^1\ q_z^1\ r_x^2\ r_y^2\ r_z^2\ v_x^2\ v_y^2\ v_z^2\ q_t^2\ q_x^2\ q_y^2\ q_z^2]^T. \quad (5.21)$$

The monocular camera on the first body provides the position of the second body in the local frame of the first body up to a scale. The position of the second body in the local frame of the first body is given by the three components of the following imaginary quaternion:

$$p_q^1 = (q^1)^* (r_q^2 - r_q^1)\, q^1. \quad (5.22)$$

Hence, the first camera provides the quaternion p_q^1 up to a scale. For the observability analysis, it is convenient to use the ratios of its components:

$$h_1(X) = \frac{[p_q^1]_x}{[p_q^1]_z}, \quad h_2(X) = \frac{[p_q^1]_y}{[p_q^1]_z}, \quad (5.23)$$

where the subscripts x, y, and z indicate, respectively, the i, j, and k components of the corresponding quaternion. Similarly, the second camera provides

$$h_3(X) = \frac{[p_q^2]_x}{[p_q^2]_z}, \quad h_4(X) = \frac{[p_q^2]_y}{[p_q^2]_z}, \quad (5.24)$$

where p_q^2 is the imaginary quaternion whose three components are the position of the first body in the local frame of the second, namely

$$p_q^2 = (q^2)^* (r_q^1 - r_q^2)\, q^2. \quad (5.25)$$

5.7. Visual inertial sensor fusion in the cooperative case

Note that using the ratios in (5.23) and (5.24) as observations can generate problems due to singularities and, when the camera measurements are used to estimate a state, it is more preferable to adopt different quantities (e.g., the two bearing angles, i.e., the azimuth and the zenith). For the observability analysis, this problem does not arise.

Finally, we need to add the two observation functions that express the constraint that the two quaternions, q^1 and q^2, are unit quaternions. The two additional observations are

$$h_5(X) = (q_t^1)^2 + (q_x^1)^2 + (q_y^1)^2 + (q_z^1)^2, \quad h_6(X) = (q_t^2)^2 + (q_x^2)^2 + (q_y^2)^2 + (q_z^2)^2. \quad (5.26)$$

5.7.2 ▪ Observability analysis

We study the observability properties of the system characterized by the state in (5.21), the dynamics in (5.20), and the six outputs in (5.23), (5.24), and (5.26).

We start by computing the observable codistribution. By comparing the above equations with (4.1) we obtain $m = 12$, $p = 6$, $u_1 = A_x^1$, $u_2 = A_y^1$, $u_3 = A_z^1$, $u_4 = \Omega_x^1$, $u_5 = \Omega_y^1$, $u_6 = \Omega_z^1$, $u_7 = A_x^2$, $u_8 = A_y^2$, $u_9 = A_z^2$, $u_{10} = \Omega_x^2$, $u_{11} = \Omega_y^2$, $u_{12} = \Omega_z^2$,

$$f^0(x) = \begin{bmatrix} v_x^1 \\ v_y^1 \\ v_z^1 \\ 0 \\ 0 \\ -|g| \\ 0 \\ 0 \\ 0 \\ v_x^2 \\ v_y^2 \\ v_z^2 \\ 0 \\ 0 \\ -|g| \\ 0 \\ 0 \\ 0 \\ 0 \end{bmatrix}, \quad f^1(X) = \begin{bmatrix} 0 \\ 0 \\ 0 \\ (q_t^1)^2 + (q_x^1)^2 - (q_y^1)^2 - (q_z^1)^2 \\ 2q_t^1 q_z^1 + 2q_x^1 q_y^1 \\ 2q_x^1 q_z^1 - 2q_t^1 q_y^1 \\ 0 \\ 0 \\ 0 \\ 0_{10} \end{bmatrix},$$

$$f^2(X) = \begin{bmatrix} 0 \\ 0 \\ 0 \\ 2q_x^1 q_y^1 - 2q_t^1 q_z^1 \\ (q_t^1)^2 - (q_x^1)^2 + (q_y^1)^2 - (q_z^1)^2 \\ 2q_t^1 q_x^1 + 2q_y^1 q_z^1 \\ 0 \\ 0 \\ 0 \\ 0_{10} \end{bmatrix},$$

Chapter 5. Observability Analysis for Systems in the Absence of Unknown Inputs

$$f^3(X) = \begin{bmatrix} 0 \\ 0 \\ 0 \\ 2q_t^1 q_y^1 + 2q_x^1 q_z^1 \\ 2q_y^1 q_z^1 - 2q_t^1 q_x^1 \\ (q_t^1)^2 - (q_x^1)^2 - (q_y^1)^2 + (q_z^1)^2 \\ 0 \\ 0 \\ 0 \\ 0_{10} \end{bmatrix}, \quad f^4(X) = \begin{bmatrix} 0 \\ 0 \\ 0 \\ 0 \\ 0 \\ 0 \\ -q_x^1/2 \\ q_t^1/2 \\ q_z^1/2 \\ -q_y^1/2 \\ 0_{10} \end{bmatrix},$$

$$f^5(X) = \begin{bmatrix} 0 \\ 0 \\ 0 \\ 0 \\ 0 \\ 0 \\ -q_y^1/2 \\ -q_z^1/2 \\ q_t^1/2 \\ q_x^1/2 \\ 0_{10} \end{bmatrix}, \quad f^6(X) = \begin{bmatrix} 0 \\ 0 \\ 0 \\ 0 \\ 0 \\ 0 \\ -q_z^1/2 \\ q_y^1/2 \\ -q_x^1/2 \\ q_t^1/2 \\ 0_{10} \end{bmatrix}.$$

The remaining vector fields, $f_7(X)$, $f_8(X)$, $f_9(X)$, $f_{10}(X)$, $f_{11}(X)$, $f_{12}(X)$, can be easily obtained from $f_1(X)$, $f_2(X)$, $f_3(X)$, $f_4(X)$, $f_5(X)$, $f_6(X)$, respectively. In particular, the zero column vector 0_{10} moves bottom up and the apex 1 must be replaced by 2; e.g., we have

$$f^7(X) = \begin{bmatrix} 0_{10} \\ 0 \\ 0 \\ 0 \\ (q_t^2)^2 + (q_x^2)^2 - (q_y^2)^2 - (q_z^2)^2 \\ 2q_t^2 q_z^2 + 2q_x^2 q_y^2 \\ 2q_x^2 q_z^2 - 2q_t^2 q_y^2 \\ 0 \\ 0 \\ 0 \\ 0 \end{bmatrix}.$$

Finally, the first two outputs are given in (5.23), with

$$\begin{aligned}
(p_q^1)_x(X) = {} & q_t^1(q_t^1 \Delta_x^{12} + q_z^1 \Delta_y^{12} - q_y^1 \Delta_z^{12}) - q_y^1(q_t^1 \Delta_y^{12} + q_y^1 \Delta_x^{12} - q_x^1 \Delta_y^{12}) \\
& + q_z^1(q_t^1 \Delta_y^{12} - q_z^1 \Delta_x^{12} + q_x^1 \Delta_z^{12}) + q_x^1(q_x^1 \Delta_x^{12} + q_y^1 \Delta_y^{12} + q_z^1 \Delta_z^{12}),
\end{aligned}$$

$$\begin{aligned}
(p_q^1)_y(X) = {} & q_t^1(q_t^1 \Delta_y^{12} - q_z^1 \Delta_x^{12} + q_x^1 \Delta_z^{12}) + q_x^1(q_t^1 \Delta_z^{12} + q_y^1 \Delta_x^{12} - q_x^1 \Delta_y^{12}) \\
& - q_z^1(q_t^1 \Delta_x^{12} + q_z^1 \Delta_y^{12} - q_y^1 \Delta_z^{12}) + q_y^1(q_x^1 \Delta_x^{12} + q_y^1 \Delta_y^{12} + q_z^1 \Delta_z^{12}),
\end{aligned}$$

$$\begin{aligned}
(p_q^1)_z(X) = {} & q_t^1(q_t^1 \Delta_z^{12} + q_y^1 \Delta_x^{12} - q_x^1 \Delta_y^{12}) - q_x^1(q_t^1 \Delta_y^{12} - q_z^1 \Delta_x^{12} + q_x^1 \Delta_z^{12}) \\
& + q_y^1(q_t^1 \Delta_x^{12} + q_z^1 \Delta_y^{12} - q_y^1 \Delta_z^{12}) + q_z^1(q_x^1 \Delta_x^{12} + q_y^1 \Delta_y^{12} + q_z^1 \Delta_z^{12}),
\end{aligned}$$

5.7. Visual inertial sensor fusion in the cooperative case

where
$$\Delta^{12} = r^1 - r^2.$$

The second two outputs are given in (5.24), and the expression of p_q^2 is the same as given above with the substitution $1 \leftrightarrow 2$ and $\Delta^{21} = -\Delta^{12}$. The last two outputs are already given in (5.26), in terms of the components of the state.

We compute the observable codistribution by running Algorithm 4.2. We do not provide the analytic expression of the observable codistribution at each step. At the initialization we have

$$\Omega_0 = \text{span}\{\nabla h_1, \nabla h_2, \nabla h_3, \nabla h_4, \nabla h_5, \nabla h_6\}$$

and its dimension is 6. At the first iterative step we obtain

$$\Omega_1 = \text{span}\{\nabla h_1, \nabla h_2, \nabla h_3, \nabla h_4, \nabla h_5, \nabla h_6, \nabla \mathcal{L}_0 h_1, \nabla \mathcal{L}_0 h_2, \nabla \mathcal{L}_0 h_3\},$$

where, for the sake of simplicity, we set $\mathcal{L}_0 = \mathcal{L}_{f^0}$. The dimension of Ω_1 is 9. At the second iterative step we obtain

$$\Omega_2 = \text{span}\{\nabla h_1, \nabla h_2, \nabla h_3, \nabla h_4, \nabla h_5, \nabla h_6, \nabla \mathcal{L}_0 h_1, \nabla \mathcal{L}_0 h_2, \nabla \mathcal{L}_0 h_3, \nabla \mathcal{L}_{00}^2 h_1, \nabla \mathcal{L}_{01}^2 h_1\};$$

namely, all the above covectors are independent and the dimension of Ω_2 is 11. By repeating a further step, we obtain that $\Omega_3 = \Omega_2$. This means that the algorithm has converged and the observable codistribution is Ω_2 and has dimension 11.

Note that the choice of the 11 generators of Ω_2 is not unique. In particular, it is possible to avoid the Lie derivatives of the functions h_3 and h_4. Specifically, a possible choice is

$$\nabla h_1, \nabla h_2, \nabla h_5, \nabla h_6, \nabla \mathcal{L}_0 h_1, \nabla \mathcal{L}_0 h_2, \nabla \mathcal{L}_{00}^2 h_1, \nabla \mathcal{L}_{01}^2 h_1, \nabla \mathcal{L}_{07}^2 h_1, \nabla \mathcal{L}_{08}^2 h_1, \nabla \mathcal{L}_{07}^2 h_2.$$

This means that we obtain the same observability properties when only $\mathcal{B}_1$ (or only $\mathcal{B}_2$) is equipped with a camera. In other words, the presence of two cameras does not improve the observability properties with respect to the case of a single camera mounted on one of the two bodies.

Once we have obtained the observable codistribution, the next step is to obtain the observable state. This state has 11 components. Obviously, a possible choice would be the state that contains the previous 11 Lie derivatives. On the other hand, their expression is too complex and it is much more preferable to find an easier state, whose components have a clear physical meaning.

We need to compute the continuous symmetries of our system. Since the dimension of the state is 20, we have 9 ($= 20 - 11$) independent symmetries. We could obtain these symmetries by directly computing the distribution orthogonal to Ω_2. We found that, in this specific case, the computation is very onerous. It is much more convenient to obtain easier generators of the observable codistribution. In particular, we can verify that

$$\nabla (p_q^1)_x(X) \in \Omega_2.$$

From the expression of h_1 and h_2, we obtain that

$$\nabla (p_q^1)_y(X) \in \Omega_2, \quad \nabla (p_q^1)_z(X) \in \Omega_2.$$

By symmetry, we also have

$$\nabla (p_q^2)_x(X) \in \Omega_2, \quad \nabla (p_q^2)_y(X) \in \Omega_2, \quad \nabla (p_q^2)_z(X) \in \Omega_2.$$

122 Chapter 5. Observability Analysis for Systems in the Absence of Unknown Inputs

We can consider a new system, characterized by the same state in (5.21) and the same dynamics given in (5.20). The new system differs from the original one for the output. In particular, it is characterized by $p = 8$ outputs. We set

$$h_1 = (p_q^1)_x, \; h_2 = (p_q^1)_y, \; h_3 = (p_q^1)_z, \; h_4 = (p_q^2)_x, \; h_5 = (p_q^2)_y, \; h_6 = (p_q^2)_z,$$

$$h_7 = (q_t^1)^2 + (q_x^1)^2 + (q_y^1)^2 + (q_z^1)^2, \; h_8 = (q_t^2)^2 + (q_x^2)^2 + (q_y^2)^2 + (q_z^2)^2.$$

We compute the observable codistribution by running Algorithm 4.2. At the initialization we have

$$\Omega_0 = \text{span}\{\nabla h_1, \nabla h_2, \nabla h_3, \nabla h_4, \nabla h_5, \nabla h_7, \nabla h_8\}$$

and its dimension is 7. At the first iterative step, we obtain

$$\Omega_1 = \text{span}\{\nabla h_1, \nabla h_2, \nabla h_3, \nabla h_4, \nabla h_5, \nabla h_7, \nabla h_8, \nabla \mathcal{L}_0 h_1, \nabla \mathcal{L}_0 h_2, \nabla \mathcal{L}_0 h_3, \nabla \mathcal{L}_0 h_4\}.$$

Since its dimension is 11, we know that we attained the entire observable codistribution. On the other hand, the above generators are much simpler than the ones previously derived. Therefore, the computation of the nine independent symmetries becomes less demanding. By a direct computation of the null space of the matrix whose lines are the covectors that generates the above Ω_1, we obtain

$$\begin{bmatrix} 1 \\ 0 \\ 0 \\ 0 \\ 0 \\ 0 \\ 0 \\ 0 \\ 0 \\ 0 \\ 1 \\ 0 \\ 0 \\ 0 \\ 0 \\ 0 \\ 0 \\ 0 \\ 0 \\ 0 \end{bmatrix}, \begin{bmatrix} 0 \\ 1 \\ 0 \\ 0 \\ 0 \\ 0 \\ 0 \\ 0 \\ 0 \\ 0 \\ 0 \\ 1 \\ 0 \\ 0 \\ 0 \\ 0 \\ 0 \\ 0 \\ 0 \\ 0 \end{bmatrix}, \begin{bmatrix} 0 \\ 0 \\ 1 \\ 0 \\ 0 \\ 0 \\ 0 \\ 0 \\ 0 \\ 0 \\ 0 \\ 0 \\ 1 \\ 0 \\ 0 \\ 0 \\ 0 \\ 0 \\ 0 \\ 0 \end{bmatrix}, \begin{bmatrix} 0 \\ 0 \\ 0 \\ 1 \\ 0 \\ 0 \\ 0 \\ 0 \\ 0 \\ 0 \\ 0 \\ 0 \\ 0 \\ 1 \\ 0 \\ 0 \\ 0 \\ 0 \\ 0 \\ 0 \end{bmatrix}, \begin{bmatrix} 0 \\ 0 \\ 0 \\ 0 \\ 1 \\ 0 \\ 0 \\ 0 \\ 0 \\ 0 \\ 0 \\ 0 \\ 0 \\ 0 \\ 1 \\ 0 \\ 0 \\ 0 \\ 0 \\ 0 \end{bmatrix}, \begin{bmatrix} 0 \\ 0 \\ 0 \\ 0 \\ 0 \\ 1 \\ 0 \\ 0 \\ 0 \\ 0 \\ 0 \\ 0 \\ 0 \\ 0 \\ 0 \\ 1 \\ 0 \\ 0 \\ 0 \\ 0 \end{bmatrix}, \begin{bmatrix} 0 \\ -r_z^1 \\ r_y^1 \\ 0 \\ -v_z^1 \\ v_y^1 \\ -q_x^1/2 \\ q_t^1/2 \\ -q_z^1/2 \\ q_y^1/2 \\ 0 \\ -r_z^2 \\ r_y^2 \\ 0 \\ -v_z^2 \\ v_y^2 \\ -q_x^2/2 \\ q_t^2/2 \\ -q_z^2/2 \\ q_y^2/2 \end{bmatrix}, \begin{bmatrix} r_z^1 \\ 0 \\ -r_x^1 \\ v_z^1 \\ 0 \\ -v_x^1 \\ -q_y^1/2 \\ q_z^1/2 \\ q_t^1/2 \\ -q_x^1/2 \\ r_z^2 \\ 0 \\ -r_x^2 \\ v_z^2 \\ 0 \\ -v_x^2 \\ -q_y^2/2 \\ q_z^2/2 \\ q_t^2/2 \\ -q_x^2/2 \end{bmatrix}, \begin{bmatrix} -r_y^1 \\ r_x^1 \\ 0 \\ -v_y^1 \\ v_x^1 \\ 0 \\ -q_z^1/2 \\ -q_y^1/2 \\ q_x^1/2 \\ q_t^1/2 \\ -r_y^2 \\ r_x^2 \\ 0 \\ -v_y^2 \\ v_x^2 \\ 0 \\ -q_z^2/2 \\ -q_y^2/2 \\ q_x^2/2 \\ q_t^2/2 \end{bmatrix}.$$

Their physical meaning can be easily obtained by considering the infinitesimal transformation given in (4.24). The first three symmetries express the system invariance under any shift that acts simultaneously on the two bodies (or, in other words, a shift of the global frame). The second three express the system invariance under any constant speed that acts simultaneously on the two bodies (or, in other words, a constant speed of the global frame). The last three express the system invariance under any rotation that acts simultaneously on the two bodies (or, in other words, a rotation of the global frame). As mentioned at the beginning of this section, these results could have been obtained by following intuitive reasoning. It is remarkable that they can be obtained algebraically, by following an automatic procedure.

5.7. Visual inertial sensor fusion in the cooperative case

We want to describe our system with an observable state. We have many choices. Since we want to achieve a local observable subsystem, we are interested in introducing very simple observable functions that generate the observable codistribution.

We remark that all the above invariances are satisfied by the relative state, which consists of the relative position, the relative speed, and the relative orientation. In particular, we introduce the following new quantities:

- the position (R) of $\mathcal{B}_2$ in the local frame of $\mathcal{B}_1$,
- the velocity (V) of $\mathcal{B}_2$ in the local frame of $\mathcal{B}_1$, and
- the orthogonal matrix O that characterizes the rotation between the two local frames.

Note that the matrix O is characterized by three independent angles. We denote them by

$$\psi, \ \alpha, \ \beta.$$

In particular, we set

$$O = O(\psi, \ \alpha, \ \beta) = \begin{bmatrix} q_t^2 + q_x^2 - q_y^2 - q_z^2 & 2(q_x q_y - q_t q_z) & 2(q_t q_y + q_x q_z) \\ 2(q_t q_z + q_x q_y) & q_t^2 - q_x^2 + q_y^2 - q_z^2 & 2(q_y q_z - q_t q_x) \\ 2(q_x q_z - q_t q_y) & 2(q_t q_x + q_y q_z) & q_t^2 - q_x^2 - q_y^2 + q_z^2 \end{bmatrix},$$

where

$$q = q(\psi, \ \alpha, \ \beta) = q_t(\psi, \ \alpha, \ \beta) + q_x(\psi, \ \alpha, \ \beta) \, i + q_y(\psi, \ \alpha, \ \beta) \, j + q_z(\psi, \ \alpha, \ \beta) \, k$$
$$= \cos \frac{\psi}{2} + \sin \frac{\psi}{2} (\cos \alpha \sin \beta \, i + \cos \alpha \cos \beta \, j + \sin \alpha \, k).$$

Specifically, the above quaternion characterizes a rotation of ψ about the axis $[\cos \alpha \sin \beta, \cos \alpha \cos \beta, \sin \alpha]^T$ (see Appendix B).

The observable codistribution is generated by the gradients of the components of R, V and the angles ψ, α, β, together with the two functions h_7 and h_8. These are precisely 11 generators. Therefore, we can fully characterize our system by the state

$$X = \begin{bmatrix} R \\ V \\ \psi \\ \alpha \\ \beta \end{bmatrix}.$$

This state has dimension equal to 9, and it is observable.

From the above equations, we obtain the following description of the local observable subsystem:

$$\begin{bmatrix} \dot{R} &= \Omega^1 \wedge R + V, \\ \dot{V} &= \Omega^1 \wedge V + OA^2 - A^1, \\ \dot{O} &= [\Omega^1]_\times^T O + O \, [\Omega^2]_\times, \end{bmatrix}$$

where $[\Omega^{1/2}]_\times$ are the skew-symmetric matrices associated to the vectors $\Omega^{1/2}$:

$$[\Omega^i]_\times = \begin{bmatrix} 0 & \Omega_z^i & -\Omega_y^i \\ -\Omega_z^i & 0 & \Omega_x^i \\ \Omega_y^i & -\Omega_x^i & 0 \end{bmatrix}, \ i = 1, 2.$$

By some further computation, we can directly obtain the dynamics of the three angles, ψ, α, β. The dynamics of the state are

$$\begin{cases}
\dot{R} = \Omega^1 \wedge R + V, \\
\dot{V} = \Omega^1 \wedge V + OA^2 - A^1, \\
\dot{\psi} = \Omega_z^2 \sin\alpha - \Omega_z^1 \sin\alpha - \Omega_y^1 \cos\alpha\cos\beta + \Omega_y^2 \cos\alpha\cos\beta - \Omega_x^1 \cos\alpha\sin\beta \\
\qquad + \Omega_x^2 \cos\alpha\sin\beta, \\
\dot{\alpha} = \frac{1}{2}\left(\Omega_y^1 \sin\beta - \Omega_x^2 \cos\beta - \Omega_x^1 \cos\beta + \Omega_y^2 \sin\beta - \Omega_z^1 \cot\frac{\psi}{2}\cos\alpha \right. \\
\qquad + \Omega_z^2 \cot\frac{\psi}{2}\cos\alpha + \Omega_y^1 \cot\frac{\psi}{2}\cos\beta\sin\alpha - \Omega_y^2 \cot\frac{\psi}{2}\cos\beta\sin\alpha \\
\qquad \left. + \Omega_x^1 \cot\frac{\psi}{2}\sin\alpha\sin\beta - \Omega_x^2 \cot\frac{\psi}{2}\sin\alpha\sin\beta\right), \\
\dot{\beta} = -\frac{\Omega_y^1 \cos\beta\sin\alpha - \Omega_z^2 \cos\alpha - \Omega_z^1 \cos\alpha + \Omega_y^2 \cos\beta\sin\alpha + \Omega_x^1 \sin\alpha\sin\beta + \Omega_x^2 \sin\alpha\sin\beta}{2\cos\alpha} \\
\qquad - \frac{\Omega_x^1 \cos\frac{\psi}{2}\cos\beta - \Omega_x^2 \cos\frac{\psi}{2}\cos\beta - \Omega_y^1 \cos\frac{\psi}{2}\sin\beta + \Omega_y^2 \cos\frac{\psi}{2}\sin\beta}{2\sin\frac{\psi}{2}\cos\alpha}.
\end{cases} \quad (5.27)$$

Finally, the two cameras provide the two vectors, R and $-O^T R$, up to a scale.

We conclude this section by remarking that, as for the visual inertial sensor fusion problem in the noncooperative case (studied in the previous sections), we cannot obtain automatically an exhaustive analysis of all the singularities, degeneracies, and minimal cases, by only executing an observability analysis. As we remarked at the end of section 5.5, in many computer vision problems, the key to such analyses is the establishment of an equivalence between the problem at hand and a polynomial equation system (PES). In [36], we established that the cooperative visual inertial sensor fusion problem, in the case of two agents, is equivalent to a PES that consists of several linear equations and three polynomial equations of second degree (in the noncooperative case, the number of polynomial equations of second degree was 1). This PES was solved by using the method based on the Macaulay resultant matrices [43]. This is the analytic solution of the cooperative visual inertial sensor fusion problem in the case of two agents. The equivalence between the cooperative visual inertial sensor fusion problem in the case of two agents, and the aforementioned PES allows us to obtain all the structural properties of the problem (e.g., how the minimal cases, singularities, and degeneracies depend on the trajectories and on the number of camera images). This exhaustive analysis is available in [37]. It is shown that the problem, when nonsingular, can have up to eight distinct solutions.

Exercise: Prove that, when the relative acceleration between the two bodies vanishes, the absolute scales becomes unobservable.

5.8 • Visual inertial sensor fusion with virtual point features

The example discussed in this section has both a practical and an academic importance. We study the visual inertial sensor fusion problem with a single agent, but, in contrast to the cases investigated in the previous sections, we contemplate the case when the motion occurs in the surrounding of a planar surface, and, because of a limited computation capability, we assume that we cannot extract natural point features from the camera images. Most of visual based navigation approaches require us to extract natural features from the images provided by the camera and in particular to detect the same features in different images. The feature matching task becomes critical in the outdoor environment because of possible illumination changes. In order to significantly reduce the computational burden required to perform these tasks and to

5.8. Visual inertial sensor fusion with virtual point features

make the feature matching more robust, we introduce a *virtual* feature by equipping our vehicle with a laser pointer. The laser beam produces a laser spot on the planar surface. This laser spot is observed by the monocular camera, and it is the unique available point feature. Compared to classical vision and IMU data fusion problems, the feature is moving in the environment, but we exploit the hypothesis that it moves on a planar surface. The first question which arises is to understand which is the observable state, i.e, the state that includes all the physical quantities that can be determined by only using the inertial data and the camera measurements of the laser spot during a short time interval. Then the second question is to analytically determine the link between this state and the system inputs and output. In other words, we aim at obtaining a local observable subsystem, as we discussed in section 4.6.1.

5.8.1 ▪ The system

Let us consider an aerial vehicle equipped with a monocular camera and an IMU. The vehicle is also equipped with a laser pointer. The configuration of the laser pointer in the camera reference frame is known. The vehicle moves in the surrounding of a planar surface, and we assume that the laser spot produced by the laser beam belongs to this planar surface (see Figure 5.5). The position and the orientation of this planar surface are unknown. The camera measurement consists of the position of the laser spot in the camera frame up to a scale factor. The IMU consists of three orthogonal accelerometers and three orthogonal gyroscopes. We assume that the transformation between the camera frame and the IMU frame is known (we can assume that the vehicle frame coincides with the camera frame).

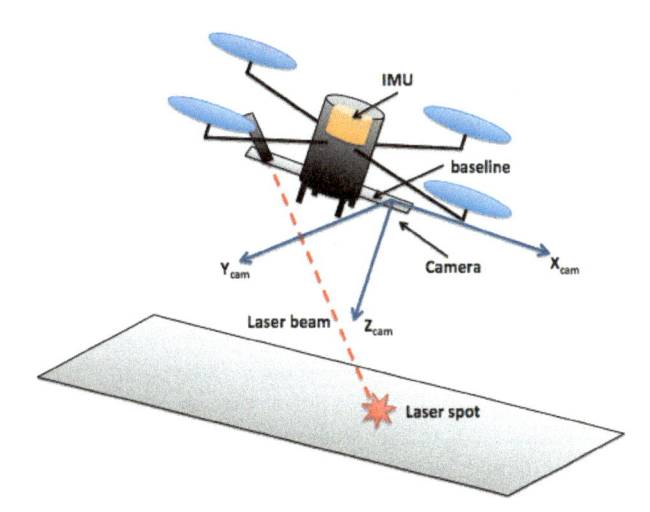

Figure 5.5. *Quadrotor equipped with a monocular camera, an IMU, and a laser pointer. The laser spot is on a planar surface, and its position in the camera frame is obtained by the camera up to a scale factor.* © 2017 IEEE. Reprinted, with permission, from [35].

We adopt the same notation adopted in section 5.5 to denote the position, speed, and orientation of the vehicle. Their dynamics are given by (5.12), i.e.,

$$\left[\begin{array}{ll} \dot{r}_q & = v_q, \\ \dot{v}_q & = qA_qq^* - |g|k, \\ \dot{q} & = \frac{1}{2}q\Omega_q. \end{array} \right.$$

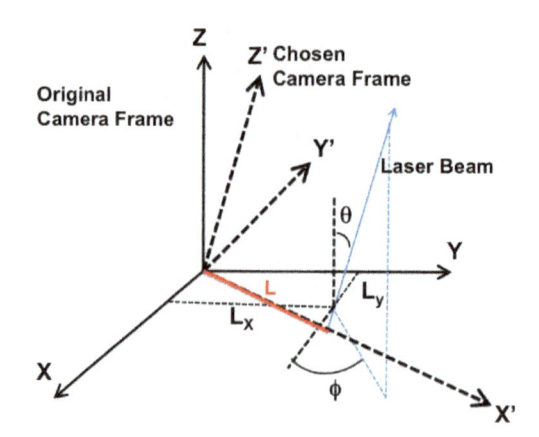

Figure 5.6. *The original camera frame XYZ, the chosen camera frame $X'Y'Z'$, and the laser module at the position $[L_x, L_y, 0]$ and the direction (θ, ϕ) (in the original frame) and the position $[L, 0, 0]$ and the direction $(0, 0)$ (in the chosen camera frame). © 2017 IEEE. Reprinted, with permission, from [35].*

We derive the expression of the output (camera measurement, which consists of the position of the laser spot in the camera frame up to a scale factor). The laser spot is on a planar surface whose configuration is unknown. First of all, we redefine the camera frame as follows. We choose the camera frame with the z-axis parallel to the laser pointer (see Figure 5.6). In addition, the camera frame is such that the laser beam intercepts the xy-plane in $[L, 0, 0]$. This choice of the camera frame can be considered a calibration of the camera-laser module. We describe this calibration task.

Camera-laser module calibration

In Figure 5.6, we display the position and the direction of the laser pointer in the original camera frame. The calibration consists in estimating the four parameters L_x, L_y, θ, ϕ. In other words, it consists in estimating the line made by the laser beam in the original camera frame. This line is determined starting from the position of the laser spot in the original camera frame for at least two spots. To have an accurate estimate, the two spots must be as far as possible from each other. The precision can be further improved by considering more than two spots and by finding the best line fit. In order to have the Cartesian coordinates of a single spot in the original camera frame, it suffices to project the spot on a checkerboard. By using the Camera Calibration Toolbox for MATLAB [6], it is possible to get the equation of the plane containing the checkerboard in the original camera frame and, knowing the direction of the spot from the camera measurement, the 3D position is finally obtained.

The chosen camera frame is obtained by rotating the original frame such that in the new frame the z-axis has the same orientation of the laser beam. Additionally, we also require that the laser beam intersects the new x-axis. In other words, we require that the laser beam intersects the new xy-plane in the point $[L, 0, 0]^T$ for a given L. We want to obtain the quaternion q which characterizes this rotation. This will allow us to express the vectors provided by the camera in the chosen frame. Note that, since the two frames only differ by a rotation (i.e., they share the same origin), we are allowed to express the vectors provided by the camera in the new frame, even if these vectors are defined up to a scale. Finally, in this section we want to determine the value

5.8. Visual inertial sensor fusion with virtual point features

of L. As we will see, both the quaternion q and the parameter L only depend on the calibration parameters: L_x, L_y, θ, ϕ.

We start by rotating the original frame of ϕ about its z-axis. The quaternion characterizing this rotation is $q_{z\text{-axis}}(\phi) = \cos\left(\frac{\phi}{2}\right) + k \sin\left(\frac{\phi}{2}\right)$ (see Appendix B). Then we rotate the frame obtained with this rotation of θ about its y-axis. The quaternion characterizing this rotation is $q_{y\text{-axis}}(\theta) = \cos\left(\frac{\theta}{2}\right) + j \sin\left(\frac{\theta}{2}\right)$. Hence, the previous two rotations are characterized by the quaternion $q_{zy} \equiv q_{z\text{-axis}}(\phi) q_{y\text{-axis}}(\theta)$. The obtained frame has the z-axis aligned with the laser beam. On the other hand, the laser beam does not intersect necessarily the x-axis. In order to obtain this, we have to rotate again the frame about its current z-axis. Let us compute the intersection of the laser beam with the xy-plane. We compute this point in the original frame. By a direct computation we obtain $r^{\text{inters}} = [L_x - \tau_x^2 L_x - \tau_x \tau_y L_y, \ L_y - \tau_x \tau_y L_x - \tau_y^2 L_y, \ -\tau_x \tau_z L_x - \tau_y \tau_z L_y]^T$, where $\tau_x = \sin(\theta)\cos(\phi)$, $\tau_y = \sin(\theta)\sin(\phi)$, $\tau_z = \cos(\theta)$. We then compute this vector in the rotated frame by doing the quaternion product: $R_q^{\text{inters}} = q_{zy}^* r_q^{\text{inters}} q_{zy}$. Let us denote R^{inters} with $[L'_x, L'_y, 0]^T$. We finally have $L = \sqrt{L_x'^2 + L_y'^2}$ and $q = q_{zy} q_{z\text{-axis}}(\alpha)$, where $q_{z\text{-axis}}(\alpha) = \cos\left(\frac{\alpha}{2}\right) + k \sin\left(\frac{\alpha}{2}\right)$ and $\alpha = \arctan\left(\frac{L'_y}{L'_x}\right)$.

Now, to obtain the expression of the output, we need to characterize the configuration of the planar surface in the global frame. We introduce a new parameter (Γ), and we describe the planar surface by the equation

$$z = \Gamma y$$

without loss of generality, and we are assuming that the x-axis of the global frame belongs to the planar surface. We introduce the following state to characterize our system:

$$x = [r_x, \ r_y, \ r_z, \ v_x, \ v_y, \ v_z, \ q_t, \ q_x, \ q_y, \ q_z, \ \Gamma]^T. \tag{5.28}$$

For the sake of simplicity, in this example we assume that the magnitude of the gravity ($|g|$) is known. The dynamics of the above state are

$$\begin{bmatrix} \dot{r}_q &= v_q, \\ \dot{v}_q &= q A_q q^* - |g|k, \\ \dot{q} &= \frac{1}{2} q \Omega_q, \\ \dot{\Gamma} &= 0. \end{bmatrix} \tag{5.29}$$

In this setting, by carrying out analytical computation (which uses the basic quaternion rules) we obtain the analytical expression of the position $[X_s, Y_s, Z_s]$ of the laser spot in the camera reference frame. We have

$$\begin{bmatrix} X_s &= L, \\ Y_s &= 0, \\ Z_s &= \frac{r_z + 2q_z q_x L - 2q_y \Gamma q_x L - 2q_y q_t L - 2q_t L \Gamma q_z - \Gamma r_y}{2\Gamma q_z q_y - 2\Gamma q_t q_x - q_z^2 - q_t^2 + q_y^2 + q_x^2}. \end{bmatrix} \tag{5.30}$$

The camera provides the vector $[X_s, Y_s, Z_s]$ up to a scale factor. This is equivalent to the two ratios $\frac{X_s}{Z_s}$ and $\frac{Y_s}{Z_s}$. Hence, since the latter is identically zero, the camera observation is given by $h_{\text{cam}} = \frac{X_s}{Z_s}$. Since $L = X_s$ is known from the camera-laser module calibration, we can set the system output equal to Z_s. Hence,

$$h_1 = \frac{r_z + 2q_z q_x L - 2q_y \Gamma q_x L - 2q_y q_t L - 2q_t L \Gamma q_z - \Gamma r_y}{2\Gamma q_z q_y - 2\Gamma q_t q_x - q_z^2 - q_t^2 + q_y^2 + q_x^2}. \tag{5.31}$$

In addition, we must include the following output, which expresses the unitary of the quaternion q:

$$h_2 = q_t^2 + q_x^2 + q_y^2 + q_z^2. \tag{5.32}$$

5.8.2 ▪ Observability analysis

We study the observability properties of the system characterized by the state in (5.28), the dynamics in (5.29), and the outputs in (5.31) and (5.32).

We start by computing the observable codistribution. By comparing the above equations with (4.1) we obtain $m = 6$, $p = 2$, $u_1 = A_x$, $u_2 = A_y$, $u_3 = A_z$, $u_4 = \Omega_x$, $u_5 = \Omega_y$, $u_6 = \Omega_z$. In addition, all the vector fields, $f^0, f^1, \ldots, f^6$, have 11 entries, and the first 10 entries coincide with the entries of the vectors $f^0, f^1, \ldots, f^6$ in section 5.5, and the last entry is zero for all the vectors. Finally, $h_1(x)$ is given in (5.31), and $h_2(x)$ is given in (5.32).

We compute the observable codistribution by running Algorithm 4.2. We do not provide the analytic expression of the observable codistribution at each step. At the initialization we have

$$\Omega_0 = \text{span}\{\nabla h_1, \nabla h_2\}$$

and its dimension is 2. At the first iterative step we obtain

$$\Omega_1 = \text{span}\{\nabla h_1, \nabla h_2, \nabla \mathcal{L}_{f^0} h_1, \nabla \mathcal{L}_{f^5} h_1, \nabla \mathcal{L}_{f^6} h_1\};$$

namely, all the above covectors are independent and the dimension of Ω_1 is 5. At the second iterative step we obtain

$$\Omega_2 = \text{span}\{\nabla h_1, \nabla h_2, \nabla \mathcal{L}_{f^0} h_1, \nabla \mathcal{L}_{f^5} h_1, \nabla \mathcal{L}_{f^6} h_1, \nabla \mathcal{L}_{f^0 f^0}^2 h_1\};$$

namely, all the above covectors are independent and the dimension of Ω_2 is 6. At the third iterative step we obtain that Ω_3 remains the same. Hence, we conclude that the observable codistribution is $\Omega = \Omega_2$. Since its dimension is 6, the system has $5(= 11-6)$ independent symmetries (11 is the state dimension). We obtain these symmetries by computing the distribution orthogonal to Ω. A possible set of generators of this distribution is given by the following five vectors:

$$\xi_1 = \begin{bmatrix} 1 \\ 0 \\ 0 \\ 0 \\ 0 \\ 0 \\ 0 \\ 0 \\ 0 \\ 0 \\ 0 \end{bmatrix}, \xi_2 = \begin{bmatrix} 0 \\ 1 \\ \Gamma \\ 0 \\ 0 \\ 0 \\ 0 \\ 0 \\ 0 \\ 0 \\ 0 \end{bmatrix}, \xi_3 = \begin{bmatrix} 0 \\ 0 \\ 0 \\ 1 \\ 0 \\ 0 \\ 0 \\ 0 \\ 0 \\ 0 \\ 0 \end{bmatrix}, \xi_4 = \begin{bmatrix} 0 \\ 0 \\ 0 \\ 0 \\ 1 \\ \Gamma \\ 0 \\ 0 \\ 0 \\ 0 \\ 0 \end{bmatrix}, \xi_5 = \begin{bmatrix} 0 \\ 0 \\ 0 \\ 0 \\ 0 \\ 0 \\ -q_z + \Gamma q_y \\ -q_y - \Gamma q_z \\ q_x - \Gamma q_t \\ q_t + \Gamma q_x \\ 0 \end{bmatrix}.$$

By using the infinitesimal transformation given in (4.24), it is immediate to obtain the physical meaning of the first four symmetries:

1. ξ_1 characterizes a shift along the x-axis.

2. ξ_2 characterizes a shift along the intersection of the yz-plane and the planar surface.

5.8. Visual inertial sensor fusion with virtual point features

3. ξ_3 characterizes a variation of the speed along the x-axis.

4. ξ_4 characterizes a variation of the speed along the intersection of the yz-plane and the planar surface.

The presence of the above four symmetries is intuitive. A generic shift on the planar surface is in the span of ξ_1 and ξ_2. Similarly, any variation of the speed on a direction that belongs to the planar surface is characterized by a linear combination of ξ_3 and ξ_4.

Let us study the meaning of the last symmetry, ξ_5. First of all, we remark that it only involves the four components of the quaternion.

By using the infinitesimal transformation given in (4.24), we have, for the quaternion,

$$q \rightarrow q' = q + \epsilon \begin{bmatrix} -q_z + \Gamma q_y \\ -q_y - \Gamma q_z \\ q_x - \Gamma q_t \\ q_t + \Gamma q_x \end{bmatrix}. \quad (5.33)$$

Let us consider the third equation in (5.29), which provides

$$\dot{q} = \frac{1}{2}\omega_q q,$$

where ω is the angular speed in the global frame and ω_q is the imaginary quaternion associated to ω. In terms of differentials, we have

$$dq = \frac{1}{2}\omega_q q dt.$$

By multiplying on the right by q^* we obtain

$$dq q^* = \frac{1}{2}\omega_q dt,$$

q being a unit quaternion. By using the expression in (5.33) for dq, we obtain

$$\epsilon \begin{bmatrix} -q_z + \Gamma q_y \\ -q_y - \Gamma q_z \\ q_x - \Gamma q_t \\ q_t + \Gamma q_x \end{bmatrix} q^* = \frac{1}{2}\omega_q dt$$

and, by setting $\epsilon = \frac{1}{2}dt$ and by carrying out the quaternion product on the left side, we obtain

$$\omega_q = \begin{bmatrix} 0 \\ 0 \\ -\Gamma \\ 1 \end{bmatrix}.$$

We remark that the corresponding ω is a vector orthogonal to the planar surface. Therefore, the transformation in (5.33) describes a rotation around the axis orthogonal to the planar surface. We note that a rotation around the above axis must also impact the first six components of the state (position and speed). On the other hand, these changes are contemplated by the first four symmetries. Note that the above five symmetries are the generators of a distribution and we are free to choose any set of five vectors to generate this distribution. In other words, a rotation around the axis orthogonal to the planar surface is obtained by a suitable linear combination of all the above five symmetries.

We want to describe our system with an observable state. We have many choices. Since we want to achieve a local observable subsystem, we are interested in introducing very simple observable functions that generate the observable codistribution. We know that we have $11 - 5 = 6$ independent observable functions. They are invariant under all the above five symmetries simultaneously. Based on the above discussion about the physical meaning of the five symmetries, it is immediate to realize that four possible independent observable functions are

- the distance from the surface,
- the component of the speed along the direction orthogonal to the planar surface, and
- the roll and pitch angles with respect to the planar surface.

Finally, the last two independent observable functions are trivially the module of the quaternion ($qq^* = 1$) and the parameter Γ. This last quantity is observable since all the symmetries have the last entry equal to zero: this makes the gradient of Γ with respect to the state, i.e., the covector

$$\nabla \Gamma = [\underbrace{0, \ldots, 0}_{10 \text{ times}}, 1],$$

orthogonal to all the symmetries.

To obtain a local observable subsystem, as we discussed in section 4.6.1, we need the analytical expressions of the observable functions in terms of the state. Then, starting from the dynamics of the state given by (5.29), we obtain the dynamics of these observable functions in terms of them and we also express the outputs only in terms of these observable functions.

We start by defining a new global frame. Its xy-plane coincides with the planar surface. We denote by p the quaternion that characterizes the rotation between the old and the new global frames. This quaternion can be obtained by requiring that it transforms the unit vector orthogonal to the surface, i.e.,

$$\tau \triangleq \frac{1}{\sqrt{1+\Gamma^2}} \begin{bmatrix} 0 \\ -\Gamma \\ 1 \end{bmatrix},$$

into the new z-axis. In other words, we require that the following quaternion equality holds (see Appendix B):

$$p^* \tau_q p = k,$$

where k is the fourth fundamental quaternion unit ($k = 0 + 0\,i + 0\,j + 1\,k$). We obtain

$$p = \cos\left(\frac{\arctan(\Gamma)}{2}\right) + \sin\left(\frac{\arctan(\Gamma)}{2}\right) i.$$

We adopt a tilde to denote the coordinates in the new global frame. We have, for the vehicle position, speed, and orientation,

$$\tilde{r}_q = p^* r_q p, \qquad \tilde{v}_q = p^* v_q p, \qquad \tilde{q} = p^* q.$$

From (5.29) we obtain

$$\begin{bmatrix} \dot{\tilde{r}}_q &= \tilde{v}_q, \\ \dot{\tilde{v}}_q &= \tilde{q} A_q \tilde{q}^* - |g| p^* k p, \\ \dot{\tilde{q}} &= \frac{1}{2} \tilde{q} \Omega_q. \end{bmatrix} \qquad (5.34)$$

5.8. Visual inertial sensor fusion with virtual point features

Regarding the last term in the second equation, by a direct computation we obtain

$$p^* k p = \sin(\arctan \Gamma) \; j + \cos(\arctan \Gamma) \; k. \tag{5.35}$$

From the above discussion, we know that the z-component of both $\tilde{r}$ and $\tilde{v}$ are observable functions. In addition, also the roll and pitch angles in the new global frame (i.e., with respect to the axis orthogonal to the planar surface) are observable. Hence, we introduce the following four quantities:

$$l_1 = \tilde{r}_z, \qquad l_2 = \tilde{v}_z, \qquad l_3 = \psi_R = \arctan\left(2 \frac{\tilde{q}_t \tilde{q}_x + \tilde{q}_y \tilde{q}_z}{1 - 2(\tilde{q}_x^2 + \tilde{q}_y^2)}\right),$$

$$l_4 = \psi_P = \arcsin\left(2(\tilde{q}_t \tilde{q}_y - \tilde{q}_x \tilde{q}_z)\right).$$

By using the first equation in (5.34) we obtain the dynamics of l_1. We have

$$\dot{l}_1 = l_2.$$

We want to compute the dynamics of l_2. We need to compute the quaternion product $\tilde{q} A_q \tilde{q}^*$. Since we only need to compute $\dot{\tilde{v}}_z$, we only need to compute the component of $\tilde{q} A_q \tilde{q}^*$ proportional to k (k is the fourth fundamental quaternion unit, $k = 0 + 0\, i + 0\, j + 1\, k$). This component only depends on the roll (l_3) and the pitch (l_4) angles. We have

$$-A_x \sin(2l_4) + A_y \cos(2l_4) \sin(2l_3) - A_z (2 \sin^2 l_3 \cos(2l_4) + \sin(2l_4)).$$

Finally, by taking into account the last term in the second equation of (5.34) and by using (5.35), we obtain

$$\dot{l}_2 = -A_x \sin(2l_4) + A_y \cos(2l_4) \sin(2l_3) - A_z (2 \sin^2 l_3 \cos(2l_4) + \sin(2l_4)) - |g| \cos(\arctan \Gamma).$$

By using the third and fourth equations in (5.16) we obtain the dynamics of l_3 and l_4. We have

$$\dot{l}_3 = \Omega_x + \Omega_y \tan l_4 \sin l_3 + \Omega_z \tan l_4 \cos l_3, \qquad \dot{l}_4 = \Omega_y \cos l_3 - \Omega_z \sin l_3.$$

Finally, we need to express the output in terms of the above observable functions. To achieve this, we start from the expression in (5.31). We express this function in terms of the vehicle position and orientation in the new global frame. We remark that, in the new frame, the planar surface is the xy-plane, i.e., it is characterized by the equation

$$z = 0.$$

Hence, it suffices to substitute $\Gamma = 0$ in (5.31). We obtain

$$h_1 = \frac{\tilde{r}_z - 2L(\tilde{q}_y \tilde{q}_t - \tilde{q}_z \tilde{q}_x)}{-\tilde{q}_z^2 - \tilde{q}_t^2 + \tilde{q}_y^2 + \tilde{q}_x^2}.$$

By a direct computation we can verify that the above function only depends on l_1, l_3, and l_4. Specifically, we obtain

$$h_1 = \frac{L \sin l_4 - l_1}{\cos(2l_3) \cos(2l_4)}.$$

The local observable subsystem is

$$\begin{bmatrix} \dot{l}_1 &= l_2, \\ \dot{l}_2 &= -A_x \sin(2l_4) + A_y \cos(2l_4)\sin(2l_3) - A_z(2\sin^2 l_3 \cos(2l_4) + \sin(2l_4)) - l_5, \\ \dot{l}_3 &= \Omega_x + \Omega_y \tan l_4 \sin l_3 + \Omega_z \tan l_4 \cos l_3, \\ \dot{l}_4 &= \Omega_y \cos l_3 - \Omega_z \sin l_3, \\ \dot{l}_5 &= 0, \\ y &= \frac{L \sin l_4 - l_1}{\cos(2l_3)\cos(2l_4)}, \end{bmatrix}$$

(5.36)

where $l_5 \triangleq |g|\cos(\arctan \Gamma)$, which is the component of the gravity vector orthogonal to the planar surface. In [44] we used a similar observable subsystem to implement an extended Kalman filter that runs on an autonomous drone, equipped with the same set of sensors described in this section.

Part III

Nonlinear Unknown Input Observability

Chapter 6

General Concepts on Nonlinear Unknown Input Observability

In the third part of this book, we study the observability properties of systems whose dynamics are also driven by one or more inputs which are unknown. An example of such a system can be a drone that operates in the presence of wind. The wind is in general unknown and acts on the dynamics as an (unknown) input. An unknown input (or disturbance) differs from a known input not only because it is unknown but also because it cannot be assigned.

A general characterization of an input-output system, when some of the inputs are unknown, is given by the following equations:

$$\begin{cases} \dot{x} &= g^0(x) + \sum_{k=1}^{m_u} f^k(x) u_k + \sum_{j=1}^{m_w} g^j(x) w_j, \\ y &= [h_1(x), \ldots, h_p(x)]^T, \end{cases} \quad (6.1)$$

where m_u is the number of inputs which are known and m_w the number of inputs which are unknown. This generalizes the systems investigated in Chapters 4 and 5, which are characterized by (4.1). They are simply obtained by setting $m_u = m$ and $m_w = 0$ in the above equation. In this chapter, we always refer to the general case of multiple unknown inputs. We denote by $w(t)$ the vector function $[w_1(t), \ldots, w_{m_w}(t)]^T$ and, in some cases, we use the diction of *unknown input* to mean the above unknown input vector. Finally, in some cases we refer to the case of single output. In this case, we set $h = h_1$.

In Chapter 4, we provided a complete theory of observability in the absence of unknown inputs. In particular, we showed that all the observability properties of a nonlinear system are determined by computing the observable codistribution. The computation of this codistribution is the core of the observability rank condition and, more broadly, it is the main step to obtain all the observability properties of a given system. In particular, it allows us to determine whether the state is weakly observable. If this is not the case, it allows us to determine all the system symmetries and to easily check whether a given scalar function is observable or not. Finally, it is the fundamental step to achieve a local observable subsystem.

In order to deal with the case of unknown inputs, we need to derive a new algorithm able to generate the observable codistribution. The only difference between the theory of observability in the presence of unknown inputs from the theory in the case without unknown inputs resides precisely in the computation of the observable codistribution. This will be the matter of the rest of this book.

In this chapter, we provide general concepts and properties that will allow us to obtain the algorithm to compute the observable codistribution in Chapters 7 and 8. In particular, in this chapter we provide an algorithm (Algorithm 6.1) that builds a codistribution that includes the observable codistribution: specifically, the codistribution provided by this algorithm is defined in an extended space that includes the original state that defines our system together with the unknown inputs and their time derivatives up to a given order. As a result, this codistribution provides mixed information about the observability of the state and the observability of the unknown inputs. In Chapters 7 and 8 we separate this information.

As in the case without unknown inputs, we introduce a constructive approach to observability by providing a definition of observability (in section 6.1) that is not based on the concept of indistinguishability. Based on this definition, in section 6.2 we introduce the augmented state (section 6.2.1) and we provide basic properties that extend the results stated by Theorems 4.3 and 4.5 to the case with unknown inputs. These are the two new Theorems 6.10 and 6.11. These results allow us to introduce Algorithm 6.1 in section 6.3. This algorithm could be a tool to obtain the observability properties. However, it has very restrictive impediments that prevent its use in practice: besides being nonconvergent, its implementation becomes computationally prohibitive after few steps. On the other hand, it is fundamental to obtain the convergent (and simple) algorithm that provides the observable codistribution directly on the space of the states (and not in the augmented space). This will be provided later for the simplified system (in Chapter 7) and for the general case (in Chapter 8).

Finally, in this chapter we also provide a definition of indistinguishability when we are in the presence of unknown inputs (section 6.4). Based on this, we introduce the standard definition of observability and we show that, starting from it, it is possible to achieve the same properties obtained starting from our constructive definition.

6.1 ▪ A constructive definition of observability

We consider a given time interval that we denote by $\mathcal{I}$. Without loss of generality, we set $\mathcal{I} = [0, T]$. We also denote the initial state by x_0 (i.e., $x(0) = x_0$).

As in the case without unknown inputs, we wish to define as observable a system such that, by suitably setting the known inputs during the time interval $\mathcal{I}$, it is possible to reconstruct the initial state x_0 by only using the values that the outputs take in $\mathcal{I}$. However, in the presence of one or more unknown inputs, the answer could depend on the values that the unknown input $w(t)$ takes in $\mathcal{I}$. Even if the unknown input is in any case unknown, a given time assignment of the unknown input in $\mathcal{I}$ could drive our system on a more informative trajectory than the one produced by a different time assignment of the unknown inputs. In section 6.1.1, we provide the definition of the *w-observability*, which depends on the specific values taken by $w(t)$ in $\mathcal{I}$. In section 6.1.2, we provide the definition of *observability*, which is independent of the specific values taken by $w(t)$ in $\mathcal{I}$.

6.1.1 ▪ Observability for a specific unknown input

We assume that, during the time interval $\mathcal{I}$, the unknown input is $w(t)$. The definitions of w-observability, w-weak observability, w-observable function, and w-weak observable function coincide with the corresponding definitions given in Chapter 4. Here, we only provide the definition of w-observable function.

6.1. A constructive definition of observability

Definition 6.1. *The scalar function $\theta(x)$ is w-observable at x_0 if we can select the inputs $u_1(t), \ldots, u_{m_u}(t)$ such that the value that θ takes at x_0 ($\theta(x_0)$) can be determined from the knowledge of the output $y(t)$ and the inputs $u_1(t), \ldots, u_{m_u}(t)$ in the time interval $\mathcal{I}$. In addition, $\theta(x)$ is w-observable on a given set $\mathcal{U} \subseteq \mathcal{M}$ if it is observable at any $x \in \mathcal{U}$.*

Note that the above definition explicitly states that we can select the known inputs to determine $\theta(x_0)$. This means that, although the observability depends on the specific $w(t)$, the choice of the known inputs is independent of $w(t)$ (which is unknown).

6.1.2 ▪ Observability independent of the unknown input

We wish to obtain a general result, independently of the specific values that the unknown inputs take in a given experiment. From one side, we want to be conservative. In other words, we want to define a system observable if we can determine x_0 even in the most unlucky situation. On the other side, we do not want the system to be unobservable if we cannot determine x_0 only for *very few* time assignments of the unknown inputs. To proceed, we need to formalize the meaning of the above "very few." The proper manner to deal with this is obtained by introducing the concept of probability. In order to do this, we must consider the infinite-dimensional Banach space:

$$\mathcal{W} \triangleq \left\{ w(t) = [w_1(t), \ldots, w_{m_w}(t)]^T, \; t \in \mathcal{I} \right\}, \tag{6.2}$$

where each $w_j(t)$ is an analytic function of t.

We consider the subset $\mathcal{W}_0 \subseteq \mathcal{W}$ such that if $w \in \mathcal{W}_0$, we cannot reconstruct x_0. If the probability that $w \in \mathcal{W}_0$ vanishes, we define the system as observable. For finite-dimensional spaces, the probability is defined by using the concept of measure. Unfortunately, the concept of measure in spaces with infinite dimensions is not trivial. In particular, there is no analogue of Lebesgue measure on an infinite-dimensional Banach space. One possibility, which is frequently adopted, is to use the concept of prevalent and shy sets [22]. The probability that a given unknown input belongs to a shy subset of $\mathcal{W}$ is 0. The probability that a given unknown input belongs to a prevalent subset of $\mathcal{W}$ is 1. In accordance with these remarks, we could define our system observable if the above set $\mathcal{W}_0$ is a shy set. This is certainly possible, and it would provide all the results that we will derive in the rest of this book. However, thanks to a result that will be proved later (Theorem 6.2) we can introduce an easier definition that avoids the concept of prevalence.

Theorem 6.2. *The observable codistribution of the system in (6.1) is independent of the values that the unknown inputs take in $\mathcal{I}$, provided that all the m_w values that they take at the initial time, i.e.,*

$$w_1(0), \; \ldots, \; w_{m_w}(0),$$

do not vanish.

Proof. The proof will be given later since it needs the introduction of the *canonic form* with respect to the unknown inputs (see Appendix C and the discussion after the proof of Theorem 6.12). ◂

The above theorem states that the possibility of reconstructing the initial state x_0 is independent of the values that the unknown inputs take in $\mathcal{I}$, provided that all the unknown inputs at the initial time do not vanish. Note that, in the m_w-dimensional space of the unknown inputs, the set that only consists of the origin (i.e., all the unknown inputs vanish) has zero Lebesgue measure.

Now we are ready to introduce the definition of the observability in the presence of unknown inputs, by avoiding the use of the concept of prevalence. We introduce the following definition.

Definition 6.3 (State Observability in the Presence of Unknown Input). *An input-output system is observable at a given $x_0 \in \mathcal{M}$ if there exists at least one choice of inputs $u_1(t), \ldots, u_{m_u}(t)$ that allows us to determine the initial state x_0 from the knowledge of the output $y(t)$ and the inputs $u_1(t), \ldots, u_{m_u}(t)$ on the time interval $\mathcal{I}$, provided that all the unknown inputs at the initial time do not vanish. In addition, the system is observable on a given set $\mathcal{U} \subseteq \mathcal{M}$ if it is observable at any $x \in \mathcal{U}$.*

In accordance with this definition, if a system is observable on a given $\mathcal{U} \subseteq \mathcal{M}$, it means that by suitably setting the m_u known inputs, we can reconstruct the initial state by only knowing these m_u inputs and the outputs in a given time interval. We do not need further knowledge.

As in the case without unknown inputs, we define the concept of weak observability.

Definition 6.4 (State Weak Observability in the Presence of Unknown Input). *An input-output system is weakly observable at a given $x_0 \in \mathcal{M}$ if there exists an open set B of x_0 such that, by knowing that $x_0 \in B$, there exists at least one choice of inputs $u_1(t), \ldots, u_{m_u}(t)$ that allows us to determine the initial state x_0 from the knowledge of the output $y(t)$ and the inputs $u_1(t), \ldots, u_{m_u}(t)$ on the time interval $\mathcal{I}$, provided that all the unknown inputs at the initial time do not vanish. In addition, the system is weakly observable on a given set $\mathcal{U} \subseteq \mathcal{M}$ if it is weakly observable at any $x \in \mathcal{U}$.*

As in the case without unknown inputs, when a system is neither observable nor weakly observable, we can wonder whether we can obtain the value that a given scalar function takes at the initial state. Let us consider a given scalar function $\theta(x)$. Obviously, if the system is observable at x_0, we can obtain the value $\theta(x_0)$. We are interested in understanding what happens when the system is not weakly observable. In particular, we introduce the following two definitions.

Definition 6.5 (Observable Function in the Presence of Unknown Input). *The scalar function $\theta(x)$ is observable at x_0 if there exists at least one choice of inputs $u_1(t), \ldots, u_{m_u}(t)$ such that the value that θ takes at x_0 ($\theta(x_0)$) can be determined from the knowledge of the output $y(t)$ and the inputs $u_1(t), \ldots, u_{m_u}(t)$ in the time interval $\mathcal{I}$, provided that all the unknown inputs at the initial time do not vanish. In addition, $\theta(x)$ is observable on a given set $\mathcal{U} \subseteq \mathcal{M}$ if it is observable at any $x \in \mathcal{U}$.*

Definition 6.6 (Weak Observable Function in the Presence of Unknown Input). *The scalar function $\theta(x)$ is weakly observable at x_0 if there exists an open neighborhood B of x_0 such that, by knowing that $x_0 \in B$, there exists at least one choice of inputs $u_1(t), \ldots, u_{m_u}(t)$ such that the value that θ takes at x_0 ($\theta(x_0)$) can be determined from the knowledge of the output $y(t)$ and the inputs $u_1(t), \ldots, u_{m_u}(t)$ in the time interval $\mathcal{I}$, provided that all the unknown inputs at the initial time do not vanish. In addition, $\theta(x)$ is weakly observable on a given set $\mathcal{U} \subseteq \mathcal{M}$ if it is weakly observable at any $x \in \mathcal{U}$.*

6.2 • Basic properties to obtain the observable codistribution

In this section we refer to a specific unknown $w(t)$, as in section 6.1.1. Our goal is to obtain the algorithm that generates the observable codistribution in the presence of $w(t)$, i.e., the span of the

6.2. Basic properties to obtain the observable codistribution

gradients of the functions which are w-weak observable. In the absence of unknown inputs, this is Algorithm 4.2. The validity of this algorithm is based on two fundamental results: Theorem 4.3 (or better Corollary 4.4) and Theorem 4.5. The former states that the values that the Lie derivatives take at x_0 provide an upper bound on the knowledge that we can gather on x_0. The latter states that we can achieve this upper bound.

We want to follow the same approach. We need to detect, also in this case, a set of scalar functions such that the values that they take at x_0 enclose the knowledge that we can acquire on x_0. From now on, we call such a set of functions an *upper bound functions' set* (note that this holds at x_0, which must be clarified by the context). In addition, we say that a given set of functions is *achievable* (at x_0) if all the functions that belong to this set are w-observable.

In the absence of unknown inputs, Theorem 4.3 guarantees that the Lie derivatives of the system constitute an upper bound functions' set. Additionally, Theorem 4.5 guarantees that this set of functions is achievable. In this case, we say that they form an *achievable upper bound functions' set*.

We immediately remark that the result stated by Theorem 4.3, i.e., (4.11), still holds if we replace $[u_1, \ldots, u_m]$ by $[u_1, \ldots, u_{m_u}, w_1, \ldots, w_{m_w}]$. This means that the Lie derivatives of the output, computed along all the vector fields that characterize the dynamics in (6.1), constitute an upper bound functions' set also in the presence of unknown inputs. On the other hand, this set of functions is not achievable. The proof of Theorem 4.5 relies on the possibility of setting all the system inputs and on their knowledge. In the presence of unknown inputs, we are not allowed to set all the system inputs to our purpose. Note also that if the above set of functions were achievable also in the presence of unknown inputs, the algorithm that provides the observable codistribution would have been Algorithm 4.2. We know that this is not the case, as is illustrated by the example discussed in Chapter 1: for a unicycle equipped with a monocular camera, the absolute scale becomes unobservable if the linear speed acts on the dynamics as an unknown input (see section 1.5.1). Therefore, the Lie derivatives of the output, computed along all the vector fields that characterize the dynamics in (6.1), constitute an upper bound functions' set that is not achievable, in general.

6.2.1 ▪ Augmented state and its properties

In order to obtain an upper bound functions' set which is achievable, we need to introduce new concepts and properties. We extend the original state by including the unknown inputs together with their time derivatives. Specifically, we denote by $^k x$ the extended state that includes the unknown inputs and their time derivatives up to the $(k-1)$-order. Since the dimension of the original state is n, the dimension of $^k x$ is $n + k m_w$ and the extended state is

$$^k x \triangleq [x_1, \ldots, x_n, \, w_1, \ldots, w_{m_w}, \, w_1^{(1)}, \ldots, w_{m_w}^{(1)}, \, \ldots, \, w_1^{(k-1)}, \ldots, w_{m_w}^{(k-1)}]^T, \quad (6.3)$$

where

$$w_j^{(m)}(\tau) \triangleq \left. \frac{d^m w_j}{dt^m} \right|_{t=\tau}, \quad j = 1, \ldots, m_w, \;\; m = 0, 1, \ldots.$$

Note that, in this notation, $w_j^{(0)} = w_j$. This same extended state was introduced in [3].

In the following, we often use the notation

$$E^k \triangleq \begin{bmatrix} w_1^{(0)} \\ \cdots \\ w_{m_w}^{(0)} \\ w_1^{(1)}(t) \\ \cdots \\ w_{m_w}^{(1)} \\ \cdots \\ w_1^{(k-1)} \\ \cdots \\ w_{m_w}^{(k-1)} \end{bmatrix} \quad (6.4)$$

and we refer to E^k as the *state extension*. From (6.1) it is immediate to obtain the dynamics of the extended state:

$$^k\dot{x} = G + \sum_{i=1}^{m_u} F^i u_i + \sum_{j=1}^{m_w} W^j w_j^{(k)}, \quad (6.5)$$

where

$$G \triangleq \begin{bmatrix} g^0 + \sum_{j=1}^{m_w} g^j w_j \\ w^{(1)} \\ w^{(2)} \\ \cdots \\ w^{(k-1)} \\ 0_{m_w} \end{bmatrix}, \quad F^i \triangleq \begin{bmatrix} f^i \\ 0_{m_w} \\ 0_{m_w} \\ \cdots \\ 0_{m_w} \\ 0_{m_w} \end{bmatrix}, \quad W^j \triangleq \begin{bmatrix} 0_n \\ 0_{m_w} \\ 0_{m_w} \\ \cdots \\ 0_{m_w} \\ e_j \end{bmatrix} \quad (6.6)$$

and we denoted by 0_m the m-dimensional zero column vector and by e_j the m_w-dimensional column vector whose entries are all zero with the exception of the one at the jth position, which is 1. We remark that the resulting system still has m_u known inputs and m_w unknown inputs. However, while the m_u known inputs coincide with the original ones, the m_w unknown inputs are now the k-order time derivatives of the original unknown inputs. The state evolution depends on the known inputs via the vector fields F^i ($i = 1, \ldots, m_u$), and it depends on the unknown inputs via the unit vectors W^j ($j = 1, \ldots, m_w$). Finally, we remark that only the vector field G depends on the new state elements.

In what follows, we will denote the extended system by $\Sigma^{(k)}$. We also denote by $\Sigma^{(0)}$ the original system, i.e., the one characterized by the state x and the equations in (6.1). The Lie derivatives of $\Sigma^{(k)}$ are the Lie derivatives of the outputs, $h_1(x), \ldots, h_p(x)$, in (6.1), along the vector fields in (6.6). Note that, when we compute such Lie derivatives, we are considering the functions $h_1(x), \ldots, h_p(x)$ as functions of the augmented state. From now on, we denote the gradient operator in the augmented space by ∇_k. In other words,

$$\nabla_k \triangleq \frac{\partial}{\partial {}^k x} = \left(\frac{\partial}{\partial x}, \frac{\partial}{\partial E^k} \right)$$

$$= \left(\frac{\partial}{\partial x_1}, \ldots, \frac{\partial}{\partial x_n}, \frac{\partial}{\partial w_1^{(0)}}, \ldots, \frac{\partial}{\partial w_{m_w}^{(0)}}, \ldots, \frac{\partial}{\partial w_1^{(k-1)}}, \ldots, \frac{\partial}{\partial w_{m_w}^{(k-1)}} \right)$$

and $\nabla \triangleq \nabla_0 = \frac{\partial}{\partial x}$.

6.2. Basic properties to obtain the observable codistribution

We start by providing simple results for $\Sigma^{(k)}$. For simplicity's sake (in particular to avoid the use of a further index), we refer to the case of a single output, i.e., $p = 1$ and $h = h_1(x)$. All the derivations can be easily extended to the multiple output case.

Lemma 6.7. *In $\Sigma^{(k)}$, the Lie derivatives of the output/outputs up to the mth order ($m \leq k$) are independent of $w_j^{(f)}$, $j = 1, \ldots, m_w$, $\forall f \geq m$.*

Proof. We proceed by induction on m for any k. When $m = 0$ we only have one zero-order Lie derivative (i.e., $h(x)$), which only depends on x; namely, it is independent of $w^{(f)}$ $\forall f \geq 0$. Let us assume that the previous assert is true for m, and let us prove that it holds for $m + 1$. If it is true for m, any Lie derivative up to the mth order is independent of $w^{(f)}$ for any $f \geq m$. In other words, the analytical expression of any Lie derivative up to the m-order is represented by a function $g(x, w, w^{(1)}, \ldots, w^{(m-1)})$. Hence,

$$\nabla_m g = \left[\frac{\partial g}{\partial x}, \frac{\partial g}{\partial w}, \frac{\partial g}{\partial w^{(1)}}, \ldots, \frac{\partial g}{\partial w^{(m-1)}}, 0_{(k-m)m_w} \right].$$

It is immediate to realize that the product of the above covector by any vector field in (6.6) depends at most on $w^{(m)}$, i.e., it is independent of $w^{(f)}$ $\forall f \geq m + 1$. ◄

A simple consequence of this lemma are the following two properties.

Proposition 6.8. *Let us consider the system $\Sigma^{(k)}$. The Lie derivatives of the output up to the kth order along at least one vector among W^j ($j = 1, \ldots, m_w$) are identically zero.*

Proof. From the previous lemma it follows that all the Lie derivatives up to the $(k-1)$-order are independent of $w^{(k-1)}$, which are the last m_w components of the extended state in (6.3). Then the proof follows from the fact that any vector among W^j ($j = 1, \ldots, m_w$) has the first $n + (k-1)m_w$ components equal to zero. ◄

Proposition 6.9. *The Lie derivatives of the output up to the kth order along any vector field G, $F^1, \ldots, F^{m_u}$ for the system $\Sigma^{(k)}$ coincide with the same Lie derivatives for the system $\Sigma^{(k+1)}$.*

Proof. We proceed by induction on m for any k. When $m = 0$ we only have one zero-order Lie derivative (i.e., $h(x)$), which is obviously the same for the two systems, $\Sigma^{(k)}$ and $\Sigma^{(k+1)}$. Let us assume that the previous assertion is true for m and let us prove that it holds for $m+1 \leq k$. If it is true for m, any Lie derivative up to the mth order is the same for the two systems. Additionally, from Lemma 6.7, we know that these Lie derivatives are independent of $w^{(f)}$ $\forall f \geq m$. The proof follows from the fact that the first $n + mm_w$ components of the vector fields $G, F^1, \ldots, F^{m_u}$ for $\Sigma^{(k)}$ coincide with the first $n + mm_w$ components of the same vector fields for $\Sigma^{(k+1)}$ when $m < k$. ◄

6.2.2 ▪ An achievable upper bound functions' set

Let us consider the system in (6.1), during the time interval $\mathcal{I} = [0, T]$. The following result guarantees that the Lie derivatives of $\Sigma^{(k)}$ form an upper bound functions' set.

Theorem 6.10. *Let us suppose that, during the time interval $\mathcal{I}$, the unknown inputs are $w_1(t), \ldots, w_{m_w}(t)$, $t \in \mathcal{I}$. The k-order Lie derivatives of $\Sigma^{(k)}$ along $G, F^1, \ldots, F^{m_u}$, for any k, computed at $x = x_0$ and at $\left.\frac{d^m w_j}{dt^m}\right|_{t=0}$ $(m = 1, \ldots, k-1, j = 1, \ldots, m_w)$ are an upper bound functions' set.*

Proof. The knowledge that we can gather on x_0 is based on the following knowledge/degrees of freedom:

- knowledge of all the values $\left.\frac{d^k y}{dt^k}\right|_{t=0}$ and $\left.\frac{d^k u_i}{dt^k}\right|_{t=0}$ for any positive integer $k = 0, 1, \ldots$ and $i = 1, \ldots, m_u$ and

- the possibility of setting all the values $\left.\frac{d^k u_i}{dt^k}\right|_{t=0}$.

Let us consider a given k, and let us refer to the extended system $\Sigma^{(k)}$ (whose dynamics are given by (6.5) and (6.6)). We can express the k-order time derivative of the output as follows:

$$\frac{d^k y}{dt^k} = F_k(u_1^{(0)}, \ldots, u_{m_u}^{(k-1)}, L^0, \ldots, L^k_{\alpha_1 \cdots \alpha_k}), \tag{6.7}$$

where $L^0, \ldots, L^k_{\alpha_1 \cdots \alpha_k}$ are the Lie derivatives of $\Sigma^{(k)}$ up to the k-order. In other words, the k-order time derivative of the output can be expressed in terms of the time derivatives of the known inputs (up to the $(k-1)$-order) and the aforementioned Lie derivatives. The above equality can be easily proved by induction, and the proof is similar to the proof of the equality in (4.13). Because of Proposition 6.8, we can ignore all the Lie derivatives along W^j, $j = 1, \ldots, m_w$. Hence, the Lie derivatives of $\Sigma^{(k)}$ along $G, F^1, \ldots, F^{m_u}$ (for any k) constitute an upper bound functions' set. ◂

Our next question is to understand whether, by a suitable choice of the known inputs, we can achieve this upper bound, i.e., whether the above Lie derivatives form an achievable upper bound functions' set. To answer to this question, we must determine whether, for any integer k, the Lie derivatives of $\Sigma^{(k)}$ up to the k-order are w-observable. Namely, we must determine whether we can obtain the values that all the Lie derivatives of $\Sigma^{(k)}$ up to the k-order take at $x = x_0$ and at $\left.\frac{d^m w_j}{dt^m}\right|_{t=0}$ $(m = 1, \ldots, k-1, j = 1, \ldots, m_w)$. Answering to this question is very hard. It is possible to prove that the answer is positive.

Specifically, we have the following result.

Theorem 6.11. *For any integer k, there exists a suitable choice of the known inputs such that we can obtain the values that all the Lie derivatives of $\Sigma^{(k)}$, up to the k-order, take at $x = x_0$ and at $\left.\frac{d^m w_j}{dt^m}\right|_{t=0}$ $(m = 1, \ldots, k-1, j = 1, \ldots, m_w)$ from the knowledge of the output in $\mathcal{I}$.*

Proof. We do not provide the proof of this theorem. We provide, later, the proof of its analogue in the standard approach based on the indistinguishability (see Theorem 6.20). ◂

6.3 ▪ A partial tool to investigate the observability properties in the presence of unknown inputs

The results stated by Theorems 6.10 and 6.11 tell us that the Lie derivatives of $\Sigma^{(k)}$ up to the k-order, and for any integer k, form an achievable upper bound functions' set. We can easily

6.3. Partial tool to investigate observability properties in presence of unknown inputs

build the codistribution that contains the gradients of all these functions. It is obtained by the following algorithm.

ALGORITHM 6.1. Observable codistribution in the augmented space that includes the unknown inputs.

Set $k = 0$
Set $\overline{\Omega}_k = \text{span}\{\nabla_k h_1, \ldots, \nabla_k h_p\}$
loop
    Set $\overline{\Omega}_k = [\overline{\Omega}_k, 0_{m_w}]$
    Set $k = k + 1$
    Build $\Sigma^{(k)}$, i.e., $G, F^1, \ldots, F^{m_u}$ by using (6.6)
    Set $\overline{\Omega}_k = \overline{\Omega}_{k-1} \oplus \mathcal{L}_G \overline{\Omega}_{k-1} \oplus \bigoplus_{i=1}^{m_u} \mathcal{L}_{F^i} \overline{\Omega}_{k-1}$
end loop

where the first line in the loop, i.e., $\overline{\Omega}_k = [\overline{\Omega}_k, 0_{m_w}]$, consists in embedding the codistribution $\overline{\Omega}_k$ in the new extended space (the one that also includes the new m_w axes, $w_1^{(k)}, \ldots, w_{m_w}^{(k)}$). In practice, each covector $\omega \in \overline{\Omega}_k$ that is represented by a row-vector of dimension $n + km_w$ is extended as follows:

$$\omega \to [\omega, \underbrace{0, \ldots, 0}_{m_w}].$$

For a given k, the codistribution $\overline{\Omega}_k$ contains mixed information on both the original initial state x_0 and the values that the unknown inputs (together with their time derivatives up to the $(k-1)$-order) take at the initial time. One of the most important results provided in Chapters 7 and 8 is that it is possible to separate the information on x_0 from the rest. Before providing this result and before discussing the main features of Algorithm 6.1, we provide the extension of Theorem 4.8 to the unknown inputs case. This result, together with Algorithm 6.1, provides a first analytic and automatic (but incomplete) procedure to determine the observability properties of the system defined by (6.1). This procedure will be illustrated in section 6.3.1, by discussing simple examples that illustrate its practical importance and its limits of applicability.

Theorem 6.12. *Let us assume that, for any integer k, the codistribution $\overline{\Omega}_k$ returned by Algorithm 4.2 is nonsingular on an open set of the augmented space. Let us suppose that $\theta(x)$ is a scalar function that only depends on the original state. Then there exists an integer k such that $\nabla_k \theta \in \overline{\Omega}_k$ if and only if $\theta(x)$ is weakly observable.*

Proof. For any integer k, we consider the system $\Sigma^{(k)}$, which is defined in the augmented space with dimension $n^k \triangleq n + km_w$. Let us denote by s^k the dimension of $\overline{\Omega}_k$. We introduce the following local coordinates' change:

$$(x_1, \ldots, x_n, w_1^{(0)}, \ldots, w_{m_w}^{(k-1)}) \to (\lambda_1, \ldots, \lambda_{s^k}, \lambda_{s^k+1}, \ldots, \lambda_{n^k}),$$

where the functions $\lambda_1, \ldots, \lambda_{s^k}$ are s^k independent Lie derivatives of $\Sigma^{(k)}$ up to the k-order and the remaining $\lambda_{s^k+1}, \ldots, \lambda_{n^k}$ are chosen from the initial coordinates in such a way that all the gradients $\nabla_k \lambda_i$, $i = 1, \ldots, n^k$, form a set of n^k independent covectors in a given open set.

Let us suppose that there exists an integer k such that $\nabla_k \theta \in \overline{\Omega}_k$. This means that, in the new coordinates, θ can be expressed in terms of $\lambda_1, \ldots, \lambda_{s^k}$. From Theorem 6.11 we know that we can determine the initial values of $\lambda_1, \ldots, \lambda_{s^k}$. Hence, the same holds for θ.

Vice versa, let us suppose that θ is weakly observable. Let us proceed by contradiction. For any integer k, $\nabla_k \theta \notin \overline{\Omega}_k$. This means that, in the new coordinates, θ also depends on some of the last $n^k - s^k$ coordinates and this holds for any k. On the other hand, from Theorem 6.10 we know that the first s^k coordinates, for all the k, constitute an upper bound functions' set. As a result, the initial value of θ cannot be determined, in contrast with the assumption. ◀

As we mentioned above, this theorem, together with Algorithm 6.1, can be used to investigate the observability properties of a system characterized by unknown inputs. A first issue that needs to be discussed is the fact that Algorithm 6.1 provides functions which are w-observable (i.e., they are observable under the action of a specific unknown input). In contrast, we want to obtain a more general result, i.e., independent of the specific unknown input. We note that Theorem 6.12 states that the same result holds at any extended state that belongs to an open set where the codistribution $\overline{\Omega}_k$ is nonsingular. In other words, all the state extensions (i.e., all the unknown inputs) that belong to this open set provide the same observability properties. We need to better understand the structure of this open set. In particular, given an extended state

$$^k x = \begin{bmatrix} x \\ E^k \end{bmatrix}$$

that belongs to the above open set, we need to understand which points that differ from it only for the state extension, i.e.,

$$^k x' = \begin{bmatrix} x \\ E'^k \end{bmatrix},$$

belong to this open set. In Appendix C, we introduce the concept of canonic form of a given input-output system, where some of the inputs are unknown. In the next chapters, we will see that, to obtain the state observability, it is necessary to set the system in canonic form (this can be achieved by following an automatic procedure that is provided by Algorithm C.2). From the derivations provided in Appendix C, it is possible to obtain a complete answer to the above question. Specifically, as we summarize in Appendix C.2.3, a necessary condition for the nonsingularity of $\overline{\Omega}_k$ is that $w_1, \ldots, w_k$ (i.e., the first m_w components of E^k) do not vanish. Additionally, the codistribution $\overline{\Omega}_k$ is nonsingular at any extended state such that the first m_w components of E'^k do not vanish. In other words, it is sufficient to consider any unknown input that does not vanish.

The results obtained so far tell us that the codistribution generated by Algorithm 6.1 consists of the gradients of observable functions. On the other hand, these functions depend on the state together with the unknown inputs and their time derivatives up to a given order. However, given a scalar function that only depends on the original state x, i.e., $\theta(x)$, we have, for any $i = 1, 2, \ldots, m_u$,

$$\mathcal{L}_{F^i} \theta = \nabla_k \theta \cdot F^i = \nabla \theta \cdot f^i = \mathcal{L}_{f^i} \theta,$$

which is still a function that only depends on the original state. We introduce the following functional space.

Definition 6.13. *Given the input-output system characterized by* (6.1)*, we denote by $\mathcal{F}$ the smallest functional space that includes all the outputs together with their Lie derivatives up to any order only along $f^1, \ldots, f^{m_u}$.*

This space (or better the codistribution that consists of the gradients of the functions in $\mathcal{F}$) can be automatically generated by Algorithm 4.2, with $m = m_u$ and without the drift f^0.

6.3. Partial tool to investigate observability properties in presence of unknown inputs 145

On the basis of the above discussion, we remark the following property.

Remark 6.14. *All the functions that belong to $\mathcal{F}$ are weakly observable.*

On the other hand, the space $\mathcal{F}$ does not contain *all* the observable functions. Algorithm 6.1 generates additional functions which depend on the entire extended state. This fact makes the implementation of this algorithm prohibitive after few steps. Before discussing this fundamental issue, together with the fact that Algorithm 6.1 does not converge automatically, we use this algorithm to investigate the observability of two simple systems.

6.3.1 ▪ Examples

We apply Algorithm 6.1 to the unicycle discussed in Chapter 1. Specifically, we consider systems 1 and 2 in section 1.2, which are characterized by a single output that is the distance of the origin (system 1) and the bearing angle of the origin in the local frame (system 2). In addition, in both cases, we assume that the linear speed v is unknown. We show that Algorithm 6.1, together with Theorem 6.12, provides results that agree with the results that we obtained in section 1.5.1 by following a heuristic procedure.

These two systems were also investigated in section 5.1 by assuming that both the inputs (v and ω) are known. In particular, we obtained that these systems can be described by the two-dimensional state $x = [\rho,\ \theta]^T$, where $\theta = l_2 = \theta_v - \phi$ (see section 5.1 and Figure 1.1 for an illustration). The dynamics of this state are (see section 5.1)

$$\begin{bmatrix} \dot{\rho} &= v\cos\theta, \\ \dot{\theta} &= \omega - \frac{v}{\rho}\sin\theta. \end{bmatrix} \tag{6.8}$$

In the following, we use Algorithm 6.1 to obtain the observability properties of the systems characterized by the above dynamics, where v acts as an unknown input, and characterized by the output $y = h(x) = \rho$ for system 1, and $y = h(x) = \pi - \theta$ for system 2. By comparing (6.8) with (6.1) we obtain

$$g^0 = \begin{bmatrix} 0 \\ 0 \end{bmatrix}, \quad f \triangleq f^1 = \begin{bmatrix} 0 \\ 1 \end{bmatrix}, \quad g \triangleq g^1 = \begin{bmatrix} \cos\theta \\ -\frac{1}{\rho}\sin\theta \end{bmatrix}.$$

We proceed as follows. Since Algorithm 6.1 does not converge, we check, at each step, whether the components of the original state (i.e., ρ and θ) are weakly observable. Specifically, in accordance with Theorem 6.12, we check, for each $k = 0, 1, \ldots$, whether $\nabla_k \rho \in \overline{\Omega}_k$ and whether $\nabla_k \theta \in \overline{\Omega}_k$. In addition, at each step, we also compute the distribution $\overline{\Delta}_k$ orthogonal to $\overline{\Omega}_k$. Since $\overline{\Omega}_k \subset \lim_{k\to\infty} \overline{\Omega}_k$, for any finite k, $\overline{\Delta}_k$ includes the symmetries of the system.

System 1

The output is $y = h(x) = \rho$. We follow the steps of Algorithm 6.1. We set $k = 0$. The state is

$$^0x = [\rho,\ \theta]^T$$

and $\nabla_0 = \left(\frac{\partial}{\partial \rho},\ \frac{\partial}{\partial \theta}\right)$. We have

$$\overline{\Omega}_0 = \text{span}\{\nabla_0 \rho\} = \text{span}\{[1,\ 0]\}.$$

146   Chapter 6. General Concepts on Nonlinear Unknown Input Observability

We have $\nabla_0 \rho \in \overline{\Omega}_0$, while $\nabla_0 \theta \notin \overline{\Omega}_0$. The distribution orthogonal to $\overline{\Omega}_0$ is

$$\overline{\Delta}_0 = \text{span}\left\{[0, 1]^T\right\}.$$

We continue to run Algorithm 6.1 in order to check whether also θ is observable. We set $k = 1$. The state is

$$^1x = [\rho, \theta, v^{(0)}]^T.$$

Additionally, $\nabla_1 = \left(\frac{\partial}{\partial \rho}, \frac{\partial}{\partial \theta}, \frac{\partial}{\partial v^{(0)}}\right)$, $\overline{\Omega}_0 = \text{span}\{[1, 0, 0]\}$,

$$F \triangleq F^1 = \begin{bmatrix} 0 \\ 1 \\ 0 \end{bmatrix}, \quad G = \begin{bmatrix} v^{(0)} \cos\theta \\ -\frac{v^{(0)}}{\rho} \sin\theta \\ 0 \end{bmatrix},$$

and $\mathcal{L}_F \overline{\Omega}_0 = \{0\}$ and $\mathcal{L}_G \overline{\Omega}_0 = \text{span}\{[0, -v^{(0)} \sin\theta, \cos\theta]\}$.
Therefore,

$$\overline{\Omega}_1 = \text{span}\left\{[1, 0, 0], [0, -v^{(0)} \sin\theta, \cos\theta]\right\}.$$

We still obtain $\nabla_1 \theta \notin \overline{\Omega}_1$. The distribution orthogonal to $\overline{\Omega}_1$ is

$$\overline{\Delta}_1 = \text{span}\left\{[0, \cos\theta, v^{(0)} \sin\theta]^T\right\}.$$

We continue to run Algorithm 6.1 in order to check whether θ is observable. We set $k = 2$. The state is

$$^2x = [\rho, \theta, v^{(0)}, v^{(1)}]^T.$$

Also $\nabla_2 = \left(\frac{\partial}{\partial \rho}, \frac{\partial}{\partial \theta}, \frac{\partial}{\partial v^{(0)}}, \frac{\partial}{\partial v^{(1)}}\right)$, $\overline{\Omega}_1 = \text{span}\{[1, 0, 0, 0], [0, -v^{(0)} \sin\theta, \cos\theta, 0]\}$,

$$F = \begin{bmatrix} 0 \\ 1 \\ 0 \\ 0 \end{bmatrix}, \quad G = \begin{bmatrix} v^{(0)} \cos\theta \\ -\frac{v^{(0)}}{\rho} \sin\theta \\ v^{(1)} \\ 0 \end{bmatrix}$$

and

$$\mathcal{L}_F \overline{\Omega}_1 = \text{span}\left\{\nabla_2\left(-v^{(0)} \sin\theta\right)\right\} = \text{span}\left\{[0, -v^{(0)} \cos\theta, -\sin\theta, 0]\right\},$$

$$\mathcal{L}_G \overline{\Omega}_1 = \text{span}\left\{\nabla_2\left(\frac{(v^{(0)} \sin\theta)^2}{\rho} + v^{(1)} \cos\theta\right)\right\}.$$

Therefore,

$$\overline{\Omega}_2 = \text{span}\left\{[1, 0, 0, 0], [0, -v^{(0)} \sin\theta, \cos\theta, 0], [0, -v^{(0)} \cos\theta, -\sin\theta, 0], \right.$$
$$\left. \nabla_2\left(\frac{(v^{(0)} \sin\theta)^2}{\rho} + v^{(1)} \cos\theta\right)\right\}.$$

This time we obtain $\nabla_2 \theta \in \overline{\Omega}_2$ and we conclude that also θ is observable (actually, weakly observable). We do not need to continue to run the algorithm.

Note that we obtained the same result obtained in section 1.5.1 by following a heuristic procedure.

System 2

The output is $y = h(x) = \pi - \theta$. We follow the steps of Algorithm 6.1. We set $k = 0$. The state is
$$^0x = [\rho,\ \theta]^T$$
and $\nabla_0 = \left(\frac{\partial}{\partial \rho},\ \frac{\partial}{\partial \theta}\right)$. We have
$$\overline{\Omega}_0 = \mathrm{span}\left\{\nabla_0(\pi - \theta) =\right\} = \mathrm{span}\left\{[0,\ -1]\right\}.$$

We have $\nabla_0 \rho \notin \overline{\Omega}_0$, while $\nabla_0 \theta \in \overline{\Omega}_0$. The distribution orthogonal to $\overline{\Omega}_0$ is
$$\overline{\Delta}_0 = \mathrm{span}\left\{[1,\ 0]^T\right\}.$$

We continue to run Algorithm 6.1 in order to check whether also ρ is observable. We set $k = 1$. The state is
$$^1x = [\rho,\ \theta,\ v^{(0)}]^T.$$
Additionally, $\nabla_1 = \left(\frac{\partial}{\partial \rho},\ \frac{\partial}{\partial \theta},\ \frac{\partial}{\partial v^{(0)}}\right)$, $\overline{\Omega}_0 = \mathrm{span}\left\{[0,\ -1,\ 0]\right\}$, and F and G coincide with the ones of the previous example at the same step $k = 1$.
We have $\mathcal{L}_F \overline{\Omega}_0 = \{0\}$ and $\mathcal{L}_G \overline{\Omega}_0 = \mathrm{span}\left\{\left[-\frac{v^{(0)} \sin \theta}{\rho^2},\ \frac{v^{(0)} \cos \theta}{\rho},\ \frac{\sin \theta}{\rho}\right]\right\}$.
Therefore,
$$\overline{\Omega}_1 = \mathrm{span}\left\{[0,\ -1,\ 0],\ \left[-\frac{v^{(0)} \sin \theta}{\rho^2},\ \frac{v^{(0)} \cos \theta}{\rho},\ \frac{\sin \theta}{\rho}\right]\right\}.$$

We still obtain $\nabla_1 \rho \notin \overline{\Omega}_1$. The distribution orthogonal to $\overline{\Omega}_1$ is
$$\overline{\Delta}_1 = \mathrm{span}\left\{[\rho,\ 0,\ v^{(0)}]^T\right\}.$$

Note that the above vector is the generator of a scale transform ($\rho \to (1+\lambda)\rho$, $v^{(0)} \to (1+\lambda)v^{(0)}$, $\lambda \in \mathbb{R}^+$).
We continue to run Algorithm 6.1 in order to check whether ρ is observable. We set $k = 2$. The state ^2x, the operator ∇_2, F, and G coincide with the ones of the previous example at the same step $k = 2$. Additionally, $\overline{\Omega}_1 = \mathrm{span}\left\{[0,\ -1,\ 0,\ 0],\ \left[-\frac{v^{(0)} \sin \theta}{\rho^2},\ \frac{v^{(0)} \cos \theta}{\rho},\ \frac{\sin \theta}{\rho},\ 0\right]\right\}$.
We have
$$\mathcal{L}_F \overline{\Omega}_1 = \mathrm{span}\left\{\nabla_2 \left(\frac{v^{(0)} \cos \theta}{\rho}\right)\right\},\quad \mathcal{L}_G \overline{\Omega}_1 = \mathrm{span}\left\{\nabla_2 \left(-\frac{((v^{(0)})^2 \sin(2\theta)}{\rho^2} + \frac{v^{(1)} \sin \theta}{\rho}\right)\right\}.$$

We obtain
$$\overline{\Omega}_2 = \mathrm{span}\left\{[0,\ -1,\ 0,\ 0],\ \left[-\frac{v^{(0)} \sin \theta}{\rho^2},\ \frac{v^{(0)} \cos \theta}{\rho},\ \frac{\sin \theta}{\rho},\ 0\right],\right.$$
$$\left.\nabla_2\left(-\frac{(v^{(0)})^2 \sin(2\theta)}{\rho^2} + \frac{v^{(1)} \sin \theta}{\rho}\right)\right\}.$$

We still obtain $\nabla_2 \rho \notin \overline{\Omega}_2$. The distribution orthogonal to $\overline{\Omega}_2$ is
$$\overline{\Delta}_2 = \mathrm{span}\left\{[\rho,\ 0,\ v^{(0)},\ v^{(1)}]^T\right\},$$

which is still the generator of a scale transform. We should continue to run the algorithm to check whether ρ is observable or not. However, as we remarked in section 1.5.1, we know that ρ is not observable because the system does not contain any metric information: both the output (the bearing of the origin in the local frame) and the known input (the vehicle angular speed) consist of angular measurements. Hence, the absolute scale will remain unobservable at any step and we do not continue to run the algorithm.

6.3.2 • Main features and limits of applicability

In the previous examples, we obtained the observability properties of two systems by following an automatic procedure that consists of the implementation of a certain number of steps of Algorithm 6.1 together with the test to check whether the components of the original state are observable or not (Theorem 6.12). We wonder whether we can use the above procedure to analyze the observability of more complex systems, e.g., by using a tool for symbolic calculation. We immediately remark the following fundamental aspects:

- Algorithm 6.1 does not converge automatically. This means that if for a given step k the gradients of the components of the original state belong to $\overline{\Omega}_k$, we can conclude that the original state is weakly observable (as in the case of system 1). On the other hand, if this is not true, we cannot exclude that it is true for a larger k. To this regard, note that for system 2 we stopped to run the algorithm by exploiting some knowledge that is not inherent to the algorithm.

- The codistribution returned by Algorithm 6.1 is defined in the extended space; i.e., its covectors are the gradients of scalar functions with respect to the augmented state. As a result, the computation dramatically increases with the state augmentation and it becomes prohibitive after few steps. In addition, this makes it hard to determine the physical meaning of the vectors in $\overline{\Delta}_k$ and, in particular, to understand whether they are really symmetries for the original system.

The goal of the next chapters is precisely to deal with the above fundamental issues.

We conclude this section with the following remark. We go back to system 1. In section 1.5.1 we showed that we can retrieve the initial θ by using the system of two equations given in (1.16) and (1.17). On the other hand, this is not possible when v_0 vanishes. Note that the codistribution $\overline{\Omega}_2$ loses one dimension when v_0 vanishes. This agrees with Theorem 6.2.

6.4 • Theory based on the standard approach

Exactly as in the case without unknown inputs, we provide the theory of observability based on the concept of indistinguishability. In this section we still refer to the system defined on the manifold $\mathcal{M}$ by the equations in (6.1) during the time interval $\mathcal{I} = [0, T]$ and we introduce the concept of indistinguishability. We actually refer to the case of a single output ($p = 1$) for the sake of simplicity. The extension to the multiple outputs case is trivial. Based on the concept of indistinguishability we provide the standard definition of observability and we show that it is equivalent to our previous definition.

6.4.1 • Indistinguishability and observability

Let us refer to the system defined by (6.1). To be conservative, we want to define as indistinguishable two states that cannot be distinguished even if this occurs only in the most unlucky

6.4. Theory based on the standard approach

situation. On the other side, we do not want to consider these states indistinguishable if the aforementioned unlucky situation can occur with zero probability. This brings us to the same considerations carried out in section 6.1 and, in particular, as for the constructive definition of observability, to adopt the concepts of prevalent and shy set.

Let us denote by $x(t; u; w; x_0)$ the state of the system at time $t \in \mathcal{I}$ when the initial state (i.e., the state at time $t = 0$) was x_0 and the system was driven by the known input $u = [u_1(\tau), \ldots, u_{m_u}(\tau)]^T$ and the unknown input $w = [w_1(\tau), \ldots, w_{m_w}(\tau)]^T$, $\tau \in \mathcal{I}$. We introduce the following definition.

Definition 6.15 (Indistinguishable States in the Presence of UI). x_a and x_b are indistinguishable if, for every time assignment of the m_u known inputs in $\mathcal{I}$, there exists a nonshy subset $\mathcal{W}_s \subseteq \mathcal{W}$ such that, for any unknown input function $w_a \in \mathcal{W}_s$ there exists an unknown input function w_b (in general, but not necessarily, $w_b = w_a$) such that $h(x(t; u; w_a; x_a)) = h(x(t; u; w_b; x_b))$ $\forall t \in \mathcal{I}$.

Based on this definition, we introduce the observability by proceeding as in section 4.5. We denote by $I(x)$ the subset of $\mathcal{M}$ of all the points which are indistinguishable from x. In what follows, we often refer to $I(x)$ as the indistinguishable set. The following definitions are exactly the same as Definitions 4.10–4.13.

Definition 6.16. *An input-output system is observable at a given $x_0 \in \mathcal{M}$ if $I(x_0) = \{x_0\}$. In addition, it is observable on a given set $\mathcal{U} \subseteq \mathcal{M}$ if it is observable at any $x \in \mathcal{U}$.*

Definition 6.17. *An input-output system is weakly observable at a given $x_0 \in \mathcal{M}$ if there exists an open neighborhood B of x_0 such that $I(x_0) \cap B = \{x_0\}$. In addition, it is weakly observable on a given set $\mathcal{U} \subseteq \mathcal{M}$ if it is weakly observable at any $x \in \mathcal{U}$.*

Definition 6.18. *The scalar function $\theta(x)$ is observable at x_0 if it is constant on $I(x_0)$. In addition, $\theta(x)$ is observable on a given set $\mathcal{U} \subseteq \mathcal{M}$ if it is observable at any $x \in \mathcal{U}$.*

Definition 6.19. *The scalar function $\theta(x)$ is weakly observable at x_0 if there exists an open neighborhood B of x_0 such that $\theta(x)$ is constant on $I(x_0) \cap B$. In addition, $\theta(x)$ is weakly observable on a given set $\mathcal{U} \subseteq \mathcal{M}$ if it is weakly observable at any $x \in \mathcal{U}$.*

6.4.2 ▪ Equivalence of the approaches

In the following, we will always assume that the codistribution returned by Algorithm 6.1 is nonsingular on a given open set and we restrict our analysis to this open set. This means that its dimension is constant on this open set. Under this assumption, we show the equivalence of the definitions of observability provided in the constructive approach and the above definition based on the concept of indistinguishability.

The result stated by Theorem 4.14 becomes the following.

Theorem 6.20. x_a and x_b are indistinguishable if and only if, for any integer k, there exist E_a^k and E_b^k such that, in $\Sigma^{(k)}$, the Lie derivatives of the output up to the kth order, along all the vector fields that characterize the dynamics of $\Sigma^{(k)}$, take the same values at $[x_a, E_a^k]$ and $[x_b, E_b^k]$.

Proof. Let us assume that x_a and x_b are indistinguishable.

We consider a piecewise-constant input $\tilde{u}$ as follows ($i = 1, \ldots, m_u$):

$$\tilde{u}_i(t) = \begin{cases} u_i^1, & t \in [0, t_1), \\ u_i^2, & t \in [t_1, t_1 + t_2), \\ \vdots \\ u_i^g, & t \in [t_1 + t_2 + \cdots + t_{g-1}, t_1 + t_2 + \cdots + t_{g-1} + t_g) \end{cases} \quad (6.9)$$

$\forall t \in [0, t_1 + t_2 + \cdots + t_{g-1} + t_g) \subset \mathcal{I}$.

Since x_a and x_b are indistinguishable, there exists a nonshy subset $\mathcal{W}_s \subseteq \mathcal{W}$ such that, for any unknown input function $w_a \in \mathcal{W}_s$, there exists an unknown input function w_b such that

$$h(x(t; u; w_a; x_a)) = h(x(t; u; w_b; x_b)) \; \forall t \in \mathcal{I}.$$

Let us consider an unknown input w_a and the correspondent w_b. For any integer k, we consider the extended system $\Sigma^{(k)}$ and we set

$$E_a^k \triangleq \begin{bmatrix} w_a^{(0)}(0) \\ w_a^{(1)}(0) \\ \vdots \\ w_a^{(k-1)}(0) \end{bmatrix}, \quad E_b^k \triangleq \begin{bmatrix} w_b^{(0)}(0) \\ w_b^{(1)}(0) \\ \vdots \\ w_b^{(k-1)}(0) \end{bmatrix}.$$

In $\Sigma^{(k)}$ the unknown input function is the k-order time derivative of the unknown input of the original system, namely $w^{(k)}(t)$. We have

$$h(x(t; \tilde{u}; w_a^{(k)}; [x_a, E_a^k];)) = h(x(t; \tilde{u}; w_b^{(k)}; [x_b, E_b^k])). \quad (6.10)$$

On the other hand, by taking the two quantities in (6.10) at $t = t_1 + t_2 + \cdots + t_{g-1} + t_g$, we can consider them as functions of the g arguments $t_1, t_2, \ldots, t_g$. Hence, by differentiating with respect to all these variables, we also have

$$\frac{\partial^g h(x(t_1 + \cdots + t_g; \tilde{u}; w_a^{(k)}; [x_a, E_a^k]))}{\partial t_1 \partial t_2 \cdots \partial t_g} = \frac{\partial^g h(x(t_1 + \cdots + t_g; \tilde{u}; w_b^{(k)}; [x_b, E_b^k]))}{\partial t_1 \partial t_2 \cdots \partial t_g}. \quad (6.11)$$

By computing the previous derivatives at $t_1 = t_2 = \cdots = t_g = 0$ and by using Proposition 6.8 we obtain, if $g \leq k$,

$$\left. \mathcal{L}^g_{\theta_1 \theta_2 \cdots \theta_g} h \right|_{\substack{x = x_a \\ E^k = E_a^k}} = \left. \mathcal{L}^g_{\theta_1 \theta_2 \cdots \theta_g} h \right|_{\substack{x = x_b \\ E^k = E_b^k}}, \quad (6.12)$$

where $\theta_h = G + \sum_{i=1}^{m_u} F^i u_i^h$, $h = 1, \ldots, g$. The equality in (6.12) must hold for all possible choices of $u_1^h, \ldots, u_{m_u}^h$. By appropriately selecting these $u_1^h, \ldots, u_{m_u}^h$, we finally obtain

$$\left. \mathcal{L}^g_{v_1 v_2 \cdots v_g} h \right|_{\substack{x = x_a \\ E^k = E_a^k}} = \left. \mathcal{L}^g_{v_1 v_2 \cdots v_g} h \right|_{\substack{x = x_b \\ E^k = E_b^k}}, \quad (6.13)$$

6.4. Theory based on the standard approach

where $v_1 v_2 \cdots v_g$ are vector fields belonging to the set $\{G, F^1, \ldots, F^{m_u}\}$. Since w_a was chosen in an open set, we can find a w_a such that E_a^k does not vanish.

Proving the reverse is immediate. From Theorem 6.10 we know that the Lie derivatives of $\Sigma^{(k)}$ are an upper bound functions' set. ◄

Theorem 6.21. *Let us assume that, for any integer k, the codistribution $\overline{\Omega}_k$ returned by Algorithm 6.1 is nonsingular on a given open set. Then the following hold:*

- *Definition 6.3 is equivalent to Definition 6.16.*
- *Definition 6.4 is equivalent to Definition 6.17.*
- *Definition 6.5 is equivalent to Definition 6.18.*
- *Definition 6.6 is equivalent to Definition 6.19.*

Proof. This proof is very similar to the one of Theorem 4.15, by exploiting, in this case, the coordinates' change adopted in the proof of Theorem 6.12. ◄

Chapter 7

Unknown Input Observability for Driftless Systems with a Single Unknown Input

In Chapter 6 we provided several results to investigate the observability properties of an input-output system, when some of the inputs are unknown. In particular, we provided an analytic procedure that consists of Algorithm 6.1 together with the result stated by Theorem 6.12. However, as we mentioned in section 6.3.2, this procedure presents two fundamental obstacles:

- Algorithm 6.1 does not converge automatically.

- The computation requested to run Algorithm 6.1 dramatically increases with the state augmentation, and it becomes prohibitive after few steps.

The goal of this and the next chapter is precisely to address both the above issues. This will be very complex. It requires the introduction of several key tensorial fields, with respect to the groups of invariance introduced in Chapter 3. In particular, we will introduce tensors with respect to the $\mathcal{SUIO}$ (Simultaneous Unknown Input-Output transformations' group). On the other hand, before introducing these tensorial fields and dealing with the general case, it is very convenient to deal with the simplified system introduced in section 3.3, which was obtained by requiring that the $\mathcal{SUIO}$ is Abelian.

The simplified system is characterized by the following equations:

$$\begin{cases} \dot{x} &= \sum_{k=1}^{m_u} f^k(x) u_k + g(x) w, \\ y &= h(x), \end{cases} \tag{7.1}$$

which are obtained by setting, in (6.1), $g^0(x) = 0_n$ (n being the dimension of the state), $m_w = p = 1$. Note that the case of multiple outputs ($p > 1$) has the same complexity and the solution provided in this chapter also includes this case (see Algorithm 7.3).

This chapter is organized as follows. We start by directly providing the solution in section 7.1. This consists of Algorithm 7.3, which returns the observable codistribution for systems characterized by (7.1). Then, in section 7.2, we provide several applications. Finally, in section 7.3, we provide all the analytic derivations that prove the validity of the solution.

7.1 ▪ Extension of the observability rank condition for the simplified systems

This section[24] provides the algorithm that returns the observable codistribution together with

---
[24]© 2019 IEEE. Reprinted, with permission, from [38].

its convergence criterion (section 7.1.1). Then, based on this algorithm, in section 7.1.2 we summarize the analytic method to determine the observability properties of systems described by (7.1).

7.1.1 ▪ Observable codistribution

We first need to compute the first-order Lie derivative of the function $h(x)$ along the vector field $g(x)$. Let us denote it by $L_g^1 = L_g^1(x)$. We have

$$L_g^1 = L_g^1(x) \triangleq \mathcal{L}_g h = \nabla h \cdot g. \qquad (7.2)$$

The analytic computation of the observable codistribution is based on the assumption that $L_g^1 \neq 0$ in a given neighborhood of x_0. In Appendix C, we introduce the concept of *canonic form* with respect to the unknown inputs. For the case $m_w = 1$ (dealt with in section C.1), we show that the system is in canonic form with respect to the unknown input w, if either $L_g^1 \neq 0$ or it is possible to redefine the output, without altering the system observability properties, such that the Lie derivative of the new output along g does not vanish.[25] Finally, if a system characterized by $m_w = 1$ is not in canonic form, it means that the unknown input is spurious (i.e., it does not affect the observability properties). For these reasons, we can assume that $L_g^1 \neq 0$.

Before introducing the new algorithm that generates the entire observable codistribution, we introduce a new set of vector fields ${}^i\phi_k$ ($i = 1, \ldots, m_u$ and for any integer k). They are obtained recursively by the following algorithm.

ALGORITHM 7.1. Vectors ${}^i\phi_k$ for driftless systems with a single unknown input.

Set $m = 0$
Set ${}^i\phi_m = f^i$ for $i = 1, \ldots, m_u$
while $m < k$ **do**
    Set $m = m + 1$
    Set ${}^i\phi_m = \frac{[{}^i\phi_{m-1},\, g]}{L_g^1}$ for $i = 1, \ldots, m_u$
end while

where $[\cdot, \cdot]$ denotes the Lie bracket of vector fields, defined in (2.15), namely

$$[a,\, b] \triangleq \frac{\partial b}{\partial x} a(x) - \frac{\partial a}{\partial x} b(x).$$

In other words, for each $i = 1, \ldots, m_u$, we have one new vector field at every step of the algorithm. Throughout this book, in the case when $m_u = 1$, we denote by ϕ_m the vector field ${}^1\phi_m$.

We are now ready to provide the entire observable codistribution. It is obtained by running the following algorithm for a given value of the integer k (that will be provided in the following).

---

[25] The new output is selected from the space of functions $\mathcal{F}$, defined, in this case, as the space that contains h and its Lie derivative up to any order along the vector fields $f^1, \ldots, f^{m_u}$.

7.1. Extension of the observability rank condition for the simplified systems

ALGORITHM 7.2. Observable codistribution for driftless systems with a single unknown input.

Set $m = 0$
Set $\Omega_m = \text{span}\{\nabla h\}$
while $m < k$ **do**
    Set $m = m + 1$
    Set $\Omega_m = \Omega_{m-1} \oplus \bigoplus_{i=1}^{m_u} \mathcal{L}_{f^i}\Omega_{m-1} \oplus \mathcal{L}_{\frac{g}{L_g^1}}\Omega_{m-1} \oplus \bigoplus_{i=1}^{m_u} \text{span}\{\mathcal{L}_{\phi_{m-1}^i}\nabla h\}$
end while

In the presence of multiple outputs, we only need to add to the codistribution Ω_0 the span of the gradients of the remaining outputs. Note that, in the presence of multiple outputs, the function L_g^1 is still a scalar function since it is still defined by using a single output. The result is independent of the chosen output (provided that L_g^1 does not vanish[26]).

In section 7.3.2 we investigate the convergence properties of Algorithm 7.2. We consider first the case of a single known input (i.e., $m_u = 1$), and then the results are easily extended to the case of multiple inputs ($m_u > 1$) in section 7.3.3. We prove that Algorithm 7.2 converges in at most $k = n + 2$ steps, and we also provide the analytic criterion to check that the convergence has been attained. This proof and the convergence criterion cannot be the same that hold for Algorithm 4.2, because of the last term that appears in the recursive step,[27] i.e., the term $\bigoplus_{i=1}^{m_u} \text{span}\{\mathcal{L}_{\phi_{m-1}^i}\nabla h\}$ (the special case when the contribution due to this last term is included in the other terms is considered separately by Lemma 7.8). In general, the criterion to establish that the convergence has been attained is not simply obtained by checking if $\Omega_{m+1} = \Omega_m$. Deriving the new analytic criterion is not immediate. It requires us to derive the analytic expression that describes the behavior of the last term in the recursive step. This fundamental equation is provided in section 7.3, and it is (7.20). The analytic derivation of this equation allows us to detect the key quantity that governs the convergence of Algorithm 7.2, in particular regarding the contribution due to the last term in the recursive step. This key quantity is the following scalar:

$$\tau \triangleq \frac{\mathcal{L}_g^2 h}{(L_g^1)^2}. \tag{7.3}$$

We prove (see Lemma 7.13 in Chapter 7.3) that, in general, there exists m' such that $\nabla \tau \in \Omega_{m'}$ (and therefore $\nabla \tau \in \Omega_m \ \forall m \geq m'$). Additionally, we prove that the convergence of the algorithm has been reached when $\Omega_{m+1} = \Omega_m$, $m \geq m'$, and $m \geq 2$ (Theorem 7.14). We also prove that the required number of steps is at most $n + 2$.

In section 7.3.1 it is also shown that the computed codistribution is the entire observable codistribution. Also in this case, the proof is given by first considering the case of a single known input (see Theorem 7.7), and then its validity is extended to the case of multiple inputs in section 7.3.3. Note that this proof is based on the assumption that the unknown input (w) is a differentiable function of time up to a given order (the order depends on the specific case).

---

[26] The case when $L_g^1 = 0$ for all the outputs is dealt with in Appendix C.

[27] The convergence criterion of Algorithm 4.2 is a consequence of the fact that all the terms that appear in the recursive step of Algorithm 4.2 are the Lie derivative of the codistribution at the previous step, along fixed vector fields (i.e., vector fields that remain the same at each step of the algorithm). This is not the case for the last term in the recursive step of Algorithm 7.2 (see also section 2.3 and Algorithm 2.2).

Below we give Algorithm 7.3 which fuses Algorithm 7.2 with Algorithm 7.1. It also includes the convergence criterion, and it accounts for multiple outputs, i.e., when $y = [h_1, \ldots, h_p]^T$. We denote by h the output that was chosen to define L_g^1 in (7.2) and τ in (7.3).

ALGORITHM 7.3. Fusion of Algorithm 7.1 and Algorithm 7.2.

Set $m = 0$
Set $\Omega_m = \text{span}\{\nabla h_1, \ldots, \nabla h_p\}$
Set ${}^i\phi_m = f^i$ for $i = 1, \ldots, m_u$
Set $m = m + 1$
Set $\Omega_m = \Omega_{m-1} \oplus \bigoplus_{i=1}^{m_u} \mathcal{L}_{f^i}\Omega_{m-1} \oplus \mathcal{L}_{\frac{g}{L_g^1}}\Omega_{m-1}$
Set ${}^i\phi_m = \frac{[{}^i\phi_{m-1},\, g]}{L_g^1}$ for $i = 1, \ldots, m_u$
Set $m = m + 1$
Set $\Omega_m = \Omega_{m-1} \oplus \bigoplus_{i=1}^{m_u} \mathcal{L}_{f^i}\Omega_{m-1} \oplus \mathcal{L}_{\frac{g}{L_g^1}}\Omega_{m-1} \oplus \bigoplus_{i=1}^{m_u} \text{span}\{\mathcal{L}_{{}^i\phi_{m-1}}\nabla h\}$
Set ${}^i\phi_m = \frac{[{}^i\phi_{m-1},\, g]}{L_g^1}$ for $i = 1, \ldots, m_u$
while $\dim(\Omega_m) > \dim(\Omega_{m-1})$ OR $\nabla\tau \notin \Omega_m$ **do**
  Set $m = m + 1$
  Set $\Omega_m = \Omega_{m-1} \oplus \bigoplus_{i=1}^{m_u} \mathcal{L}_{f^i}\Omega_{m-1} \oplus \mathcal{L}_{\frac{g}{L_g^1}}\Omega_{m-1} \oplus \bigoplus_{i=1}^{m_u} \text{span}\{\mathcal{L}_{{}^i\phi_{m-1}}\nabla h\}$
  Set ${}^i\phi_m = \frac{[{}^i\phi_{m-1},\, g]}{L_g^1}$ for $i = 1, \ldots, m_u$
end while
Set $\Omega = \Omega_m$ and $s = \dim(\Omega)$

Note that, when at the fifth line we set $\Omega_m = \Omega_{m-1} \oplus \bigoplus_{i=1}^{m_u} \mathcal{L}_{f^i}\Omega_{m-1} \oplus \mathcal{L}_{\frac{g}{L_g^1}}\Omega_{m-1}$, we are automatically also including the last term of the recursive step ($\oplus \bigoplus_{i=1}^{m_u} \text{span}\{\mathcal{L}_{{}^i\phi_{m-1}}\nabla h\}$) because ${}^i\phi_0 = f^i$ and $\nabla h \in \Omega$ from the initialization. In addition, in the case of a single output, and always at the fifth line, the term $\mathcal{L}_{\frac{g}{L_g^1}}\Omega$ vanishes because $\mathcal{L}_{\frac{g}{L_g^1}}\nabla h = \nabla \mathcal{L}_{\frac{g}{L_g^1}} h = \nabla \frac{\mathcal{L}_g h}{L_g^1} = \nabla \frac{L_g^1}{L_g^1} = \nabla 1$, which is the zero covector.

Algorithm 7.3 differs from the algorithm that returns the observable codistribution in the absence of unknown inputs (i.e., Algorithm 4.2) for the following reasons:

- In the recursive step, the vector field that corresponds to the unknown input (i.e., g) must be rescaled by dividing by L_g^1.

- The recursive step also contains the sum of the contributions $\bigoplus_{i=1}^{m_u} \mathcal{L}_{{}^i\phi}\nabla h$. In other words, we need to compute the Lie derivatives of the gradient of the chosen output along the vector fields obtained through the recursive Algorithm 7.1.

- The convergence of Algorithm 7.3 is achieved in at most $n+2$ steps, instead of $n-1$ steps (in the special case contemplated by Lemma 7.8, this upper bound is $n-1$ for both cases).

- When $\Omega_m = \Omega_{m-1}$ Algorithm 4.2 has converged. For Algorithm 7.2, we also need to check that $\nabla\tau \in \Omega_m$ and $m \geq 2$ (with the exception of the special case contemplated by Lemma 7.8).

7.1.2 ▪ The analytic procedure

In this section, we outline all the steps to investigate the weak observability at a given point x_0 of a nonlinear system characterized by (7.1). Basically, these steps are the steps necessary to compute the observable codistribution (i.e., the steps of Algorithm 7.3) and to prove that the gradient of a given state component belongs to this codistribution.

Note that, in the trivial case analyzed by Lemma 7.8, the method provided below simplifies, since we do not need to compute the quantity $\tau \left(= \frac{\mathcal{L}_g^2 h}{(\mathcal{L}_g^1 h)^2} \right)$, and we do not need to check that its gradient belongs to the codistribution computed at every step of the algorithm. In practice, we skip steps 4 and 5 in the procedure below.

1. For the chosen x_0, compute $L_g^1 (= \mathcal{L}_g^1 h)$. In the case when $L_g^1 = 0$, choose another function in the space of functions $\mathcal{F}$ (defined as the space that contains h and its Lie derivatives up to any order along the vector fields $f^1, \ldots, f^{m_u}$) such that its Lie derivative along g does not vanish.[28]

2. Compute the codistribution Ω_0 and Ω_1 (at x_0) by using Algorithm 7.3.

3. Compute the vector fields ${}^i\phi_m$ ($i = 1, \ldots, m_u$) by using Algorithm 7.1, starting from $m = 0$, to check whether the considered system is in the special case contemplated by Lemma 7.8. In this trivial case, set $m' = 0$, use the recursive step of Algorithm 7.3 to build the codistribution Ω_m for $m \geq 2$, and skip to step 6.

4. Compute $\tau \left(= \frac{\mathcal{L}_g^2 h}{(L_g^1)^2} \right)$ and $\nabla \tau$.

5. Use the recursive step of Algorithm 7.3 to build the codistribution Ω_m for $m \geq 2$, and, for each m, check whether $\nabla \tau \in \Omega_m$. Denote by m' the smallest m such that $\nabla \tau \in \Omega_m$.

6. For each $m \geq m'$, check whether $\Omega_{m+1} = \Omega_m$ and denote by $\Omega = \Omega_m$, where m is the smallest integer such that $m \geq m'$ and $\Omega_{m+1} = \Omega_m$ (note that $m \leq n + 2$).

7. If the gradient of a given state component (x_j, $j = 1, \ldots, n$) belongs to Ω (namely if $\nabla x_j \in \Omega$) on a given neighborhood of x_0, then x_j is weakly observable at x_0. If this holds for all the state components, the state x is weakly observable at x_0. Finally, if the dimension of Ω is smaller than n on a given neighborhood of x_0, then the state is not weakly observable at x_0.

In the presence of multiple outputs, the procedure remains the same, with the exception of the first step and the initialization of Ω. Ω_0 includes the gradients of all the outputs (second line of Algorithm 7.3). In addition, at the first step of the procedure, one of the outputs is selected and denoted by h, and it is used to define L_g^1 in (7.2) and τ in (7.3). The result is independent of the chosen output, provided that its Lie derivative along g does not vanish at x_0. If no output satisfies this condition, the function h must be selected from the space of functions $\mathcal{F}$ (which now includes all the outputs and their Lie derivatives up to any order along the vector fields $f^1, \ldots, f^{m_u}$) and its Lie derivative along g does not vanish. If the Lie derivative of any function in $\mathcal{F}$ vanishes, it means that the unknown input can be ignored to obtain the observability properties (the system is not canonic with respect to the unknown input, as shown in Appendix C).

---

[28] If the Lie derivative of any function in $\mathcal{F}$ vanishes, it means that the unknown input can be ignored to obtain the observability properties (the system is not canonic with respect to the unknown input, as shown in Appendix C).

7.2 ▪ Applications

We apply the analytic method described in section 7.1.2 in order to investigate the observability properties of several nonlinear systems characterized by the equations given in (7.1).

For educational purposes, the examples studied in this section are deliberately very trivial in order to allow us to compare the analytic results provided by the analytic method with what we can expect by following intuitive reasoning. In addition, this allows easily following the steps of the automatic procedure in section 7.1.2.

Note that the method is very powerful and can be used to automatically obtain the observability properties of any system that satisfies (7.1), i.e., independently of the state dimension (intuitive reasoning often becomes prohibitive for systems characterized by high-dimensional states), independently of the type of nonlinearity, and in general independently of the system complexity. In this regard, note that the method can be implemented automatically by using a simple symbolic computation tool (in our examples, we executed the computation manually).

7.2.1 ▪ Unicycle with one input unknown

The system

As in section 4.3.3, in this section[29] we work in polar coordinates. The state is

$$x = [\rho, \ \phi, \ \theta_v]^T.$$

Its dynamics are given by (4.19), i.e.,

$$\begin{bmatrix} \dot{\rho} &= v\cos(\theta_v - \phi), \\ \dot{\phi} &= \frac{v}{\rho}\sin(\theta_v - \phi), \\ \dot{\theta}_v &= \omega. \end{bmatrix} \quad (7.4)$$

We consider the same three outputs (separately) considered in section 4.3.3 (see also Figure 1.1 for an illustration), i.e.,

$$y = \rho, \qquad y = \theta_v - \phi, \qquad y = \phi.$$

Note that the second output differs from the one considered in sections 4.3.3 and 6.3.1, $y = \beta = \pi - (\theta_v - \phi)$. This does not change the observability properties.

For each of these three outputs, we consider the following two cases:

- v is known, ω is unknown.

- v is unknown, ω is known.

Hence, we have six cases. Note that the last two outputs when v is unknown were already investigated by applying the procedure introduced in section 6.3 (see section 6.3.1).

Intuitive procedure to obtain the observability properties

From the analysis carried out in section 4.3.3, we know that, when both the inputs are known, the dimension of the observable codistribution is 2 for the first two outputs ($y = \rho$ and $y = \theta_v - \phi$) and 3 for the last one ($y = \phi$). In particular, for the first two outputs all the initial states rotated

---
[29] © 2019 IEEE. Reprinted, with permission, from [38].

7.2. Applications

Table 7.1. *Dimension of the observable codistribution (Ω) obtained by following intuitive reasoning*

Case	Dimension of Ω
1st: $y = \rho$, $u = \omega$, $w = v$	2
2nd: $y = \rho$, $w = \omega$, $u = v$	2
3rd: $y = \theta_v - \phi$, $u = \omega$, $w = v$	1
4th: $y = \theta_v - \phi$, $w = \omega$, $u = v$	2
5th: $y = \phi$, $u = \omega$, $w = v$	2
6th: $y = \phi$, $w = \omega$, $u = v$	3

around the vertical axis are indistinguishable. When one of the inputs misses, this unobservable degree of freedom obviously remains. On the other hand, when the linear speed is unknown (i.e., it acts as an unknown input ($w = v$)) and the output is an angle (second and third outputs, i.e., $y = \theta_v - \phi$ and $y = \phi$, respectively), we lose a further degree of freedom, which corresponds to the absolute scale. This result was already confirmed by applying the procedure introduced in section 6.3 (see section 6.3.1). In Table 7.1 we provide the dimension of the observable codistribution obtained by following this intuitive reasoning for the six considered cases.

Analytic results

We now derive the observability properties by applying the analytic criterion described in section 7.1.2. For all the cases we have $m_u = 1$. Hence, we adopt the following notation: $f \triangleq f^1$ (for the vector field in (7.1)) and $\phi_m \triangleq {}^1\phi_m$ (for the vectors defined by Algorithm 7.1). We consider the six cases defined in section 7.2.1 separately.

First Case: $y = \rho$, $u = \omega$, $w = v$

We have

$$f = \begin{bmatrix} 0 \\ 0 \\ 1 \end{bmatrix}, \quad g = \begin{bmatrix} \cos(\theta_v - \phi) \\ \frac{\sin(\theta_v - \phi)}{\rho} \\ 0 \end{bmatrix}.$$

We apply the analytic criterion in section 7.1.2. We obtain the following:

Step 1

We have $L_g^1 = \cos(\theta_v - \phi)$, which does not vanish, in general.

Step 2

We have $\Omega_0 = \text{span}\{[1, 0, 0]\}$. Additionally, $\Omega_1 = \Omega_0$.

Step 3

We have $\mathcal{L}_{\phi_0} L_g^1 = \mathcal{L}_f L_g^1 = -\sin(\theta_v - \phi)$, which does not vanish, in general. This suffices to conclude that the considered system is not in the special case contemplated by Lemma 7.8, and we need to continue with step 4.

Step 4

We have $\tau \triangleq \frac{L_g^2}{(L_g^1)^2} = \frac{\tan^2(\theta_v - \phi)}{\rho}$ and $\nabla \tau = \frac{\tan(\theta_v - \phi)}{\rho} \left[-\frac{\tan(\theta_v - \phi)}{\rho}, -\frac{2}{\cos^2(\theta_v - \phi)}, \frac{2}{\cos^2(\theta_v - \phi)} \right]$.

Step 5

We need to compute Ω_2, and, in order to do this, we need to compute ϕ_1. We obtain

$$\phi_1 = \begin{bmatrix} -\tan(\theta_v - \phi) \\ \frac{1}{\rho} \\ 0 \end{bmatrix}$$

and

$$\Omega_2 = \text{span}\left\{ [1, 0, 0], \left[0, \frac{1}{\cos^2(\theta_v - \phi)}, -\frac{1}{\cos^2(\theta_v - \phi)} \right] \right\}.$$

It is immediate to check that $\nabla \tau \in \Omega_2$, meaning that $m' = 2$.

Step 6

By a direct computation, it is possible to check that $\Omega_3 = \Omega_2$, meaning that $\Omega = \Omega_2$

Step 7

The dimension of the observable codistribution is 2. We conclude that the state is not weakly observable. This result agrees with the one in Table 7.1 (second line).

Second Case: $y = \rho$, $u = v$, $w = \omega$

We have

$$f = \begin{bmatrix} \cos(\theta_v - \phi) \\ \frac{\sin(\theta_v - \phi)}{\rho} \\ 0 \end{bmatrix}, \quad g = \begin{bmatrix} 0 \\ 0 \\ 1 \end{bmatrix}.$$

We apply the analytic criterion in section 7.1.2. We obtain the following:

Step 1

We easily obtain $\mathcal{L}_g h = 0$. We consider the function $\mathcal{L}_f h \in \mathcal{F}$. We have $\mathcal{L}_f h = \cos(\theta_v - \phi)$ and $\mathcal{L}_g \mathcal{L}_f h = -\sin(\theta_v - \phi)$, which does not vanish, in general. Hence, we can proceed with the steps in section 7.1.2 by setting

$$h = \cos(\theta_v - \phi)$$

and

$$L_g^1 = -\sin(\theta_v - \phi).$$

Step 2

We have

$$\Omega_0 = \text{span}\{[1, 0, 0], [0, \sin(\theta_v - \phi), -\sin(\theta_v - \phi)]\},$$

as long as the function ρ is also a system output. Additionally, $\Omega_1 = \Omega_0$.

7.2. Applications

Step 3

We have $\mathcal{L}_{\phi_0} L_g^1 = \mathcal{L}_f L_g^1 = \frac{\sin(\theta_v - \phi)\cos(\theta_v - \phi)}{\rho}$, which does not vanish, in general. This suffices to conclude that the considered system is not in the special case contemplated by Lemma 7.8, and we need to continue with step 4.

Step 4

We have
$$\tau = -\frac{\cos(\theta_v - \phi)}{\sin^2(\theta_v - \phi)}.$$

Step 5

By a direct computation, we obtain $\Omega_2 = \Omega_1$. Additionally, it is immediate to check that $\nabla \tau \in \Omega_2$, meaning that $m' = 2$.

Step 6

By a direct computation, it is possible to check that $\Omega_3 = \Omega_2$, meaning that $\Omega = \Omega_2$.

Step 7

The dimension of the observable codistribution is 2. We conclude that the state is not weakly observable. This result agrees with the one in Table 7.1 (third line).

Third Case: $y = \theta_v - \phi,\ u = \omega,\ w = v$

We have
$$f = \begin{bmatrix} 0 \\ 0 \\ 1 \end{bmatrix}, \quad g = \begin{bmatrix} \cos(\theta_v - \phi) \\ \frac{\sin(\theta_v - \phi)}{\rho} \\ 0 \end{bmatrix}.$$

We apply the analytic criterion in section 7.1.2. We obtain the following:

Step 1

We have $L_g^1 = -\frac{\sin(\theta_v - \phi)}{\rho}$, which does not vanish, in general.

Step 2

We have $\Omega_0 = \text{span}\{[0, -1, 1]\}$ and $\Omega_1 = \Omega_0$.

Step 3

We have $\mathcal{L}_{\phi_0} L_g^1 = \mathcal{L}_f L_g^1 = -\frac{\cos(\theta_v - \phi)}{\rho}$, which does not vanish, in general. This suffices to conclude that the considered system is not in the special case contemplated by Lemma 7.8, and we need to continue with step 4.

Step 4

We have $\tau = 2\cot(\theta_v - \phi)$, and

$$\nabla\tau = \frac{2}{\sin^2(\theta_v - \phi)}[0, 1, -1].$$

Step 5

By a direct computation, we obtain $\Omega_2 = \Omega_1$. Additionally, it is immediate to check that $\nabla\tau \in \Omega_2$, meaning that $m' = 2$.

Step 6

By a direct computation, it is possible to check that $\Omega_3 = \Omega_2$, meaning that $\Omega = \Omega_2$.

Step 7

The dimension of the observable codistribution is 1. We conclude that the state is not weakly observable. This result agrees with the one in Table 7.1 (fourth line). Note that the new unobservable direction with respect to the case when both inputs are known is precisely the absolute scale. Indeed, the new symmetry (i.e., the new vector orthogonal to the observable codistribution) is the vector

$$\begin{bmatrix} 1 \\ 0 \\ 0 \end{bmatrix}.$$

The infinitesimal transformation given in (4.24) becomes

$$\begin{bmatrix} \rho \\ \phi \\ \theta_v \end{bmatrix} \rightarrow \begin{bmatrix} \rho' \\ \phi' \\ \theta_v' \end{bmatrix} = \begin{bmatrix} \rho \\ \phi \\ \theta_v \end{bmatrix} + \epsilon \begin{bmatrix} 1 \\ 0 \\ 0 \end{bmatrix}.$$

This means that the system is invariant with respect to the scale ($\rho \rightarrow (1+\epsilon)\rho$).

Fourth Case: $y = \theta_v - \phi$, $u = v$, $w = \omega$

We have

$$f = \begin{bmatrix} \cos(\theta_v - \phi) \\ \frac{\sin(\theta_v - \phi)}{\rho} \\ 0 \end{bmatrix}, \quad g = \begin{bmatrix} 0 \\ 0 \\ 1 \end{bmatrix}.$$

We apply the analytic criterion in section 7.1.2. We obtain the following:

Step 1

We have $L_g^1 = 1 \neq 0$.

Step 2

By a direct computation, we obtain $\Omega_0 = \text{span}\{[0, -1, 1]\}$ and

$$\Omega_1 = \text{span}\left\{ [0, -1, 1], \left[-\frac{\sin(\theta_v - \phi)}{\rho^2}, -\frac{\cos(\theta_v - \phi)}{\rho}, \frac{\cos(\theta_v - \phi)}{\rho} \right] \right\}.$$

7.2. Applications

Step 3

Since $L_g^1 = 1$, it is immediate to realize that $\mathcal{L}_{\phi_j} L_g^1 = 0$ for any integer $j \geq 0$. Hence, the considered system is in the special case contemplated by Lemma 7.8 and we skip to step 6 by setting $m' = 0$.

Step 6

By a direct computation, it is possible to check that $\Omega_2 = \Omega_1$, meaning that $\Omega = \Omega_1$.

Step 7

The dimension of the observable codistribution is 2. We conclude that the state is not weakly observable. This result agrees with the one in Table 7.1 (fifth line).

Fifth Case: $y = \phi$, $u = \omega$, $w = v$

We have

$$f = \begin{bmatrix} 0 \\ 0 \\ 1 \end{bmatrix}, \quad g = \begin{bmatrix} \cos(\theta_v - \phi) \\ \frac{\sin(\theta_v - \phi)}{\rho} \\ 0 \end{bmatrix}.$$

We apply the analytic criterion in section 7.1.2. We obtain the following:

Step 1

We have $L_g^1 = \frac{\sin(\theta_v - \phi)}{\rho}$, which does not vanish, in general.

Step 2

We easily obtain $\Omega_0 = \text{span}\{[0, 1, 0]\}$ and $\Omega_1 = \Omega_0$.

Step 3

We have $\mathcal{L}_{\phi_0} L_g^1 = \mathcal{L}_f L_g^1 = \frac{\cos(\theta_v - \phi)}{\rho}$, which does not vanish, in general. This suffices to conclude that the considered system is not in the special case contemplated by Lemma 7.8, and we need to continue with step 4.

Step 4

We have $\tau = -2 \cot(\theta_v - \phi)$ and

$$\nabla \tau = \frac{2}{\sin^2(\theta_v - \phi)} [0, -1, 1].$$

Step 5

To compute Ω_2 we need to compute ϕ_1. We obtain

$$\phi_1 = \begin{bmatrix} -\rho \\ \cot(\theta_v - \phi) \\ 0 \end{bmatrix}$$

and

$$\Omega_2 = \text{span}\left\{[0,1,0], \frac{1}{\sin^2(\theta_v - \phi)}[0,1,-1]\right\}.$$

It is immediate to check that $\nabla \tau \in \Omega_2$, meaning that $m' = 2$.

Step 6

By a direct computation, we obtain $\Omega_3 = \Omega_2$, meaning that $\Omega = \Omega_2$, whose dimension is 2.

Step 7

The dimension of the observable codistribution is 2. We conclude that the state is not weakly observable. This result agrees with the one in Table 7.1 (sixth line). Note that the new unobservable direction with respect to the case when both inputs are known is precisely the absolute scale (as in the third case studied above).

Sixth Case: $y = \phi$, $u = v$, $w = \omega$

We have

$$f = \begin{bmatrix} \cos(\theta_v - \phi) \\ \frac{\sin(\theta_v - \phi)}{\rho} \\ 0 \end{bmatrix}, \quad g = \begin{bmatrix} 0 \\ 0 \\ 1 \end{bmatrix}.$$

We apply the analytic criterion in section 7.1.2. We obtain the following:

Step 1

We easily obtain $\mathcal{L}_g h = 0$. We consider the function $\mathcal{L}_f h \in \mathcal{F}$. We have $\mathcal{L}_f h = \frac{\sin(\theta_v - \phi)}{\rho}$ and $\mathcal{L}_g \mathcal{L}_f h = \frac{\cos(\theta_v - \phi)}{\rho}$, which does not vanish, in general. Hence, we can proceed with the steps in section 7.1.2 by setting

$$h = \frac{\sin(\theta_v - \phi)}{\rho}, \quad L_g^1 = \frac{\cos(\theta_v - \phi)}{\rho}.$$

Step 2

We have

$$\Omega_0 = \text{span}\left\{[0,1,0], \left[-\frac{\sin(\theta_v - \phi)}{\rho^2}, -\frac{\cos(\theta_v - \phi)}{\rho}, \frac{\cos(\theta_v - \phi)}{\rho}\right]\right\},$$

as long as the function ϕ is also a system output. We compute Ω_1. By a direct computation, we obtain that its dimension is 3. Hence, we do not need to proceed with the remaining steps since we can directly conclude that the entire state is weakly observable. This result agrees with the one in Table 7.1 (seventh line).

7.2. Applications

7.2.2 • Unicycle in the presence of an external disturbance

The system

We consider the same vehicle considered in section 7.2.1, and we adopt the Cartesian coordinates to characterize the state, i.e.,

$$x = [x_v, y_v, \theta_v]^T.$$

We assume that the vehicle motion is also affected by an unknown input that produces an additional (and unknown) robot speed (denoted by w) along a fixed direction (denoted by γ). Hence, the dynamics are characterized by the following differential equations:

$$\begin{bmatrix} \dot{x}_v &= v\cos\theta_v + w\cos\gamma, \\ \dot{y}_v &= v\sin\theta_v + w\sin\gamma, \\ \dot{\theta}_v &= \omega, \end{bmatrix} \quad (7.5)$$

where v and ω are the linear and the rotational speeds, respectively, in the absence of the unknown input. We assume that these two speeds are known (we refer to them as the known inputs), w is unknown (we refer to it as the unknown input or disturbance), and γ is constant in time. See also Figure 7.1 for an illustration.

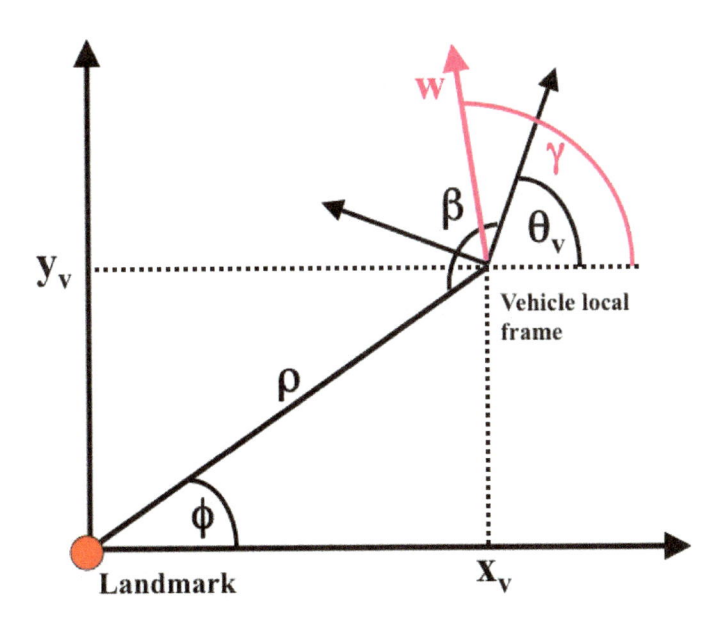

Figure 7.1. *The vehicle state together with the three considered outputs. Reprinted from* [34].

We consider the same three cases of output considered in the previous section. Additionally, we deal with both the cases when γ is known and unknown.

Observability properties when the disturbance direction is known

The state is $[x_v, y_v, \theta_v]^T$, and its dynamics are provided by the three equations in (7.5), where γ is a known parameter. These equations are a special case of (7.1). From (7.5) and (7.1) we easily obtain $m_u = 2$, $u_1 = v$, $u_2 = \omega$,

$$f^1 = \begin{bmatrix} \cos\theta_v \\ \sin\theta_v \\ 0 \end{bmatrix}, \quad f^2 = \begin{bmatrix} 0 \\ 0 \\ 1 \end{bmatrix}, \quad g = \begin{bmatrix} \cos\gamma \\ \sin\gamma \\ 0 \end{bmatrix}.$$

We consider the three outputs separately. For simplicity's sake, we actually consider the following three outputs: $y = \rho^2 = x_v^2 + y_v^2$ instead of $y = \rho$, $y = \tan\beta = \frac{y_v - x_v \tan\theta_v}{x_v + y_v \tan\theta_v}$ instead of $y = \beta$, and $y = \tan\phi = \frac{y_v}{x_v}$ instead of $y = \phi$. Obviously, the result of the observability analysis does not change.

First Case: $y = \rho^2$

We apply the analytic criterion in section 7.1.2. We obtain the following:

Step 1

We have $L_g^1 = 2(x_v \cos\gamma + y_v \sin\gamma)$, which does not vanish, in general.

Step 2

We have $\Omega_0 = \mathrm{span}\{[x_v, y_v, 0]\}$ and $\Omega_1 = \mathrm{span}\{[x_v, y_v, 0], [\cos\theta_v, \sin\theta_v, y_v\cos\theta_v - x_v\sin\theta_v]\}$.

Step 3

We have $\mathcal{L}_{1\phi_0} L_g^1 = \mathcal{L}_{f^1} L_g^1 = 2\cos(\gamma - \theta_v)$, which does not vanish, in general. This suffices to conclude that the considered system is not in the special case contemplated by Lemma 7.8, and we need to continue with step 4.

Step 4

We have $\tau = \frac{1}{2(x_v \cos\gamma + y_v \sin\gamma)^2}$ and $\nabla\tau = -\frac{1}{(x_v\cos\gamma + y_v\sin\gamma)^3}[\cos\gamma, \sin\gamma, 0]$.

Step 5

We need to compute Ω_2, and, in order to do this, we need to compute ${}^1\phi_1$ and ${}^2\phi_1$ through Algorithm 7.1. We obtain

$${}^1\phi_1 = {}^2\phi_1 = \begin{bmatrix} 0 \\ 0 \\ 0 \end{bmatrix}.$$

On the other hand, we obtain that $\mathcal{L}_{\frac{g}{L_g^1}} \nabla \mathcal{L}_{f^1} h \notin \Omega_1$. Hence, by using Algorithm 7.3 we obtain that Ω_2 has dimension equal to 3. As a result, we do not need to proceed with the remaining steps, since we can directly conclude that the entire state is weakly observable.

Second Case: $y = \tan\beta$

We apply the analytic criterion in section 7.1.2. We obtain the following:

7.2. Applications

Step 1

We have $L_g^1 = -\frac{y_v \cos\gamma - x_v \sin\gamma}{x_v^2 \cos^2\theta_v + 2\sin\theta_v \cos\theta_v x_v y_v - y_v^2 \cos^2\theta_v + y_v^2}$, which does not vanish, in general.

Step 2

By an explicit computation (by using Algorithm 7.3), we obtain that the dimension of Ω_0 is 1 and the dimension of Ω_1 is 2.

Step 3

We have $\mathcal{L}_{^2\phi_0} L_g^1 = \mathcal{L}_{f^2} L_g^1 \neq 0$, in general. This suffices to conclude that the considered system is not in the special case contemplated by Lemma 7.8, and we need to continue with step 4.

Step 4

We have $\tau = \frac{x_v \cos\gamma + y_v \sin\gamma + x_v \cos(\gamma - 2\theta_v) - y_v \sin(\gamma - 2\theta_v)}{y_v \cos\gamma - x_v \sin\gamma}$.

Step 5

We need to compute Ω_2. Also in this case, we obtain that $\mathcal{L}_{\frac{g}{L_g^1}} \nabla \mathcal{L}_{f^1} h \notin \Omega_1$. Hence, by using Algorithm 7.3 we obtain that Ω_2 has dimension equal to 3. As a result, we do not need to proceed with the remaining steps, since we can directly conclude that the entire state is weakly observable.

Third Case: $y = \tan\phi$

We apply the analytic criterion in section 7.1.2. We obtain the following:

Step 1

We have $L_g^1 = -\frac{y_v \cos\gamma - x_v \sin\gamma}{x_v^2}$, which does not vanish, in general.

Step 2

We have $\Omega_0 = \text{span}\{[-y_v, x_v, 0]\}$. In addition, by an explicit computation (by using Algorithm 7.3), we obtain that the dimension of Ω_1 is 2.

Step 3

We have $\mathcal{L}_{^2\phi_0} L_g^1 = \mathcal{L}_{f^2} L_g^1 \neq 0$, in general. This suffices to conclude that the considered system is not in the special case contemplated by Lemma 7.8, and we need to continue with step 4.

Step 4

We have $\tau = -\frac{2x_v^4 \cos\gamma}{x_v^4 \sin\gamma - x_v^3 y_v \cos\gamma}$.

… 168 Chapter 7. Unknown Input Observability for the Simplified Systems

Step 5

We need to compute Ω_2. Also in this case, we obtain that $\mathcal{L}_{\frac{g}{L_g^1}} \nabla \mathcal{L}_{f^1} h \notin \Omega_1$. Hence, by using Algorithm 7.3 we obtain that Ω_2 has dimension equal to 3. As a result, we do not need to proceed with the remaining steps, since we can directly conclude that the entire state is weakly observable.

Observability properties when the disturbance direction is unknown

The state is $[x_v, y_v, \theta_v, \gamma]^T$, and its dynamics are provided by the following four equations:

$$\begin{bmatrix} \dot{x}_v &= v \cos \theta_v + w \cos \gamma, \\ \dot{y}_v &= v \sin \theta_v + w \sin \gamma, \\ \dot{\theta}_v &= \omega, \\ \dot{\gamma} &= 0. \end{bmatrix} \quad (7.6)$$

From (7.6) and (7.1) we easily obtain $m_u = 2$, $u_1 = v$, $u_2 = \omega$,

$$f^1 = \begin{bmatrix} \cos \theta_v \\ \sin \theta_v \\ 0 \\ 0 \end{bmatrix}, \quad f^2 = \begin{bmatrix} 0 \\ 0 \\ 1 \\ 0 \end{bmatrix}, \quad g = \begin{bmatrix} \cos \gamma \\ \sin \gamma \\ 0 \\ 0 \end{bmatrix}.$$

We consider the three outputs separately.

First Case: $y = \rho^2$

We apply the analytic criterion in section 7.1.2. We obtain the following:

Step 1

We obviously obtain the same expression as in the case $y = \rho^2$ of the previous subsection, i.e., $L_g^1 = 2(x_v \cos \gamma + y_v \sin \gamma)$, which does not vanish, in general.

Step 2

We have $\Omega_0 = \text{span}\{[x_v, y_v, 0, 0]\}$ and $\Omega_1 = \text{span}\{[x_v, y_v, 0, 0], [\cos \theta_v, \sin \theta_v, y_v \cos \theta_v - x_v \sin \theta_v, 0]\}$.

Step 3

We have $\mathcal{L}_{1\phi_0} L_g^1 = \mathcal{L}_{f^1} L_g^1 = 2 \cos(\gamma - \theta_v)$, which does not vanish, in general. This suffices to conclude that the considered system is not in the special case contemplated by Lemma 7.8, and we need to continue with step 4.

Step 4

We have $\tau = \frac{1}{2(x_v \cos \gamma + y_v \sin \gamma)^2}$, as in the case $y = \rho^2$ of the previous subsection. On the other hand, the gradient of τ also includes the derivative with respect to γ, namely

$$\nabla \tau = \frac{-1}{(x_v \cos \gamma + y_v \sin \gamma)^3} [\cos \gamma, \ \sin \gamma, \ 0, \ y_v \cos \gamma - x_v \sin \gamma].$$

7.2. Applications

Step 5

We need to compute Ω_2, and, in order to do this, we need to compute $^1\phi_1$ and $^2\phi_1$ through Algorithm 7.1. We obtain

$$^1\phi_1 = {}^2\phi_1 = \begin{bmatrix} 0 \\ 0 \\ 0 \\ 0 \end{bmatrix}.$$

By using Algorithm 7.3 we compute Ω_2, and we obtain that its dimension is 3. Additionally, it is possible to verify that $\nabla\tau \in \Omega_2$, meaning that $m' = 2$.

Step 6

By a direct computation, it is possible to check that $\Omega_3 = \Omega_2$, meaning that $\Omega = \Omega_2$.

Step 7

We conclude that the dimension of the observable codistribution is equal to $3(< 4)$ and the state is not weakly observable.

We compute the system symmetries by computing the distribution orthogonal to the observable codistribution. This has dimension 1. Hence, we have one single symmetry, which is

$$\begin{bmatrix} -y_v \\ x_v \\ 1 \\ 1 \end{bmatrix}.$$

The infinitesimal transformation given in (4.24) becomes

$$\begin{bmatrix} x_v \\ y_v \\ \theta_v \\ \gamma \end{bmatrix} \rightarrow \begin{bmatrix} x'_v \\ y'_v \\ \theta'_v \\ \gamma' \end{bmatrix} = \begin{bmatrix} x_v \\ y_v \\ \theta_v \\ \gamma \end{bmatrix} + \epsilon \begin{bmatrix} -y_v \\ x_v \\ 1 \\ 1 \end{bmatrix} \simeq \begin{bmatrix} \begin{bmatrix} \cos\epsilon & -\sin\epsilon \\ \sin\epsilon & \cos\epsilon \end{bmatrix} \begin{bmatrix} x_v \\ y_v \end{bmatrix} \\ \theta_v + \epsilon \\ \gamma + \epsilon \end{bmatrix},$$

which is an infinitesimal rotation around the vertical axis.

Second Case: $y = \tan\beta$

We apply the analytic criterion in section 7.1.2. We obtain the following:

Step 1

We obviously obtain the same expression, as in the case $y = \tan\beta$ of the previous subsection, i.e., $L_g^1 = -\frac{y_v \cos\gamma - x_v \sin\gamma}{x_v^2 \cos^2\theta_v + 2\sin\theta_v \cos\theta_v x_v y_v - y_v^2 \cos^2\theta_v + y_v^2}$, which does not vanish, in general.

Step 2

We compute Ω_0 and Ω_1: their dimensions are 1 and 2, respectively.

Step 3

We have $\mathcal{L}^2_{\phi_0} L^1_g = \mathcal{L}_{f^2} L^1_g \neq 0$, in general. This suffices to conclude that the considered system is not in the special case contemplated by Lemma 7.8, and we need to continue with step 4.

Step 4

We have $\tau = \frac{x_v \cos\gamma + y_v \sin\gamma + x_v \cos(\gamma - 2\theta_v) - y_v \sin(\gamma - 2\theta_v)}{y_v \cos\gamma - x_v \sin\gamma}$, as in the case $y = \tan\beta$ of the previous subsection. On the other hand, the gradient of τ also includes the derivative with respect to γ.

Step 5

By using Algorithm 7.3 we compute Ω_2 and we obtain that its dimension is 3. Additionally, it is possible to verify that $\nabla \tau \in \Omega_2$, meaning that $m' = 2$.

Step 6

By a direct computation, it is possible to check that $\Omega_3 = \Omega_2$, meaning that $\Omega = \Omega_2$.

Step 7

We conclude that the dimension of the observable codistribution is equal to $3(< 4)$ and the state is not weakly observable. In particular, the system has the same symmetry as that of the previous case.

Third Case: $y = \tan\phi$

We apply the analytic criterion in section 7.1.2. We obtain the following:

Step 1

We obviously obtain the same expression, as in the case $y = \tan\phi$ of the previous subsection, i.e., $L^1_g = -\frac{y_v \cos\gamma - x_v \sin\gamma}{x_v^2}$, which does not vanish, in general.

Step 2

We compute Ω_0 and Ω_1: their dimensions are 1 and 2, respectively.

Step 3

We have $\mathcal{L}^2_{\phi_0} L^1_g = \mathcal{L}_{f^2} L^1_g \neq 0$, in general. This suffices to conclude that the considered system is not in the special case contemplated by Lemma 7.8, and we need to continue with step 4.

Step 4

We have $\tau = -\frac{2x_v^4 \cos\gamma}{x_v^4 \sin\gamma - x_v^3 y_v \cos\gamma}$, as in the case $y = \tan\phi$ of the previous subsection. On the other hand, the gradient of τ also includes the derivative with respect to γ.

7.2. Applications

Table 7.2. *Weak observability of the state in all the considered scenarios. Reprinted from* [34].

γ	Output	State observability
known	$y = \rho$	yes
known	$y = \beta$	yes
known	$y = \phi$	yes
unknown	$y = \rho$	no
unknown	$y = \beta$	no
unknown	$y = \phi$	yes

Step 5

By using Algorithm 7.3, we compute Ω_2 and we obtain that its dimension is 4. As a result, we do not need to proceed with the remaining steps, since we can directly conclude that the entire state is weakly observable.

Table 7.2 summarizes the results of the observability analysis carried out in this section. The reader can find in [34] the results of extensive simulations for the system studied in this section. These results clearly validate the analytic results of the previous observability analysis. In addition, we remark that the analytic results provided by our observability analysis are also understandable by following intuitive reasoning. From the analysis carried out in section 4.3.3, we know that, without the unknown input, the dimension of the observable codistribution is 2 for the first two outputs and 3 for the last one. In particular, for the first two outputs all the initial states rotated around the vertical axis are indistinguishable. In other words, in these two cases, the system exhibits a continuous symmetry. In the presence of the unknown input, when γ is known, the aforementioned system invariance is broken and the entire state becomes observable. When γ is unknown, the symmetry still remains (and obviously also concerns the new state component γ).

We conclude by remarking a very important aspect. The presence of an unknown input improves the observability properties of a system (this regards the case when γ is known). In particular, if $w = 0$ (absence of unknown input), the state becomes unobservable despite the knowledge of the unknown input (we know that it is zero), while, when $w \neq 0$, the state is observable even if w is unknown. Note that having an unknown input equal to zero is an event that occurs with zero probability and our theory accounts for this fact since it is based on Definition 6.3. In this regard, note also that the validity of Theorem 7.7, which allows us to introduce Algorithm 7.3, holds when the unknown input is different from 0.

7.2.3 ▪ Vehicle moving in 3D in the presence of a disturbance

The system

In this section,[30] we consider a vehicle that moves in a 3D environment. We assume that the dynamics of the vehicle are affected by the presence of a disturbance (e.g., this could be an aerial vehicle in the presence of wind). We assume that the direction of the disturbance is constant in time and a priori known. Conversely, the disturbance magnitude is unknown and time dependent. The vehicle is equipped with speed sensors (e.g., airspeed sensors in the case of an aerial vehicle), gyroscopes, and a bearing sensor (e.g., monocular camera). We assume that all the sensors share

---
[30] © 2017 IEEE. Reprinted, with permission, from [35].

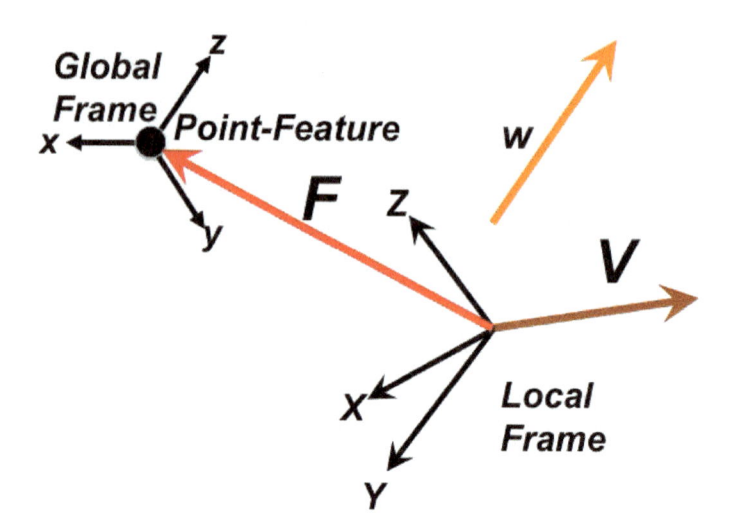

Figure 7.2. *Local and global frame for the considered problem. The z-axis of the latter is aligned with the direction of the disturbance (assumed to be known and constant in time). The speed V is the vehicle speed with respect to the air, which differs from the ground speed because of the disturbance (w).* © 2017 IEEE. Reprinted, with permission, from [35].

the same frame (in other words, they are extrinsically calibrated). Without loss of generality, we define the vehicle local frame as this common frame. The airspeed sensors measure the vehicle speed with respect to the air in the local frame. The gyroscopes provide the angular speed in the local frame. Finally, the bearing sensor provides the bearing angles of the features in the environment expressed in its own local frame. We consider the extreme case of a single point feature and, without loss of generality, we set the origin of the global frame at this point feature (see Figure 7.2 for an illustration). Additionally, we assume that the z-axis of the global frame is aligned with the direction of the disturbance.

Our system can be characterized by the following state:

$$X \triangleq [r_x,\ r_y,\ r_z,\ q_t,\ q_x,\ q_y,\ q_z]^T, \qquad (7.7)$$

where $r = [r_x,\ r_y,\ r_z]$ is the position of the vehicle in the global frame and $q = q_t + q_x i + q_y j + q_z k$ is the unit quaternion that describes the transformation change between the global and the local frames (see also Appendix B). The dynamics are affected by the presence of the disturbance. The disturbance is characterized by the following vector (in the global frame):

$$\bar{w} = w \begin{bmatrix} 0 \\ 0 \\ 1 \end{bmatrix}, \qquad (7.8)$$

where w is its unknown magnitude.

As in section 5.5, for each vector defined in the 3D space, the subscript q will be adopted to denote the corresponding imaginary quaternion. For instance, regarding the vehicle position, we have $r_q = 0 + r_x i + r_y j + r_z k$. Additionally, we denote by V and Ω the following physical quantities:

7.2. Applications

- $V = [V_x, V_y, V_z]$ is the vehicle speed with respect to the air expressed in the local frame (hence, $w\,k + qV_q q^*$ is the vehicle speed with respect to the ground expressed in the global frame).

- $\Omega \triangleq [\Omega_x\,\Omega_y\,\Omega_z]$ is the angular speed (and $\Omega_q = 0 + \Omega_x\,i + \Omega_y\,j + \Omega_z\,k$).

The dynamics of the state are

$$\begin{bmatrix} \dot{r}_q & = w\,k + qV_q q^*, \\ \dot{q} & = \tfrac{1}{2} q\Omega_q. \end{bmatrix} \tag{7.9}$$

The monocular camera provides the position of the feature in the local frame ($F_q = -q^* r_q q$) up to a scale. Hence, it provides the ratios of the components of F. We have the following output functions:

$$h_1(X) = \frac{(q^* r_q q)_x}{(q^* r_q q)_z}, \quad h_2(X) = \frac{(q^* r_q q)_y}{(q^* r_q q)_z}, \tag{7.10}$$

where the subscripts x, y, and z indicate, respectively, the i, j, and k components of the corresponding quaternion. We have also to consider the constraint $q^* q = 1$. This provides the further output

$$h_3(X) \triangleq q^* q. \tag{7.11}$$

Our system is characterized by the state in (7.7), the dynamics in (7.9), and the three outputs h_1, h_2, and h_3 in (7.10) and (7.11).

Observability in the absence of disturbance

Our system is characterized by the state in (7.7), the dynamics in (7.9) with $w = 0$, and the three outputs h_1, h_2, and h_3 in (7.10) and (7.11).

By comparing (7.9) with (7.1) we obtain that our system is characterized by six known inputs ($m_u = 6$) that are $u_1 = \Omega_x$, $u_2 = \Omega_y$, $u_3 = \Omega_z$, $u_4 = V_x$, $u_5 = V_y$, and $u_6 = V_z$. Additionally, we obtain

$$f^1 = \frac{1}{2}\begin{bmatrix} 0 \\ 0 \\ 0 \\ -q_x \\ q_t \\ q_z \\ -q_y \end{bmatrix},\ f^2 = \frac{1}{2}\begin{bmatrix} 0 \\ 0 \\ 0 \\ -q_y \\ -q_z \\ q_t \\ q_x \end{bmatrix},\ f^3 = \frac{1}{2}\begin{bmatrix} 0 \\ 0 \\ 0 \\ -q_z \\ q_y \\ -q_x \\ q_t \end{bmatrix},\ f^4 = \begin{bmatrix} q_t^2 + q_x^2 - q_y^2 - q_z^2 \\ 2q_t q_z + 2q_x q_y \\ 2q_x q_z - 2q_t q_y \\ 0 \\ 0 \\ 0 \\ 0 \end{bmatrix},$$

$$f^5 = \begin{bmatrix} 2q_x q_y - 2q_t q_z \\ q_t^2 - q_x^2 + q_y^2 - q_z^2 \\ 2q_t q_x + 2q_y q_z \\ 0 \\ 0 \\ 0 \\ 0 \end{bmatrix},\ f^6 = \begin{bmatrix} 2q_t q_y + 2q_x q_z \\ 2q_y q_z - 2q_t q_x \\ q_t^2 - q_x^2 - q_y^2 + q_z^2 \\ 0 \\ 0 \\ 0 \\ 0 \end{bmatrix}.$$

Finally, in the absence of disturbance we have

$$g = [0,\,0,\,0,\,0,\,0,\,0,\,0]^T.$$

In this case we can apply the observability rank condition, i.e., Algorithm 4.2, to obtain the observable codistribution. We compute the codistribution Ω_0 by computing the gradients of the three functions h_1, h_2, and h_3. We obtain that this codistribution has dimension equal to 3. We use Algorithm 4.2 to compute Ω_1. We obtain that its dimension is 4. In particular, the additional covector is obtained by the gradient of the following Lie derivative:

$$\mathcal{L}_{f^4} h_1.$$

In other words,

$$\Omega_1 = \text{span}\left\{\nabla h_1,\ \nabla h_2,\ \nabla h_3,\ \nabla \mathcal{L}_{f^4} h_1\right\}.$$

All the remaining first-order Lie derivatives have a gradient that is in the above codistribution. Additionally, by an explicit computation, it is easy to realize that $\Omega_2 = \Omega_1$. This means that Algorithm 4.2 has converged and the observable codistribution is Ω_1.

By an explicit computation, it is possible to check that the gradients of the components of the vector F belong to Ω_1. This means that all the independent observable functions are the components of F, i.e., the position of the feature in the local frame (obviously, the fourth observable function is the norm of the quaternion). In particular, no component of the vehicle orientation is observable.

Observability in the presence of the disturbance

We now consider the case when the dynamics are affected by the presence of the disturbance. By comparing (7.9) with (7.1) we obtain that the vector fields that characterize the dynamics are the same that characterize the dynamics in the absence of disturbance with the exception of the last one, which becomes

$$g = [0,\ 0,\ 1,\ 0,\ 0,\ 0,\ 0]^T.$$

To obtain the observability properties we apply the analytic method described in section 7.1.2 by following its seven steps.

First step

We start by computing the Lie derivatives of the outputs h_1, h_2, and h_3 along the vector field g. We find that the result differs from zero for the first two outputs. Hence, we use the first output (h_1) to define L_g^1 (we could choose also the second output h_2). In particular, we obtain

$$L_g^1 \triangleq \mathcal{L}_g h_1 = \frac{-r_y(2q_t q_z - 2q_x q_y) - r_x(q_t^2 - q_x^2 + q_y^2 - q_z^2)}{[r_z(q_t^2 - q_x^2 - q_y^2 + q_z^2) + 2r_x(q_t q_y + q_x q_z) + 2r_y(q_y q_z - q_t q_x)]^2}.$$

Second step

We compute the codistribution Ω_0 by computing the gradients of the three functions h_1, h_2, and h_3. This coincides with the case without disturbance, and we obtain that this codistribution has dimension equal to 3.

We use Algorithm 7.3 to compute Ω_1. We obtain that its dimension is 5. In particular, the additional two independent covectors are obtained by the gradients of the following two Lie derivatives:

$$\mathcal{L}_{f^4} h_1,\quad \mathcal{L}_{\frac{g}{L_g^1}} h_2.$$

In other words,

$$\Omega_1 = \text{span}\left\{\nabla h_1,\ \nabla h_2,\ \nabla h_3,\ \nabla \mathcal{L}_{f^4} h_1,\ \nabla \mathcal{L}_{\frac{g}{L_g^1}} h_2\right\}.$$

7.3. Analytic derivations

All the remaining first-order Lie derivatives have a gradient that is in the above codistribution.

Third step
We compute $^1\phi_1$, $^2\phi_1$, $^3\phi_1$, $^4\phi_1$, $^5\phi_1$, and $^6\phi_1$ by using Algorithm 7.1. We obtain that all these vectors vanish. As a result, all the subsequent steps of Algorithm 7.1 provide null vectors. Therefore, the assumptions of Lemma 7.8 are trivially met. We set $m' = 0$, and we skip to the sixth step.

Sixth step
We use Algorithm 7.3 to compute Ω_2, and we obtain

$$\Omega_2 = \Omega_1 \oplus \text{span}\left\{ \nabla \mathcal{L}_{f^4} \mathcal{L}_{\frac{g}{L_g^1}} h_2 \right\}.$$

Hence, its dimension is 6. Finally, by using again Algorithm 7.3 it is possible to compute Ω_3 and to check that $\Omega_3 = \Omega_2$. This means that the algorithm has converged and the observable codistribution is $\Omega = \Omega_2$.

Seventh step
By computing the distribution orthogonal to the codistribution Ω we can find the continuous symmetry that characterizes the unobservable space. By an explicit computation we obtain the following vector:

$$\left[-r_y,\ r_x,\ 0,\ -\frac{q_z}{2},\ -\frac{q_y}{2},\ \frac{q_x}{2},\ \frac{q_t}{2} \right]^T.$$

This symmetry corresponds to an invariance with respect to a rotation around the z-axis of the global frame. This means that we have a single unobservable mode that is the yaw in the global frame.[31] We conclude by remarking that the presence of the disturbance, even if its magnitude is unknown and is not constant, makes observable the roll and the pitch angles. This result is similar to the result that we obtained in the case of visual and inertial sensor fusion in the presence of gravity (section 5.6). The presence of gravity makes observable the roll and the pitch angles, even if its magnitude is unknown. What it is nonintuitive in the case now investigated is that not only is the magnitude of the disturbance unknown, but it is also time dependent.

The reader can find in [35] the results of simulations for the system studied in this section. These results clearly validate the analytic results of the previous observability analysis.

7.3 ▪ Analytic derivations

In this section,[32] we provide all the analytic derivations to obtain the analytic solution presented in section 7.1. In other words, we prove that the entire observable codistribution for the simplified systems introduced in section 3.3 (which are the systems characterized by (7.1)) is generated by Algorithm 7.3. This analytic derivation starts from the results obtained in Chapter 6, in particular from Algorithm 6.1 and Theorem 6.12 introduced in section 6.3, which together provide a partial tool to determine the observability properties of systems characterized by (6.1). Hence, the first step of this derivation consists in specifying the results obtained in Chapter 6 for the simplified systems.

---
[31] Note that the chosen global frame is aligned with the direction of the disturbance (Figure 7.2). Hence, what is unobservable is a rotation around the direction of the disturbance.
[32] © 2019 IEEE. Reprinted, with permission, from [38].

The simplified systems (characterized by (7.1)) can be obtained from the general systems (characterized by (6.1)) by setting g^0 to the null vector (driftless systems) and $m_w = 1$ (i.e., single unknown input).

When $m_w = 1$, the augmented state introduced in section 6.2.1 becomes

$$^k x \triangleq [x_1, \ldots, x_n, w, w^{(1)}, \ldots, w^{(k-1)}]^T. \tag{7.12}$$

Its dimension is $n + k$.

We introduce a further simplification. All the results of this section will be obtained for systems also characterized by a single known input (i.e., $m_u = 1$). Then, in section 7.3.3, all these results will be extended to the case $m_u > 1$.

Under these settings, $\Sigma^{(k)}$ is characterized by the following equations (see (6.5) and (6.6)):

$$^k\dot{x} = G(^k x) + F(x)u + W w^{(k)}, \tag{7.13}$$

where

$$F \triangleq \begin{bmatrix} f(x) \\ 0 \\ 0 \\ \ldots \\ 0 \\ 0 \end{bmatrix}, \quad G \triangleq \begin{bmatrix} g(x)w \\ w^{(1)} \\ w^{(2)} \\ \ldots \\ w^{(k-1)} \\ 0 \end{bmatrix}, \quad W \triangleq \begin{bmatrix} 0 \\ 0 \\ 0 \\ \ldots \\ 0 \\ 1 \end{bmatrix} \tag{7.14}$$

and we set $f(x) \triangleq f^1(x)$ and $u \triangleq u_1$.

Algorithm 6.1 becomes the following.

ALGORITHM 7.4. Observable codistribution in the augmented space for driftless systems with $m_w = m_u = 1$.

Set $k = 0$
Set $\overline{\Omega}_k = \text{span}\{\nabla_k h\}$
loop
    Set $\overline{\Omega}_k = [\overline{\Omega}_k, 0]$
    Set $k = k + 1$
    Build $\Sigma^{(k)}$, i.e., G, F, by using (7.14)
    Set $\overline{\Omega}_k = \overline{\Omega}_{k-1} \oplus \mathcal{L}_G \overline{\Omega}_{k-1} \oplus \mathcal{L}_F \overline{\Omega}_{k-1}$
end loop

where the first line in the loop, i.e., $\overline{\Omega}_k = [\overline{\Omega}_k, 0]$, consists in embedding the codistribution $\overline{\Omega}_k$ in the new extended space (the one that also includes the new axis, $w^{(k)}$). In practice, each covector $\omega \in \overline{\Omega}_k$ that is represented by a row-vector of dimension $n + k$ is extended as follows:

$$\omega \to [\omega, 0].$$

In section 6.3, we provided an analytic procedure to obtain the observability properties in the presence of unknown inputs. This procedure is based on Algorithm 6.1 and Theorem 6.12. In particular, it consists of the implementation of a certain number of steps of Algorithm 6.1 together with the test to check whether the components of the original state are observable or not (Theorem 6.12). In section 6.3.1 we applied this procedure on two specific systems, and in section 6.3.2 we discussed its limits of applicability. These are basically the following two:

7.3. Analytic derivations

- Algorithm 6.1, and also Algorithm 7.4, do not converge, automatically. This means that, if for a given step k the gradients of the components of the original state belong to $\overline{\Omega}_k$, we can conclude that the original state is weakly observable. On the other hand, if this is not true, we cannot exclude that it is true for a larger k.

- The codistribution returned by Algorithm 6.1, and also by Algorithm 7.4, is defined in the extended space; i.e., its covectors are the gradients of scalar functions with respect to the augmented state. As a result, the computation dramatically increases with the state augmentation and it becomes prohibitive after few steps.

The goal of the next two subsections is precisely to deal with the above fundamental issues for the simplified systems, i.e., the systems characterized by (7.1)). Specifically, we prove (Theorem 7.7) that the codistribution returned by Algorithm 7.4 can be split into two codistributions: the former is the codistribution generated by Algorithm 7.3, once embedded in the extended space, and the latter is the codistribution L^m defined at the beginning of section 7.3.1 . We prove (Lemma 7.2) that the second codistribution (L^m) can be ignored when deriving the observability properties of the original state. Additionally, we prove that Algorithm 7.3 converges in at most $n + 2$ steps and we provide the convergence criterion (section 7.3.2). All these results are first obtained in the case with a single known input ($m_u = 1$), and then they will be extended to the case when $m_u > 1$ (section 7.3.3).

7.3.1 ▪ Separation

We consider the system $\Sigma^{(k)}$, i.e., the input-output system characterized by (7.13), (7.14), and the single output $y = h(x)$. For each integer $m \leq k$, we generate the codistribution Ω_m by using Algorithm 7.3 (note that here we are considering $m_u = 1$). By construction, the generators of Ω_m are the gradients of scalar functions that only depend on the original state (x) and not on its extension. In what follows, we need to embed this codistribution in the augmented space. We will denote by $[\Omega_m, 0_k]$ the codistribution made by covectors whose first n components are covectors in Ω_m and the last components are all zero. Additionally, we will introduce a new codistribution L^m, which is defined in the augmented space, as follows:

$$L^m \triangleq \text{span}\{\mathcal{L}_G^1 \nabla_k h, \mathcal{L}_G^2 \nabla_k h, \ldots, \mathcal{L}_G^m \nabla_k h\}. \tag{7.15}$$

In other words, L^m is the span of the Lie derivatives of $\nabla_k h$ up to the order m along the vector G. We finally introduce the following codistribution.

Definition 7.1 ($\widetilde{\Omega}$ Codistribution). *This codistribution is defined as follows:* $\widetilde{\Omega}_m \triangleq [\Omega_m, 0_k] \oplus L^m$.

The codistribution $\widetilde{\Omega}_m$ consists of two parts. Specifically, we can select a basis that consists of exact differentials that are the gradients of functions that only depend on the original state (x) and not on its extension (these are the generators of $[\Omega_m, 0_k]$) and the gradients $\mathcal{L}_G^1 \nabla_k h, \mathcal{L}_G^2 \nabla_k h, \ldots, \mathcal{L}_G^m \nabla_k h$. The second set of generators, i.e., the gradients $\mathcal{L}_G^1 \nabla_k h, \mathcal{L}_G^2 \nabla_k h, \ldots, \mathcal{L}_G^m \nabla_k h$, are m and, with respect to the first set, they are gradients of functions that also depend on the state extension $[w, w^{(1)}, \ldots, w^{(m-1)}]^T$. We have the following fundamental result.

Lemma 7.2. $\nabla x_j \in \Omega_m$ *if and only if* $\nabla_k x_j \in \widetilde{\Omega}_m$, *where* x_j *is the jth component of the state* ($j = 1, \ldots, n$).

Proof. The fact that $\nabla x_j \in \Omega_m$ implies that $\nabla_k x_j \in \widetilde{\Omega}_m$ is obvious since $[\Omega_m, 0_k] \subseteq \widetilde{\Omega}_m$ by definition. Let us prove that also the contrary holds, i.e., that if $\nabla_k x_j \in \widetilde{\Omega}_m$, then $\nabla x_j \in \Omega_m$. Since $\nabla_k x_j \in \widetilde{\Omega}_m$, we have $\nabla_k x_j = \sum_{i=1}^{N_1} c_i^1 \omega_i^1 + \sum_{i=1}^{N_2} c_i^2 \omega_i^2$, where $\omega_1^1, \omega_2^1, \ldots, \omega_{N_1}^1$ are N_1 generators of $[\Omega_m, 0_k]$, $\omega_1^2, \omega_2^2, \ldots, \omega_{N_2}^2$ are N_2 generators of L^m, and $c_1^1, \ldots, c_{N_1}^1, c_1^2, \ldots, c_{N_2}^2$ are suitable coefficients. We want to prove that $N_2 = 0$.

We proceed by contradiction. Let us suppose that $N_2 \geq 1$. We remark that the first set of generators have the last k entries equal to zero, as for $\nabla_k x_j$. The second set of generators consists of the Lie derivatives of $\nabla_k h$ along G up to the m order. Let us select the one that is the highest-order Lie derivative, and let us denote by j' this highest order. We have $1 \leq N_2 \leq j' \leq m$. By a direct computation, it is immediate to realize that this is the only generator that depends on $w^{(j'-1)}$. Specifically, the dependence is linear by the product $L_g^1 w^{(j'-1)}$ (we remind the reader that $L_g^1 \neq 0$). But this means that $\nabla_k x_j$ has the $(n+j')$th entry equal to $L_g^1 \neq 0$ and this is not possible since $\nabla_k x_j = [\nabla x_j, 0_k]$. ◄

A fundamental consequence of this lemma is that if we are able to prove that $\widetilde{\Omega}_m = \overline{\Omega}_m$, the weak observability of the original state x can be investigated by only considering the codistribution Ω_m. In the rest of this section we prove this fundamental theorem, stating that $\widetilde{\Omega}_m = \overline{\Omega}_m$.

For a given $m \leq k$ we define the vector Φ_m by the following algorithm:

1. $\Phi_0 = F$;

2. $\Phi_m = [\Phi_{m-1}, G]$,

where now the Lie brackets $[\cdot, \cdot]$ are computed with respect to the extended state, i.e.,

$$[F, G] \triangleq \frac{\partial G}{\partial^k x} F(^k x) - \frac{\partial F}{\partial^k x} G(^k x).$$

By a direct computation it is easy to realize that Φ_m has the last k components identically null. In what follows, we will denote by $\check{\Phi}_m$ the vector that contains the first n components of Φ_m. In other words, $\Phi_m \triangleq [\check{\Phi}_m^T, 0_k^T]^T$. Additionally, we set $\hat{\phi}_m \triangleq \begin{bmatrix} \phi_m \\ 0_k \end{bmatrix}$ (ϕ_m is defined by Algorithm 7.1).

We have the following result.

Lemma 7.3. $\mathcal{L}_G \overline{\Omega}_m \oplus \mathrm{span}\,\{\mathcal{L}_{\Phi_m} \nabla_k h\} = \mathcal{L}_G \overline{\Omega}_m \oplus \mathrm{span}\,\{\mathcal{L}_F \mathcal{L}_G^m \nabla_k h\}$.

Proof. We have $\mathcal{L}_F \mathcal{L}_G^m \nabla_k h = \mathcal{L}_G \mathcal{L}_F \mathcal{L}_G^{m-1} \nabla_k h + \mathcal{L}_{\Phi_1} \mathcal{L}_G^{m-1} \nabla_k h$.

The first term $\mathcal{L}_G \mathcal{L}_F \mathcal{L}_G^{m-1} \nabla_k h \in \mathcal{L}_G \overline{\Omega}_m$. Hence, we need to prove $\mathcal{L}_G \overline{\Omega}_m \oplus \mathrm{span}\,\{\mathcal{L}_{\Phi_m} \nabla_k h\} = \mathcal{L}_G \overline{\Omega}_m \oplus \mathrm{span}\,\{\mathcal{L}_{\Phi_1} \mathcal{L}_G^{m-1} \nabla_k h\}$. We repeat the previous procedure m times. Specifically, we use the equality $\mathcal{L}_{\Phi_j} \mathcal{L}_G^{m-j} \nabla_k h = \mathcal{L}_G \mathcal{L}_{\Phi_j} \mathcal{L}_G^{m-j-1} \nabla_k h + \mathcal{L}_{\Phi_{j+1}} \mathcal{L}_G^{m-j-1} \nabla_k h$, for $j = 1, \ldots, m$, and we remove the first term since $\mathcal{L}_G \mathcal{L}_{\Phi_j} \mathcal{L}_G^{m-j-1} \nabla_k h \in \mathcal{L}_G \overline{\Omega}_m$. ◄

Lemma 7.4. $\check{\Phi}_m = \sum_{j=1}^m c_j^n (\mathcal{L}_G h, \mathcal{L}_G^2 h, \ldots, \mathcal{L}_G^m h) \phi_j$; i.e., the vector $\check{\Phi}_m$ is a linear combination of the vectors ϕ_j ($j = 1, \ldots, m$), where the coefficients (c_j^n) depend on the state only through the functions that generate the codistribution L^m.

7.3. Analytic derivations

Proof. We proceed by induction. By a direct computation it is immediate to obtain $\check{\Phi}_1 = \phi_1 \mathcal{L}_G h$.
Inductive step: Let us assume that $\check{\Phi}_{m-1} = \sum_{j=1}^{m-1} c_j (\mathcal{L}_G h, \mathcal{L}_G^2 h, \ldots, \mathcal{L}_G^{m-1} h) \phi_j$. We have

$$\Phi_m = [\Phi_{m-1}, G] = \sum_{j=1}^{m-1} \left[c_j \begin{bmatrix} \phi_j \\ 0_k \end{bmatrix}, G \right]$$

$$= \sum_{j=1}^{m-1} c_j \left[\begin{bmatrix} \phi_j \\ 0_k \end{bmatrix}, G \right] - \sum_{j=1}^{m-1} \mathcal{L}_G c_j \begin{bmatrix} \phi_j \\ 0_k \end{bmatrix}.$$

We directly compute the Lie bracket in the sum (note that ϕ_j is independent of the unknown input w and its time derivatives):

$$\left[\begin{bmatrix} \phi_j \\ 0_k \end{bmatrix}, G \right] = \begin{bmatrix} [\phi_j, g]w \\ 0_k \end{bmatrix} = \begin{bmatrix} \phi_{j+1} \mathcal{L}_G^1 h \\ 0_k \end{bmatrix}.$$

Regarding the second term, we remark that $\mathcal{L}_G c_j = \sum_{i=1}^{m-1} \frac{\partial c_j}{\partial (\mathcal{L}_G^i h)} \mathcal{L}_G^{i+1} h$. By setting $\widetilde{c}_j = c_{j-1} \mathcal{L}_G^1 h$ for $j = 2, \ldots, m$ and $\widetilde{c}_1 = 0$, and by setting $\overline{c}_j = -\sum_{i=1}^{m-1} \frac{\partial c_j}{\partial (\mathcal{L}_G^i h)} \mathcal{L}_G^{i+1} h$ for $j = 1, \ldots, m-1$ and $\overline{c}_m = 0$, we obtain $\check{\Phi}_m = \sum_{j=1}^{m} (\widetilde{c}_j + \overline{c}_j) \phi_j$, which proves our assertion since $c_j^n (\triangleq \widetilde{c}_j + \overline{c}_j)$ is a function of $\mathcal{L}_G h, \mathcal{L}_G^2 h, \ldots, \mathcal{L}_G^m h$. ◂

The following result also holds.

Lemma 7.5. *If $w \neq 0$, $\hat{\phi}_m = \sum_{j=1}^{m} b_j^n (\mathcal{L}_G h, \mathcal{L}_G^2 h, \ldots, \mathcal{L}_G^m h) \Phi_j$, i.e., the vector $\hat{\phi}_m$ is a linear combination of the vectors Φ_j ($j = 1, \ldots, m$), where the coefficients (b_j^n) depend on the state only through the functions that generate the codistribution L^m.*

Proof. We proceed by induction. By a direct computation it is immediate to obtain $\hat{\phi}_1 = \Phi_1 \frac{1}{\mathcal{L}_G h}$ (note that $\mathcal{L}_G h = L_g^1 w \neq 0$).
Inductive step: Let us assume that $\hat{\phi}_{m-1} = \sum_{j=1}^{m-1} b_j (\mathcal{L}_G h, \mathcal{L}_G^2 h, \ldots, \mathcal{L}_G^{m-1} h) \Phi_j$. We need to prove that $\hat{\phi}_m = \sum_{j=1}^{m} b_j^n (\mathcal{L}_G h, \mathcal{L}_G^2 h, \ldots, \mathcal{L}_G^m h) \Phi_j$. We start by applying on both members of the equality $\hat{\phi}_{m-1} = \sum_{j=1}^{m-1} b_j (\mathcal{L}_G h, \mathcal{L}_G^2 h, \ldots, \mathcal{L}_G^{m-1} h) \Phi_j$ the Lie bracket with respect to G. We obtain for the first member $[\hat{\phi}_{m-1}, G] = \hat{\phi}_m \mathcal{L}_G^1 h$. For the second member we have

$$\sum_{j=1}^{m-1} [b_j \Phi_j, G] = \sum_{j=1}^{m-1} b_j [\Phi_j, G] - \sum_{j=1}^{m-1} \mathcal{L}_G b_j \Phi_j = \sum_{j=1}^{m-1} b_j \Phi_{j+1} - \sum_{j=1}^{m-1} \sum_{i=1}^{m-1} \frac{\partial b_j}{\partial (\mathcal{L}_G^i h)} \mathcal{L}_G^{i+1} h \Phi_j.$$

Since $\mathcal{L}_G h = L_g^1 w \neq 0$, by setting $\widetilde{b}_j = \frac{b_{j-1}}{\mathcal{L}_G^1 h}$ for $j = 2, \ldots, m$ and $\widetilde{b}_1 = 0$, and by setting $\overline{b}_j = -\sum_{i=1}^{m-1} \frac{\partial b_j}{\partial (\mathcal{L}_G^i h)} \frac{\mathcal{L}_G^{i+1} h}{\mathcal{L}_G^1 h}$ for $j = 1, \ldots, m-1$ and $\overline{b}_m = 0$, we obtain $\hat{\phi}_m = \sum_{j=1}^{m} (\widetilde{b}_j + \overline{b}_j) \Phi_j$, which proves our assertion since $b_j^n (\triangleq \widetilde{b}_j + \overline{b}_j)$ is a function of $\mathcal{L}_G h, \mathcal{L}_G^2 h, \ldots, \mathcal{L}_G^m h$. ◂

An important consequence of the previous two lemmas is the following result.

Proposition 7.6. *If $w \neq 0$, the following two codistributions coincide:*

1. $\text{span}\{\mathcal{L}_{\Phi_0}\nabla_k h, \mathcal{L}_{\Phi_1}\nabla_k h, \ldots, \mathcal{L}_{\Phi_m}\nabla_k h, \mathcal{L}_G^1\nabla_k h, \ldots, \mathcal{L}_G^m\nabla_k h\}$;

2. $\text{span}\{\mathcal{L}_{\hat{\phi}_0}\nabla_k h, \mathcal{L}_{\hat{\phi}_1}\nabla_k h, \ldots, \mathcal{L}_{\hat{\phi}_m}\nabla_k h, \mathcal{L}_G^1\nabla_k h, \ldots, \mathcal{L}_G^m\nabla_k h\}$.

We are now ready to prove the following fundamental result.

Theorem 7.7 (Separation). *If $w \neq 0$, then $\overline{\Omega}_m = \widetilde{\Omega}_m \triangleq [\Omega_m, 0_k] \oplus L^m$.*

Proof. We proceed by induction. By definition, $\overline{\Omega}_0 = \widetilde{\Omega}_0$ since they are both the span of $\nabla_k h$.
Inductive step: Let us assume that $\overline{\Omega}_{m-1} = \widetilde{\Omega}_{m-1}$. We have $\overline{\Omega}_m = \overline{\Omega}_{m-1} \oplus \mathcal{L}_F\overline{\Omega}_{m-1} \oplus \mathcal{L}_G\overline{\Omega}_{m-1} = \overline{\Omega}_{m-1} \oplus \mathcal{L}_F\widetilde{\Omega}_{m-1} \oplus \mathcal{L}_G\overline{\Omega}_{m-1} = \overline{\Omega}_{m-1} \oplus [\mathcal{L}_f\Omega_{m-1}, 0_k] \oplus \mathcal{L}_F L^{m-1} \oplus \mathcal{L}_G\overline{\Omega}_{m-1}$.
On the other hand,

$$\mathcal{L}_F L^{m-1} = \text{span}\left\{\mathcal{L}_F\mathcal{L}_G^1\nabla_k h\right\} \oplus \cdots \oplus \text{span}\left\{\mathcal{L}_F\mathcal{L}_G^{m-2}\nabla_k h\right\} \oplus \text{span}\left\{\mathcal{L}_F\mathcal{L}_G^{m-1}\nabla_k h\right\}.$$

The first $m-2$ terms are in $\overline{\Omega}_{m-1}$. Hence we have

$$\overline{\Omega}_m = \overline{\Omega}_{m-1} \oplus [\mathcal{L}_f\Omega_{m-1}, 0_k] \oplus \text{span}\left\{\mathcal{L}_F\mathcal{L}_G^{m-1}\nabla_k h\right\} \oplus \mathcal{L}_G\overline{\Omega}_{m-1}.$$

By using Lemma 7.3 we obtain $\overline{\Omega}_m = \overline{\Omega}_{m-1} \oplus [\mathcal{L}_f\Omega_{m-1}, 0_k] \oplus \text{span}\left\{\mathcal{L}_{\Phi_{m-1}}\nabla_k h\right\} \oplus \mathcal{L}_G\overline{\Omega}_{m-1}$.
By using again the induction assumption we obtain $\overline{\Omega}_m = [\Omega_{m-1}, 0_k] \oplus L^{m-1} \oplus [\mathcal{L}_f\Omega_{m-1}, 0_k] \oplus \text{span}\left\{\mathcal{L}_{\Phi_{m-1}}\nabla_k h\right\} \oplus \mathcal{L}_G[\Omega_{m-1}, 0_k] \oplus \mathcal{L}_G L^{m-1} = [\Omega_{m-1}, 0_k] \oplus L^m \oplus [\mathcal{L}_f\Omega_{m-1}, 0_k] \oplus \text{span}\left\{\mathcal{L}_{\Phi_{m-1}}\nabla_k h\right\} \oplus [\mathcal{L}_{\frac{g}{L_g^1}}\Omega_{m-1}, 0_k]$, where we used $L^m \oplus \mathcal{L}_G[\Omega_{m-1}, 0_k] = L^m \oplus [\mathcal{L}_{\frac{g}{L_g^1}}\Omega_{m-1}, 0_k]$, which holds because $\mathcal{L}_G h = L_g^1 w \neq 0$. By using Proposition 7.6, we obtain
$\overline{\Omega}_m = [\Omega_{m-1}, 0_k] \oplus L^m \oplus [\mathcal{L}_f\Omega_{m-1}, 0_k] \oplus \text{span}\{\mathcal{L}_{\hat{\phi}_{m-1}}\nabla_k h\} \oplus [\mathcal{L}_{\frac{g}{L_g^1}}\Omega_{m-1}, 0_k] = \widetilde{\Omega}_m$. ◂

Theorem 7.7 is fundamental. It allows us to obtain all the observability properties of the original state by restricting the computation to the codistribution defined by Algorithm 7.3, namely a codistribution whose covectors have the same dimension of the original space. In other words, the dimension of these covectors is independent of the state augmentation.

7.3.2 ▪ Convergence

Algorithm 7.3 is recursive, and $\Omega_m \subseteq \Omega_{m+1}$. This means that if for a given m the gradients of the components of the original state belong to Ω_m, we can conclude that the original state is weakly observable. On the other hand, if this is not true, we cannot exclude that it is true for a larger m. The goal of this section is precisely to address this issue. We will show that the algorithm converges in a finite number of steps, and we will also provide the criterion to establish that the algorithm has converged (Theorem 7.14). This theorem will be proved at the end of this section since we need to introduce several important new quantities and properties.

When investigating the convergence properties of Algorithm 7.3, we remark that the main difference between Algorithms 4.2 and 7.3 is the presence of the last term in the recursive step of the latter. Without this term, the convergence criterion would simply consist of the inspection of the equality $\Omega_{m+1} = \Omega_m$, as for Algorithm 4.2.

7.3. Analytic derivations

The following result provides the convergence criterion in a very special case that basically occurs when the contribution due to the last term in the recursive step of Algorithm 7.3 is included in the other terms. In this case, we obviously obtain that the convergence criterion consists of the inspection of the equality $\Omega_{m+1} = \Omega_m$, as for Algorithm 4.2. For any integer $j \geq 0$ we define

$$\chi_j \triangleq \frac{\mathcal{L}_{\phi_j} L_g^1}{L_g^1}. \tag{7.16}$$

We have the following result.

Lemma 7.8. *Let us denote by Λ_j the distribution generated by $\phi_0, \phi_1, \ldots, \phi_j$ and by $m(\leq n-1)$ the smallest integer for which $\Lambda_{m+1} = \Lambda_m$ (n is the dimension of the state x). In the very special case when $\chi_j = 0 \ \forall j = 0, \ldots, m$, Algorithm 7.3 converges at the integer j such that $\Omega_{j+1} = \Omega_j$, and this occurs in at most $n-1$ steps.*

Proof. First of all, we remind the reader that the existence of an integer $m(\leq n-1)$ such that $\Lambda_{m+1} = \Lambda_m$ is proved in [24]. In particular, the first chapter in [24] analyzes the convergence of Λ_j with respect to j in the case when $L_g^1 = 1$. It is proved that the distribution converges to Λ^* and that the convergence is achieved at the smallest integer for which we have $\Lambda_{m+1} = \Lambda_m$. Additionally, we have $\Lambda_{m+1} = \Lambda_m = \Lambda^*$ and m cannot exceed $n-1$ (see Lemmas 1.8.2 and 1.8.3 in [24]). The same result still holds when $L_g^1 \neq 1$ and can be proved by following the same steps.

In the very special case when $\chi_j = 0 \ \forall j = 0, \ldots, m$, thanks to the aforementioned convergence of the distribution Λ_j, we easily obtain that $\mathcal{L}_{\phi_{j-1}} \mathcal{L}_g h = 0 \ \forall j \geq 1$. Now, let us consider the following equation:

$$\mathcal{L}_{\phi_j} h = \frac{1}{L_g^1} \left(\mathcal{L}_{\phi_{j-1}} \mathcal{L}_g h - \mathcal{L}_g \mathcal{L}_{\phi_{j-1}} h \right). \tag{7.17}$$

Since $\mathcal{L}_{\phi_{j-1}} \mathcal{L}_g h = 0 \ \forall j \geq 1$, we have $\mathcal{L}_{\phi_j} h = -\mathcal{L}_{\frac{g}{L_g^1}} \mathcal{L}_{\phi_{j-1}} h$ for any $j \geq 1$. Therefore, we conclude that the last term in the recursive step of Algorithm 7.3 is included in the second last term and, in this special case, Algorithm 7.3 has converged when $\Omega_{m+1} = \Omega_m$. This occurs in at most $n-1$ steps, as for Algorithm 4.2. ◀

Let us consider now the general case. To proceed we need to introduce several important new quantities and properties.

For a given positive integer j we define the vector ψ_j by the following algorithm:

1. $\psi_0 = f$;
2. $\psi_j = [\psi_{j-1}, \frac{g}{L_g^1}]$.

It is possible to find the expression that relates these vectors to the vectors ϕ_j, previously defined. Specifically we have what follows.

Lemma 7.9. *The following equation holds:*

$$\psi_j = \phi_j + \left\{ \sum_{i=0}^{j-1} (-)^{j-i} \mathcal{L}_{\frac{g}{L_g^1}}^{j-i-1} \left(\frac{\mathcal{L}_{\phi_i} L_g^1}{L_g^1} \right) \right\} \frac{g}{L_g^1}. \tag{7.18}$$

Proof. We proceed by induction. By definition, $\psi_0 = \phi_0 = f$ and (7.18) holds for $j = 0$.
Inductive step: Let us assume that it holds for a given $j - 1 \geq 0$, and let us prove its validity for j. We have

$$\psi_j = \left[\psi_{j-1}, \frac{g}{L_g^1}\right] = \left[\phi_{j-1}, \frac{g}{L_g^1}\right] + \left[\left\{\sum_{i=0}^{j-2} (-)^{j-i-1} \mathcal{L}_{\frac{g}{L_g^1}}^{j-i-2} \left(\frac{\mathcal{L}_{\phi_i} L_g^1}{L_g^1}\right)\right\} \frac{g}{L_g^1}, \frac{g}{L_g^1}\right].$$

On the other hand,

$$\left[\phi_{j-1}, \frac{g}{L_g^1}\right] = \phi_j - \frac{\mathcal{L}_{\phi_{j-1}} L_g^1}{L_g^1} \frac{g}{L_g^1}$$

and

$$\left[\left\{\sum_{i=0}^{j-2} (-)^{j-i-1} \mathcal{L}_{\frac{g}{L_g^1}}^{j-i-2} \left(\frac{\mathcal{L}_{\phi_i} L_g^1}{L_g^1}\right)\right\} \frac{g}{L_g^1}, \frac{g}{L_g^1}\right] = -\mathcal{L}_{\frac{g}{L_g^1}} \left\{\sum_{i=0}^{j-2} (-)^{j-i-1} \mathcal{L}_{\frac{g}{L_g^1}}^{j-i-2} \left(\frac{\mathcal{L}_{\phi_i} L_g^1}{L_g^1}\right)\right\} \frac{g}{L_g^1}$$

$$= \left\{\sum_{i=0}^{j-2} (-)^{j-i} \mathcal{L}_{\frac{g}{L_g^1}}^{j-i-1} \left(\frac{\mathcal{L}_{\phi_i} L_g^1}{L_g^1}\right)\right\} \frac{g}{L_g^1}.$$

Hence,

$$\psi_j = \phi_j - \frac{\mathcal{L}_{\phi_{j-1}} L_g^1}{L_g^1} \frac{g}{L_g^1} + \left\{\sum_{i=0}^{j-2} (-)^{j-i} \mathcal{L}_{\frac{g}{L_g^1}}^{j-i-1} \left(\frac{\mathcal{L}_{\phi_i} L_g^1}{L_g^1}\right)\right\} \frac{g}{L_g^1},$$

which coincides with (7.18). ◀

We have the following result.

Lemma 7.10. *For $i = 0, 1, \ldots, m - 2$, we have*

$$\nabla \frac{\mathcal{L}_{\phi_i} L_g^1}{L_g^1} \in \Omega_m. \tag{7.19}$$

Proof. By construction, $\nabla \mathcal{L}_{\phi_i} h \in \Omega_m$ for any $i = 1, \ldots, m - 1$. On the other hand, we have

$$\mathcal{L}_{\phi_i} h = \frac{1}{L_g^1}[\mathcal{L}_{\phi_{i-1}} \mathcal{L}_g h - \mathcal{L}_g \mathcal{L}_{\phi_{i-1}} h] = \frac{\mathcal{L}_{\phi_{i-1}} L_g^1}{L_g^1} - \mathcal{L}_{\frac{g}{L_g^1}} \mathcal{L}_{\phi_{i-1}} h.$$

We compute the gradient of both members of this equation. Since $\nabla \mathcal{L}_{\frac{g}{L_g^1}} \mathcal{L}_{\phi_{i-1}} h \in \Omega_m$, for any $i = 1, \ldots, m - 1$, also $\nabla \frac{\mathcal{L}_{\phi_{i-1}} L_g^1}{L_g^1} \in \Omega_m$. ◀

From Lemma 7.9 with $j = 1, \ldots, m - 1$ and Lemma 7.10 it is immediate to obtain the following result.

Proposition 7.11. *If Ω_m is invariant with respect to $\mathcal{L}_f$ and $\mathcal{L}_{\frac{g}{L_g^1}}$, then it is also invariant with respect to $\mathcal{L}_{\phi_j}$, $j = 1, \ldots, m - 1$.*

In order to obtain the convergence criterion for Algorithm 7.3 we need to substitute the expression of ϕ_j in terms of ϕ_{j-2} in the term $\mathcal{L}_{\phi_j} h$. This will allow us to detect the key quantity

7.3. Analytic derivations

that governs the convergence of Algorithm 7.3, in particular regarding the contribution due to the last term in the recursive step. For the simplified systems analyzed in this chapter (i.e., the systems characterized by (7.1)), this quantity is a scalar and it is the one provided in (7.3). For the sake of clarity, we provide (7.3) below:

$$\tau \triangleq \frac{\mathcal{L}_g^2 h}{(L_g^1)^2}.$$

The behavior of the last term in the recursive step of Algorithm 7.3 is given by the following lemma.

Lemma 7.12. *We have the following key equality:*

$$\mathcal{L}_{\phi_j} h = \mathcal{L}_{\phi_{j-2}} \tau + \tau \frac{\mathcal{L}_{\phi_{j-2}} L_g^1}{L_g^1} - \mathcal{L}_{\frac{g}{L_g^1}} \left(\frac{\mathcal{L}_{\phi_{j-2}} L_g^1}{L_g^1} + \mathcal{L}_{\phi_{j-1}} h \right), \quad j \geq 2. \qquad (7.20)$$

Proof. We will prove this equality by an explicit computation. We have

$$\mathcal{L}_{\phi_j} h = \frac{1}{L_g^1} \left(\mathcal{L}_{\phi_{j-1}} \mathcal{L}_g h - \mathcal{L}_g \mathcal{L}_{\phi_{j-1}} h \right).$$

The second term on the right-hand side simplifies with the last term in (7.20). Hence we have to prove

$$\frac{1}{L_g^1} \mathcal{L}_{\phi_{j-1}} L_g^1 = \mathcal{L}_{\phi_{j-2}} \tau + \tau \frac{\mathcal{L}_{\phi_{j-2}} L_g^1}{L_g^1} - \mathcal{L}_{\frac{g}{L_g^1}} \frac{\mathcal{L}_{\phi_{j-2}} L_g^1}{L_g^1}. \qquad (7.21)$$

We have

$$\frac{1}{L_g^1} \mathcal{L}_{\phi_{j-1}} L_g^1 = \frac{1}{(L_g^1)^2} \left(\mathcal{L}_{\phi_{j-2}} L_g^2 - \mathcal{L}_g \mathcal{L}_{\phi_{j-2}} L_g^1 \right). \qquad (7.22)$$

We remark that

$$\frac{1}{(L_g^1)^2} \mathcal{L}_{\phi_{j-2}} L_g^2 = \mathcal{L}_{\phi_{j-2}} \tau + 2\tau \frac{\mathcal{L}_{\phi_{j-2}} L_g^1}{L_g^1}$$

and

$$\frac{1}{(L_g^1)^2} \mathcal{L}_g \mathcal{L}_{\phi_{j-2}} L_g^1 = \tau \frac{\mathcal{L}_{\phi_{j-2}} L_g^1}{L_g^1} + \mathcal{L}_{\frac{g}{L_g^1}} \frac{\mathcal{L}_{\phi_{j-2}} L_g^1}{L_g^1}.$$

By substituting these two last equalities in (7.22), we immediately obtain (7.21). ◄

Lemma 7.13. *In general, there exists an integer $m \leq n + 2$ (n being the dimension of x) such that $\nabla \tau \in \Omega_m$.*

Proof. Let us introduce the following notation for a given integer j:

- $\mathcal{Z}_j \triangleq \mathcal{L}_{\phi_{j+2}} h$;
- $\mathcal{B}_j \triangleq \mathcal{L}_{\phi_j} \tau$;
- $\chi_j \triangleq \frac{\mathcal{L}_{\phi_j} L_g^1}{L_g^1}$.

By construction, $\nabla \mathcal{Z}_j \in \Omega_{j+3}$. On the other hand, from (7.20), we immediately obtain

$$\nabla \mathcal{Z}_j = \nabla \mathcal{B}_j + \chi_j \nabla \tau + \tau \nabla \chi_j - \mathcal{L}_{\frac{g}{L_g^1}}\left(\nabla \chi_j + \nabla \mathcal{L}_{\phi_{j+1}} h\right). \tag{7.23}$$

By using Lemma 7.10, we obtain the following results:

- $\tau \nabla \chi_j \in \Omega_{j+2}$;
- $\mathcal{L}_{\frac{g}{L_g^1}} \nabla \chi_j \in \Omega_{j+3}$.

Additionally, $\mathcal{L}_{\frac{g}{L_g^1}} \nabla \mathcal{L}_{\phi_{j+1}} h \in \Omega_{j+3}$. Hence, from (7.23), we obtain that the covector

$$\mathcal{Z}'_j \triangleq \nabla \mathcal{B}_j + \chi_j \nabla \tau \tag{7.24}$$

belongs to Ω_{j+3}. Let us denote by j^* the smallest integer such that

$$\nabla \mathcal{B}_{j^*} = \sum_{j=0}^{j^*-1} c_j \nabla \mathcal{B}_j + c_{-1} \nabla h. \tag{7.25}$$

Note that j^* is a finite integer and in particular $j^* \leq n-1$. Indeed, if this would not be the case, the dimension of the codistribution generated by $\nabla h, \nabla \mathcal{B}_0, \nabla \mathcal{B}_1, \ldots, \nabla \mathcal{B}_{n-1}$ would be $n+1$, i.e., larger than n. From (7.25) and (7.24) we obtain

$$\mathcal{Z}'_{j^*} = \sum_{j=0}^{j^*-1} c_j \nabla \mathcal{B}_j + c_{-1} \nabla h + \chi_{j^*} \nabla \tau. \tag{7.26}$$

From (7.24), for $j = 0, \ldots, j^*-1$, we obtain $\nabla \mathcal{B}_j = \mathcal{Z}'_j - \chi_j \nabla \tau$. By substituting in (7.26) we obtain

$$\mathcal{Z}'_{j^*} - \sum_{j=0}^{j^*-1} c_j \mathcal{Z}'_j - c_{-1} \nabla h = \left(-\sum_{j=0}^{j^*-1} c_j \chi_j + \chi_{j^*}\right) \nabla \tau. \tag{7.27}$$

We remark that the left-hand side consists of the sum of covectors that belong to Ω_{j^*+3}. Since in general $\chi_{j^*} \neq \sum_{j=0}^{j^*-1} c_j \chi_j$, we have $\nabla \tau \in \Omega_{j^*+3}$. By setting $m \triangleq j^* + 3$, we have $m \leq n+2$ and $\nabla \tau \in \Omega_m$. ◀

The previous lemma ensures that, in general, there exists a finite $m \leq n+2$ such that $\nabla \tau \in \Omega_m$. Note that the previous proof holds if the quantity $\chi_{j^*} - \sum_{j=0}^{j^*-1} c_j \chi_j$ does not vanish. This holds in general, with the exception of the trivial case considered in Lemma 7.8, in which case $\chi_j = 0\ \forall j$.

The following theorem allows us to obtain the criterion to stop Algorithm 7.3:

Theorem 7.14. *If $\nabla \tau \in \Omega_m$ and Ω_m is invariant under $\mathcal{L}_f$ and $\mathcal{L}_{\frac{g}{L_g^1}}$, then $\Omega_{m+p} = \Omega_m\ \forall p \geq 0$.*

Proof. We proceed by induction. Obviously, the equality holds for $p = 0$.
Inductive step: Let us assume that $\Omega_{m+p} = \Omega_m$, and let us prove that $\Omega_{m+p+1} = \Omega_m$. We have to prove that $\nabla \mathcal{L}_{\phi_{m+p}} h \in \Omega_m$. Indeed, from the inductive assumption, we know that $\Omega_{m+p}(=\Omega_m)$ is invariant under $\mathcal{L}_f$ and $\mathcal{L}_{\frac{g}{L_g^1}}$. Additionally, because of this invariance, by using

7.3. Analytic derivations

Proposition 7.11, we obtain that Ω_m is also invariant under $\mathcal{L}_{\phi_j}$ for $j = 1, 2, \ldots, m + p - 1$. Since $\nabla \tau \in \Omega_m$, we have $\nabla \mathcal{L}_{\phi_{m+p-2}} \tau \in \Omega_m$. Additionally, $\nabla \mathcal{L}_{\phi_{m+p-1}} h \in \Omega_m$ and, because of Lemma 7.10, we also have $\nabla \frac{\mathcal{L}_{\phi_{m+p-2}} L_g^1}{L_g^1} \in \Omega_m$. Finally, because of the invariance under $\mathcal{L}_{\frac{g}{L_g^1}}$, also the Lie derivatives along $\frac{g}{L_g^1}$ of $\nabla \mathcal{L}_{\phi_{m+p-1}} h$ and $\nabla \frac{\mathcal{L}_{\phi_{m+p-2}} L_g^1}{L_g^1}$ belong to Ω_m. Now we use (7.20) for $j = m + p$. By computing the gradient of this equation it is immediate to obtain that $\nabla \mathcal{L}_{\phi_{m+p}} h \in \Omega_m$. ◄

We conclude this section by providing an upper bound for the number of steps that are in general necessary to achieve the convergence. The dimension of Ω_{j^*+2} is at least the dimension of the span of the covectors ∇h, $\mathcal{Z}_0'$, $\mathcal{Z}_1'$, ..., $\mathcal{Z}_{j^*-1}'$. From the definition of j^*, we know that the vectors ∇h, $\nabla \mathcal{B}_0$, $\nabla \mathcal{B}_1$, ..., $\nabla \mathcal{B}_{j^*-1}$ are independent, meaning that the dimension of their span is $j^* + 1$. Hence, from (7.24), it easily follows that the dimension of the span of the vectors ∇h, $\mathcal{Z}_0'$, $\mathcal{Z}_1'$, ..., $\mathcal{Z}_{j^*-1}'$, $\nabla \tau$ is at least $j^* + 1$. Since Ω_{j^*+3} contains this span, its dimension is at least $j^* + 1$. Therefore, the condition $\Omega_{m+1} = \Omega_m$ for $m \geq j^* + 3$ is achieved for $m \leq n + 2$.

7.3.3 ▪ Extension to the case of multiple known inputs

In this section we extend all the analytic results obtained in the previous section to the case of multiple known inputs (i.e., $m_u > 1$). This extension is immediate. We extend the results in subsections 7.3.1 and 7.3.2 separately.

Separation when $m_u > 1$

It is immediate to realize that Lemma 7.2 remains the same. Regarding Lemma 7.3, Lemma 7.4, and Lemma 7.5, we need, first of all, to perform the following substitutions ($i = 1, \ldots, m_u$):

- $F \to F^i$;
- $\Phi_m \to {}^i \Phi_m \; \forall m$;
- $\phi_m \to {}^i \phi_m \; \forall m$;
- $\breve{\Phi}_m \to {}^i \breve{\Phi}_m \; \forall m$;
- $\hat{\phi}_m \to {}^i \hat{\phi}_m \; \forall m$.

Then, the results stated by these lemmas hold for each i separately and can be proved for each i separately, by following precisely the same steps. Regarding Proposition 7.6, the two codistributions become, respectively,

$$\sum_{i=1}^{m_u} \text{span}\{\mathcal{L}_{{}^i \Phi_0} \nabla h, \mathcal{L}_{{}^i \Phi_1} \nabla h, \ldots, \mathcal{L}_{{}^i \Phi_m} \nabla h\} \oplus L^m$$

and

$$\sum_{i=1}^{m_u} \text{span}\{\mathcal{L}_{{}^i \hat{\phi}_0} \nabla h, \mathcal{L}_{{}^i \hat{\phi}_1} \nabla h, \ldots, \mathcal{L}_{{}^i \hat{\phi}_m} \nabla h\} \oplus L^m.$$

Finally, the proof of Theorem 7.7 follows the same steps by using the above substitutions and by summing on all the values of i.

Convergence when $m_u > 1$

The convergence in the special case dealt by Lemma 7.8 requires us to build m_u distributions. Then, we have the same result of Lemma 7.8. Specifically, we have the following.

Lemma 7.15. *Let us denote by Λ^i_j the distribution generated by $^i\phi_0, {}^i\phi_1, \ldots, {}^i\phi_j$ (i.e., the vectors obtained by running Algorithm 7.1 for a given $i = 1, \ldots, m_u$) and by $m_i (\leq n-1)$ the smallest integer for which $\Lambda^i_{m_i+1} = \Lambda^i_{m_i}$ (n is the dimension of the state x). If $\mathcal{L}_{^i\phi_j} L^1_g = 0 \; \forall i = 1, \ldots, m_u$ and $\forall j = 0, \ldots, m_i$, the convergence of Algorithm 7.3 occurs in at most $n-1$ steps, and it occurs at the smallest integer j such that $\Omega_{j+1} = \Omega_j$.*

To proceed with the general case of convergence (i.e., when the assumptions of Lemma 7.15 are not met), we need to perform the further substitution $\psi_m \to \psi^i_m \; \forall m$.

Then, the results stated by Lemma 7.9, Lemma 7.10, and Lemma 7.12 hold for each i separately and can be proved for each i separately by following precisely the same steps. Specifically, the extension of Lemma 7.12 provides the following fundamental equation, which holds for every $i = 1, \ldots, m_u$:

$$\mathcal{L}_{^i\phi_j} h = \mathcal{L}_{^i\phi_{j-2}} \tau + \tau \frac{\mathcal{L}_{^i\phi_{j-2}} L^1_g}{L^1_g} - \mathcal{L}_{\frac{g}{L^1_g}} \left(\frac{\mathcal{L}_{^i\phi_{j-2}} L^1_g}{L^1_g} + \mathcal{L}_{^i\phi_{j-1}} h \right). \tag{7.28}$$

Regarding Lemma 7.13, it is immediate to extend its validity to the case of multiple known inputs. Indeed, the codistribution returned by Algorithm 7.3 in the case of multiple known inputs includes the codistribution in the case of a single known input.

Finally, regarding Proposition 7.11 and Theorem 7.14, we need to require that Ω_m is invariant with respect to $\mathcal{L}_{\frac{g}{L^1_g}}$ and all $\mathcal{L}_{f^i}$ simultaneously. Specifically, the extension of Theorem 7.14 is as follows.

Theorem 7.16. *If $\nabla \tau \in \Omega_m$ and Ω_m is invariant under $\mathcal{L}_{f^i} \; \forall i = 1, \ldots, m_u$ and $\mathcal{L}_{\frac{g}{L^1_g}}$, then $\Omega_{m+p} = \Omega_m \; \forall p \geq 0$*

We conclude this section with the following remark, which immediately follows from Theorem 7.7.

Remark 7.17. *If the considered system has two (or more) outputs such that their first-order Lie derivative along g does not vanish, then the codistribution $\widetilde{\Omega}_m$ (for every integer m) is independent of the output that is selected to define the scalars $h(x)$ and L^1_g in (7.2).*

Proof. Let us denote by h_{i_1} and h_{i_2} two outputs that verify the above condition. For every integer m, we compute $\widetilde{\Omega}^1_m$ by using Algorithm 7.3 with $h = h_{i_1}$ and $L^1_g = \mathcal{L}_g h_{i_1}$. Similarly, we compute $\widetilde{\Omega}^2_m$ by using Algorithm 7.3 with $h = h_{i_2}$ and $L^1_g = \mathcal{L}_g h_{i_2}$. By using Theorem 7.7, we have $\widetilde{\Omega}^1_m = \overline{\Omega}_m = \widetilde{\Omega}^2_m$. ◄

Chapter 8

Unknown Input Observability for the General Case

This chapter provides the complete analytic procedure to obtain the observability properties in the presence of unknown inputs in the general case, i.e., for systems characterized by the following equations:

$$\begin{cases} \dot{x} & = g^0(x) + \sum_{i=1}^{m_u} f^i(x)u_i + \sum_{j=1}^{m_w} g^j(x)w_j, \\ y & = [h_1(x), \ldots, h_p(x)]^T. \end{cases} \quad (8.1)$$

With respect to the simplified systems investigated in Chapter 7, we now account for a drift (i.e., a nonvanishing g^0) and for multiple unknown inputs (i.e., $m_w > 1$).

In this chapter we introduce several tensor fields with respect to the group of transformations discussed in section 3.3.2, which was denoted by $\mathcal{SUIO}$ (Simultaneous Unknown Input-Output transformations' group). In general, to avoid ambiguities, we only make explicit the indices that refer to these tensors. For tensors with respect to coordinates' changes, we adopt a matrix notation. This is the case of (8.1), where $g^j(x)$ is a vector field with respect to a coordinates change and the jth component of a vector field with respect to $\mathcal{SUIO}$. In addition, we widely use the Einstein notation (see section 2.2.2) with respect to the indices that refer to this group.

This chapter is organized in the same way as Chapter 7. We start by directly providing the solution in section 8.1. This consists of Algorithm 8.3, which returns the observable codistribution for systems characterized by (8.1). Then, in section 8.2, we provide several applications. In section 8.3, we provide all the analytic derivations that prove the validity of the solution. Finally, in section 8.4, we extend the solution in order to account for an explicit time dependency of the dynamics and/or the output.

8.1 ▪ Extension of the observability rank condition for the general case

This section provides the algorithm that returns the observable codistribution together with its convergence criterion (section 8.1.1). Then, based on this algorithm, in section 8.1.2 we summarize the analytic method to determine the observability properties of systems described by (8.1).

8.1.1 • Observable codistribution

Before providing the algorithm that returns the observable codistribution, we need to introduce several new ingredients, which are tensor fields with respect to $\mathcal{SUIO}$ (see section 3.3.2) and also a new operation (the *autobracket*, provided by Definition 8.1). On the other hand, the $\mathcal{SUIO}$ will not be defined directly on the system characterized by (8.1), but on a slightly different system, which is the one characterized by (8.4). Note that the new system has exactly the same observability properties and differs from the original system only for the outputs.

The first step consists in building this new system (by showing that it has the same observability properties).

We select m_w functions, $\widetilde{h}_1, \ldots, \widetilde{h}_{m_w}$, from the space $\mathcal{F}$ (see Definition 6.13) such that the matrix

$$\begin{bmatrix} \mathcal{L}_{g^1} \widetilde{h}_1 & \mathcal{L}_{g^1} \widetilde{h}_2 & \cdots & \mathcal{L}_{g^1} \widetilde{h}_{m_w} \\ \mathcal{L}_{g^2} \widetilde{h}_1 & \mathcal{L}_{g^2} \widetilde{h}_2 & \cdots & \mathcal{L}_{g^2} \widetilde{h}_{m_w} \\ \cdots & \cdots & \cdots & \\ \mathcal{L}_{g^{m_w}} \widetilde{h}_1 & \mathcal{L}_{g^{m_w}} \widetilde{h}_2 & \cdots & \mathcal{L}_{g^{m_w}} \widetilde{h}_{m_w} \end{bmatrix} \tag{8.2}$$

is nonsingular at x_0. In Appendix C, we introduce the concept of *canonic form* with respect to the unknown inputs. The system is in its canonic form with respect to its unknown inputs if it is possible to select m_w scalar functions from $\mathcal{F}$ such that the above matrix is nonsingular. In Appendix C, we show that, for a system that is not in its canonic form, it is possible either to find a finite number of local coordinates' changes in the space of the unknown inputs and their time derivatives up to a given order, such that the canonic form is achieved, or to show that some of the unknown inputs are spurious (i.e., they do not affect the observability properties). In this latter case, it is possible to write the dynamics in (8.1) with a number of unknown inputs smaller than m_w and in canonic form with respect to these new unknown inputs. Hence, we can assume that the above matrix is nonsingular. Note that a necessary (but not sufficient) condition for the nonsingularity of the above matrix is that the functions $\widetilde{h}_1(x), \ldots, \widetilde{h}_{m_w}(x)$ are independent, i.e., their gradients are independent covectors (at x_0).

We now focus our attention on the following input-output system:

$$\begin{cases} \dot{x} &= g^0(x) + \sum_{i=1}^{m_u} f^i(x) u_i + \sum_{j=1}^{m_w} g^j(x) w_j, \\ y &= [\widetilde{h}_1(x), \ldots, \widetilde{h}_{m_w}(x)]^T, \end{cases} \tag{8.3}$$

which coincides with the original system characterized by (8.1) but with different outputs. In particular, since the new outputs belong to $\mathcal{F}$ and since the functions in $\mathcal{F}$ are observable functions for the system characterized by (8.1) (see Remark 6.14 in section 6.3), the observable codistribution of the system defined by (8.3) is certainly included in the observable codistribution of the original system. We consider all the functions

$$\widetilde{h}_1(x), \ldots, \widetilde{h}_{m_w}(x), h_1(x), \ldots, h_p(x).$$

Let us denote by s the dimension of the codistribution generated by their gradients. Since $\widetilde{h}_1(x), \ldots, \widetilde{h}_{m_w}(x)$ are independent, $s \geq m_w$ and we set $s = m_w + s'$. By rearranging the original outputs, we set the functions

$$\widetilde{h}_1(x), \ldots, \widetilde{h}_{m_w}(x), h_1(x), \ldots, h_{s'}(x)$$

as the generators of the above codistribution.

8.1. Extension of the observability rank condition for the general case

From now on, we refer to the following input-output system, which has the same observability properties of the original system:

$$\begin{cases} \dot{x} = g^0(x) + \sum_{i=1}^{m_u} f^i(x)u_i + \sum_{j=1}^{m_w} g^j(x)w_j, \\ y = [\widetilde{h}_1(x),\ldots,\widetilde{h}_{m_w}(x), h_1(x),\ldots, h_{s'}(x)]^T. \end{cases} \quad (8.4)$$

To obtain $\mathcal{SUIO}$ for the above system we work in the chronospace by only using synchronous frames, i.e., with $x^0 = t$ and with the structure of the dynamics given by (8.7) below. The chronostate is

$$\underline{x} \triangleq \begin{bmatrix} x^0 \\ x^1 \\ \ldots \\ x^n \end{bmatrix}. \quad (8.5)$$

We have

$$d\underline{x} = \underline{g}^0(\underline{x})dt + \sum_{i=1}^{m_u} \underline{f}^i(\underline{x})dU_i + \sum_{j=1}^{m_w} \underline{g}^j(\underline{x})dW_j, \quad (8.6)$$

with

$$dU_i = u_i dt, \quad dW_j = w_j dt, \quad \underline{g}^0(\underline{x}) \triangleq \begin{bmatrix} 1 \\ g^0 \end{bmatrix}, \quad \underline{f}^i(\underline{x}) \triangleq \begin{bmatrix} 0 \\ f^i \end{bmatrix}, \quad \underline{g}^j(\underline{x}) \triangleq \begin{bmatrix} 0 \\ g^j \end{bmatrix}. \quad (8.7)$$

Regarding the outputs, we remark that we need to include a new output (see Chapter 3):

$$\widetilde{h}_0(\underline{x}) = x^0 = t.$$

Hence, a full description of our system is given by the following equations:

$$\begin{cases} d\underline{x} = \underline{g}^0(\underline{x})dt + \sum_{i=1}^{m_u} \underline{f}^i(\underline{x})dU_i + \sum_{j=1}^{m_w} \underline{g}^j(\underline{x})dW_j, \\ y = [\widetilde{h}_0(\underline{x}), \widetilde{h}_1(\underline{x}),\ldots,\widetilde{h}_{m_w}(\underline{x}), h_1(\underline{x}),\ldots, h_{s'}(\underline{x})]^T. \end{cases} \quad (8.8)$$

The $\mathcal{SUIO}$ is defined by the following three equations:

$$\begin{cases} \widetilde{h}_\alpha \to \widetilde{h}'_\alpha = S_\alpha^\beta \widetilde{h}_\beta, \\ dW_\alpha \to dW'_\alpha = S_\alpha^\beta dW_\beta, \\ \underline{g}^\alpha \to \underline{g}'^\alpha = (S^{-1})^\alpha_\beta \underline{g}^\beta, \end{cases} \quad (8.9)$$

where S is a nonsingular $(m_w + 1) \times (m_w + 1)$ matrix and all the Greek indices can take the values $0, 1, \ldots, m_w$. We remind the reader that Latin indices take the values $1, \ldots, m_w$. The index β is a dummy index in all three above equations. In accordance with the Einstein notation, this implies the sum $\sum_{\beta=0}^{m_w}$. Note that the last s' outputs (i.e., $h_1(\underline{x}),\ldots, h_{s'}(\underline{x})$) are not involved in the $\mathcal{SUIO}$.

Since we want to work only with synchronous frames, we need to impose special structure to the above matrix S, in order to maintain the same structure of the vector fields that appear in (8.7). Specifically, we need to set

$$S_0^0 = 1, \quad S_0^i = S_i^0 = 0, \quad i = 1,\ldots, m_w. \quad (8.10)$$

Note that limiting the analysis to synchronous frames is not a necessary condition but allows us to obtain more comprehensible results.

We go back to the matrix defined in (8.2). We want to build a two-index tensor of type (1, 1) with respect to the $\mathcal{SUIO}$, starting from its entries (see sections 2.2.4 and 2.5). We set

$$\mu_\beta^\alpha \triangleq \mathcal{L}_{\underline{g}^\alpha} \widetilde{h}_\beta, \quad \alpha,\ \beta = 0, 1, \ldots, m_w. \tag{8.11}$$

Note that the Lie derivative in the chronospace is defined by computing the gradient with respect to the chronostate. In other words,

$$\mu_\beta^\alpha = \mathcal{L}_{\underline{g}^\alpha} \widetilde{h}_\beta = \left[\frac{\partial}{\partial t}\widetilde{h}_\beta,\ \nabla \widetilde{h}_\beta\right] \cdot \underline{g}^\alpha.$$

Specifically, we have for $\alpha = j = 1, \ldots, m_w$

$$\mathcal{L}_{\underline{g}^j} \widetilde{h}_\beta = \left[\frac{\partial}{\partial t}\widetilde{h}_\beta,\ \nabla \widetilde{h}_\beta\right] \cdot \begin{bmatrix} 0 \\ g^j \end{bmatrix} = \mathcal{L}_{g^j} \widetilde{h}_\beta,$$

where the last Lie derivative is the usual Lie derivative; i.e., it is obtained by computing the gradient in the space of the states. In particular, we obtain for $\beta = i = 1, \ldots, m_w$

$$\mu_i^j = \mathcal{L}_{g^j} \widetilde{h}_i; \tag{8.12}$$

namely, they are the entries of the matrix in (8.2). In addition, for $\beta = 0$, since $\widetilde{h}_0 = t$, we have

$$\mu_0^j = 0, \quad j = 1, \ldots, m_w. \tag{8.13}$$

For $\alpha = 0$ we obtain

$$\mathcal{L}_{\underline{g}^0} \widetilde{h}_\beta = \left[\frac{\partial}{\partial t}\widetilde{h}_\beta,\ \nabla \widetilde{h}_\beta\right] \cdot \begin{bmatrix} 1 \\ g^0 \end{bmatrix} = \frac{\partial}{\partial t}\widetilde{h}_\beta + \mathcal{L}_{g^0} \widetilde{h}_\beta$$

and

$$\mu_0^0 = 1, \quad \mu_i^0 = \frac{\partial \widetilde{h}_i}{\partial t} + \mathcal{L}_{g^0} \widetilde{h}_i, \quad i = 1, \ldots, m_w. \tag{8.14}$$

Note that the expressions in (8.13) and (8.14) only hold for synchronous frames. In addition, if the output functions are independent of time (as in many cases), we have

$$\mu_i^0 = \mathcal{L}_{g^0} \widetilde{h}_i, \quad i = 1, \ldots, m_w.$$

It is immediate to verify that μ is a tensor of type (1, 1) with respect to the transformation defined in (8.9), i.e.,

$$\mu_\beta^\alpha \rightarrow \mu_\beta'^\alpha = (S^{-1})_\gamma^\alpha S_\beta^\eta \mu_\eta^\gamma,$$

where the two indices γ and η are dummy indices and are summed over $0, 1, \ldots, m_w$. The above equation holds in any setting (and not only for synchronous frames).

We denote by ν the inverse of μ. In other words, we have

$$\mu_\gamma^\alpha \nu_\beta^\gamma = \delta_\beta^\alpha, \quad \alpha,\ \beta = 0, 1, \ldots, m_w, \tag{8.15}$$

where, in accordance with the Einstein notation, the dummy Greek index γ is summed over $\gamma = 0, 1, \ldots, m_w$. For synchronous frames, where (8.13) and (8.14) hold, we have

$$\nu_0^0 = 1, \quad \nu_0^i = 0, \quad \nu_i^0 = -\mu_k^0 \nu_i^k, \quad \nu_k^i \mu_j^k = \mu_k^i \nu_j^k = \delta_j^i. \tag{8.16}$$

8.1. Extension of the observability rank condition for the general case

Note that Latin indices take the values $1, \ldots, m_w$. When they are dummy, they imply a sum over $1, \ldots, m_w$ (this is the case of k in the above equation).

We introduce the following $m_w + 1$ vector fields $\widehat{g}^\alpha$, $\alpha = 0, 1, \ldots, m_w$:

$$\widehat{g}^\alpha \triangleq \nu_\beta^\alpha \underline{g}^\beta, \tag{8.17}$$

where, in accordance with the Einstein notation, the dummy Greek index β is summed over $\beta = 0, 1, \ldots, m_w$. Note that $\widehat{g}^\alpha$ (as $\underline{g}^\alpha$) is a vector field with respect to a coordinates' change in the chronospace and is also a vector field (or more precisely the αth component of a vector field) with respect to $\mathcal{SUIO}$. We also denote by $\widehat{g}^\alpha$ the following vector fields (in the space of the states):

$$\widehat{g}^\alpha \triangleq \nu_\beta^\alpha g^\beta. \tag{8.18}$$

By a direct computation it is immediate to obtain the following equation, which holds for synchronous frames:

$$\underline{\widehat{g}}^0 = \begin{bmatrix} 1 \\ \widehat{g}^0 \end{bmatrix}, \quad \underline{\widehat{g}}^i = \begin{bmatrix} 0 \\ \widehat{g}^i \end{bmatrix}, \quad i = 1, \ldots, m_w. \tag{8.19}$$

Finally, given the system characterized by (8.4), we introduce the following unary operation (i.e., an operation with only one operand) which will be called the *autobracket*. The operand must be a vector field with respect to a change of coordinates and a tensor of type $(0, p)$ with respect to $\mathcal{SUIO}$. The output of the operation is still a vector field with respect to a change of coordinates and a tensor of type $(0, p+1)$ with respect to $\mathcal{SUIO}$. The operation is defined both in the chronospace and in the space of states (we use the underline on the operand to distinguish the two cases).

Definition 8.1 (Autobracket). *Given the quantity $\phi^{\alpha_1, \ldots, \alpha_p}$ (or the quantity $\underline{\phi}^{\alpha_1, \ldots, \alpha_p}$), which is a vector field with respect to a change of coordinates in the space of the states (or in the chronospace) and one component of a tensor of type $(0, p)$ with respect to $\mathcal{SUIO}$, we define its autobracket with respect to the system characterized by (8.4) as the following object:*

$$[\phi^{\alpha_1, \ldots, \alpha_p}]^{\alpha_{p+1}} \triangleq \nu_\beta^{\alpha_{p+1}} [\phi^{\alpha_1, \ldots, \alpha_p}, g^\beta] - \nu_0^{\alpha_{p+1}} \frac{\partial \phi^{\alpha_1, \ldots, \alpha_p}}{\partial t} \tag{8.20}$$

or

$$[\underline{\phi}^{\alpha_1, \ldots, \alpha_p}]^{\alpha_{p+1}} \triangleq \nu_\beta^{\alpha_{p+1}} [\underline{\phi}^{\alpha_1, \ldots, \alpha_p}, \underline{g}^\beta], \tag{8.21}$$

where the square brackets on the right-hand side of both the above equations are the Lie brackets (in the second case computed in the chronospace) and the dummy Greek index β is summed over $\beta = 0, 1, \ldots, m_w$ (in accordance with the Einstein notation).

The operation defined by (8.20) only holds for synchronous frames, i.e., when the transformations of $\mathcal{SUIO}$ satisfy (8.10). Note that the validity of (8.10) guarantees that $[\phi^{\alpha_1, \ldots, \alpha_p}]^{\alpha_{p+1}}$ is a tensor of type $(0, p+1)$ with respect to $\mathcal{SUIO}$ (in particular, it guarantees that $\nu_0^{\alpha_{p+1}}$ is a tensor of type $(0, 1)$). The second operation in the chronospace, i.e., the one defined by (8.21), holds in general, i.e., also when the structure of S given in (8.10) is not honored. When $\phi^{\alpha_1, \ldots, \alpha_p}$ does not depend explicitly on time, the operation defined by (8.20) becomes

$$[\phi^{\alpha_1, \ldots, \alpha_p}]^{\alpha_{p+1}} \triangleq \nu_\beta^{\alpha_{p+1}} [\phi^{\alpha_1, \ldots, \alpha_p}, g^\beta]. \tag{8.22}$$

$\phi^{\alpha_1, \ldots, \alpha_p}$ consists of $p \times (m_w + 1)$ vector fields in the space of states that is $p \times (m_w + 1) \times n$ real values (n is the dimension of the state). Similarly, $\underline{\phi}^{\alpha_1, \ldots, \alpha_p}$ consists of $p \times (m_w + 1)$ vector fields in the chronospace that is $p \times (m_w + 1) \times (n + 1)$ real values.

Let us consider the case (very common when we work in a synchronous frame) when the operand has the following structure:

$$\underline{\phi}^{\alpha_1,\ldots,\alpha_p} = \begin{bmatrix} 0 \\ \phi^{\alpha_1,\ldots,\alpha_p} \end{bmatrix}. \tag{8.23}$$

We have

$$[\underline{\phi}^{\alpha_1,\ldots,\alpha_p}, \underline{g}^\beta] = \frac{\partial \underline{g}^\beta}{\partial \underline{x}} \underline{\phi}^{\alpha_1,\ldots,\alpha_p} - \frac{\partial \underline{\phi}^{\alpha_1,\ldots,\alpha_p}}{\partial \underline{x}} \underline{g}^\beta$$

$$= \begin{bmatrix} 0 & 0 & \cdots & 0 \\ \frac{\partial g^\beta}{\partial t} & \frac{\partial g^\beta}{\partial x} & & \end{bmatrix} \begin{bmatrix} 0 \\ \phi^{\alpha_1,\ldots,\alpha_p} \end{bmatrix} - \begin{bmatrix} 0 & 0 & \cdots & 0 \\ \frac{\partial \phi^{\alpha_1,\ldots,\alpha_p}}{\partial t} & \frac{\partial \phi^{\alpha_1,\ldots,\alpha_p}}{\partial x} & & \end{bmatrix} \underline{g}^\beta.$$

We distinguish the case $\beta = 0$ from $\beta = j = 1,\ldots,m_w$. By using (8.7) we obtain

$$[\underline{\phi}^{\alpha_1,\ldots,\alpha_p}, \underline{g}^0] = \begin{bmatrix} 0 \\ [\phi^{\alpha_1,\ldots,\alpha_p}, g^0] - \frac{\partial \phi^{\alpha_1,\ldots,\alpha_p}}{\partial t} \end{bmatrix}$$

and

$$[\underline{\phi}^{\alpha_1,\ldots,\alpha_p}, \underline{g}^j] = \begin{bmatrix} 0 \\ [\phi^{\alpha_1,\ldots,\alpha_p}, g^j] \end{bmatrix}.$$

Hence, when (8.23) holds, we have

$$[\underline{\phi}^{\alpha_1,\ldots,\alpha_p}]^{\alpha_{p+1}} = \begin{bmatrix} 0 \\ [\phi^{\alpha_1,\ldots,\alpha_p}]^{\alpha_{p+1}} \end{bmatrix}. \tag{8.24}$$

The autobracket and the tensors μ and ν for the simplified system

To illustrate the previous objects, we compute them in the case of the simplified system, i.e., the system characterized by (7.1) (which is obtained by requiring that $\mathcal{SUIO}$ is an Abelian group). We also assume that the outputs and the vectors f^i and g are independent of time. The tensor μ has the following four components:

$$\mu_0^0 = 1, \quad \mu_0^1 = \mu_1^0 = 0, \quad \mu_1^1 = L_g^1.$$

Note that only the last one is nontrivial and was one of the key quantities that characterizes the solution (Chapter 7). Its inverse becomes

$$\nu_0^0 = 1, \quad \nu_0^1 = \nu_1^0 = 0, \quad \nu_1^1 = \frac{1}{L_g^1}.$$

In addition, (8.18) defines the single vector field

$$\widehat{g}^1 \triangleq \frac{g^1}{L_g^1},$$

which is the vector field that appears in Algorithm 7.3 (where we denoted $g \triangleq g^1$).

Finally, the autobracket (when the operand does not depend explicitly on time) reduces to $\frac{[\phi, g]}{L_g^1}$, i.e., to the Lie bracket of ϕ along g divided by L_g^1 (when $\mathcal{SUIO}$ is Abelian, we do not

8.1. Extension of the observability rank condition for the general case

have indices with respect to $\mathcal{SUIO}$). We remark that we obtain the same operation that appears in Algorithm 7.1.

Let us go back to the general case. Thanks to the autobracket, we can express in compact form the general algorithm that provides the entire observable codistribution of the system characterized by (8.4) (or (8.1)). We start by providing the analogue of Algorithm 7.1. In this case, the new algorithm generates, at the kth step, a quantity which is a vector field with respect to a coordinates' change and a $(0, k)$ tensor with respect to $\mathcal{SUIO}$. We can define the algorithm both in the space of the states and in the chronospace. In the former we have what follows.

ALGORITHM 8.1. Vectors ${}^i\phi_k^{\alpha_1,\ldots,\alpha_k}$ **in the general case with unknown inputs.**

Set $m = 0$
Set ${}^i\phi_m = f^i$ for $i = 1, \ldots, m_u$
while $m < k$ **do**
    Set $m = m + 1$
    Set ${}^i\phi_m^{\alpha_1,\ldots,\alpha_m} = [{}^i\phi_{m-1}^{\alpha_1,\ldots,\alpha_{m-1}}]^{\alpha_m}$ for $i = 1, \ldots, m_u$
end while

In the latter we only need to change the second line with

$${}^i\underline{\phi}_0 = \underline{f}^i.$$

From (8.24) we immediately obtain

$${}^i\underline{\phi}_k^{\alpha_1,\ldots,\alpha_k} = \begin{bmatrix} 0 \\ {}^i\phi_k^{\alpha_1,\ldots,\alpha_k} \end{bmatrix} \quad \forall k. \tag{8.25}$$

The above algorithm coincides with Algorithm 7.1 for the simplified system.

We have now all the ingredients to provide the entire observable codistribution. It is obtained by running the following algorithm for a given value of the integer k (that will be provided in the following).

ALGORITHM 8.2. Observable codistribution in the general case with unknown inputs.

Set $m = 0$
Set $\Omega_m = \text{span}\{\nabla h_1, \ldots, \nabla h_p\}$
while $m < k$ **do**
    Set $m = m + 1$
    Set $\Omega_m = \Omega_{m-1} \oplus \bigoplus_{i=1}^{m_u} \mathcal{L}_{f^i}\Omega_{m-1} \oplus \bigoplus_{\alpha=0}^{m_w} \mathcal{L}_{\widehat{g}^\alpha}\Omega_{m-1} \oplus$
    $\bigoplus_{i=1}^{m_u} \bigoplus_{l=1}^{m_w} \bigoplus_{\alpha_1=0}^{m_w} \cdots \bigoplus_{\alpha_{m-1}=0}^{m_w} \text{span}\left\{\mathcal{L}_{{}^i\phi_{m-1}^{\alpha_1,\ldots,\alpha_{m-1}}} \nabla \widetilde{h}_l\right\}$
end while

In section 8.3.2 we investigate the convergence properties of Algorithm 8.2. We consider first the case of a single known input (i.e., $m_u = 1$), and then the results are easily extended to the case of multiple known inputs ($m_u > 1$) in section 8.3.3. We prove that Algorithm 8.2 converges, and we also provide the analytic criterion to check that the convergence has been attained. As for Algorithm 7.3, this proof and the convergence criterion cannot be the same that hold for Algorithm 4.2, because of the last term that appears in the recursive step,[33] i.e., the term $\bigoplus_{i=1}^{m_u} \text{span}\left\{\mathcal{L}_{i\phi_{m-1}^{\alpha_1,\ldots,\alpha_{m-1}}} \nabla h_l\right\}$ (the special case when the contribution due to this last term is included in the other terms is considered separately by Lemma 8.11). In general, the criterion to establish that the convergence has been attained is not simply obtained by checking whether $\Omega_{m+1} = \Omega_m$. Deriving the new analytic criterion is a very laborious and demanding task. It requires us to derive the analytic expression that describes the behavior of the last term in the recursive step. This fundamental equation is provided in section 8.3, and it is (8.56). The analytic derivation of this equation allows us to detect the key quantity that governs the convergence of Algorithm 8.2, in particular regarding the contribution due to the last term in the recursive step. Its derivation was very troublesome since we did not a priori know that, instead of a scalar as for Algorithm 7.3, the key quantity becomes a three-index tensor with respect to $\mathcal{SUIO}$. Specifically, it is the following three-index tensor of type $(1, 2)$:

$$\mathcal{T}_\gamma^{\alpha\beta} \triangleq \nu_\eta^\beta(\mathcal{L}_{\hat{g}^\alpha} \mu_\gamma^\eta), \quad \alpha, \beta, \gamma = 0, 1, \ldots, m_w, \tag{8.26}$$

where, in accordance with the Einstein notation, the dummy Greek index η is summed over $\eta = 0, 1, \ldots, m_w$.

Note that, for the simplified system characterized by (7.1), this tensor has the single component $\mathcal{T}_1^{11}$, which coincides with the quantity τ defined in (7.3). In general, this tensor has $(m_w + 1) \times (m_w + 1) \times (m_w + 1)$ components. On the other hand, for synchronous frames it is immediate to obtain that $\mathcal{T}_0^{\alpha,\beta} = 0$ $\alpha, \beta = 0, 1, \ldots, m_w$. In other words, for these frames, we can consider the lower index as a Latin index and the components of this tensor are $(m_w + 1) \times (m_w + 1) \times m_w$. We prove (see Lemma 8.16 in section 8.3) that, in general, there exists m' such that the gradients of all these components belong to $\Omega_{m'}$ (and therefore belong to Ω_m $\forall m \geq m'$). Additionally, we prove that the convergence of the algorithm has been reached when $\Omega_{m+1} = \Omega_m$, $m \geq m'$, and $m \geq 2$. We also prove that the required number of steps is at most $n + 2$.

In section 8.3.1 it is also shown that the computed codistribution is the entire observable codistribution. Also in this case, the proof is given by first considering the case of a single known input, and then, its validity is extended to the case of multiple inputs in section 8.3.3. Note that this proof is based on the assumption that the unknown inputs are differentiable functions of time up to a given order (the order depends on the specific case).

Below we provide Algorithm 8.3 that fuses Algorithm 8.2 with Algorithm 8.1. It also includes the convergence criterion.

ALGORITHM 8.3. Fusion of Algorithm 8.1 and Algorithm 8.2.

Set $m = 0$
Set $\Omega_m = \text{span}\{\nabla h_1, \ldots, \nabla h_p\}$
Set $^i\phi_m = f^i$ for $i = 1, \ldots, m_u$
Set $m = m + 1$
Set $\Omega_m = \Omega_{m-1} \oplus \bigoplus_{i=1}^{m_u} \mathcal{L}_{f^i}\Omega_{m-1} \oplus \bigoplus_{\alpha=0}^{m_w} \mathcal{L}_{\hat{g}^\alpha}\Omega_{m-1}$
Set $^i\phi_m^{\alpha_1} = [^i\phi_{m-1}]^{\alpha_1}$ for $i = 1, \ldots, m_u$
Set $m = m + 1$

---

[33] Again, we remind the reader that the convergence criterion of Algorithm 4.2 is a consequence of the fact that all the terms that appear in the recursive step of Algorithm 4.2 are the Lie derivative of the codistribution at the previous step, along fixed vector fields (i.e., vector fields that remain the same at each step of the algorithm). This is not the case for the last term in the recursive step of Algorithm 8.2.

8.1. Extension of the observability rank condition for the general case

Set $\Omega_m = \Omega_{m-1} \oplus \bigoplus_{i=1}^{m_u} \mathcal{L}_{f^i}\Omega_{m-1} \oplus \bigoplus_{\alpha=0}^{m_w} \mathcal{L}_{\hat{g}^\alpha}\Omega_{m-1} \oplus$
$\bigoplus_{i=1}^{m_u} \bigoplus_{l=1}^{m_w} \bigoplus_{\alpha_1=0}^{m_w} \text{span}\left\{\mathcal{L}_{i\phi_{m-1}^{\alpha_1}} \nabla \widetilde{h}_l\right\}$
Set ${}^i\phi_m^{\alpha_1,\alpha_2} = [{}^i\phi_{m-1}^{\alpha_1}]^{\alpha_2}$ for $i = 1,\ldots,m_u$
while $\dim(\Omega_m) > \dim(\Omega_{m-1})$ OR $\nabla \mathcal{T} \not\in \Omega_m$ **do**
$\quad$ Set $m = m+1$
$\quad$ Set $\Omega_m = \Omega_{m-1} \oplus \bigoplus_{i=1}^{m_u} \mathcal{L}_{f^i}\Omega_{m-1} \oplus \bigoplus_{\alpha=0}^{m_w} \mathcal{L}_{\hat{g}^\alpha}\Omega_{m-1} \oplus$
$\quad \bigoplus_{i=1}^{m_u} \bigoplus_{l=1}^{m_w} \bigoplus_{\alpha_1=0}^{m_w} \cdots \bigoplus_{\alpha_{m-1}=0}^{m_w} \text{span}\left\{\mathcal{L}_{i\phi_{m-1}^{\alpha_1,\ldots,\alpha_{m-1}}} \nabla \widetilde{h}_l\right\}$
$\quad$ Set ${}^i\phi_m^{\alpha_1,\ldots,\alpha_m} = [{}^i\phi_{m-1}^{\alpha_1,\ldots,\alpha_{m-1}}]^{\alpha_m}$ for $i = 1,\ldots,m_u$
end while
Set $\Omega = \Omega_m$ and $s = \dim(\Omega)$

where with $\nabla \mathcal{T} \not\in \Omega_m$ we mean that at least one component $\nabla \mathcal{T}_\gamma^{\alpha\beta} \not\in \Omega_m$ for given α, β, γ among $0, 1, \ldots, m_w$. Note that, for synchronous frames, we can restrict ourselves to the values $\mathcal{T}_k^{\alpha\beta}$, $k = 1, \ldots, m_w$.

8.1.2 ▪ The analytic procedure

In this section, we outline all the steps to investigate the weak observability at a given point x_0 of a nonlinear system characterized by (8.1). Basically, these steps are the steps necessary to compute the observable codistribution (i.e., the steps of Algorithm 8.3) and to prove that the gradient of a given state component belongs to this codistribution.

Note that, in the trivial case analyzed by Lemma 8.11, the method provided below simplifies, since we do not need to compute the tensor $\mathcal{T}$ defined in (8.26), and we do not need to check that the gradients of its components belong to the codistribution computed at every step of Algorithm 8.3. In practice, we skip steps 5 and 6 in the procedure below.

1. For the chosen x_0, select m_w scalar functions $(\widetilde{h}_1, \widetilde{h}_2, \ldots, \widetilde{h}_{m_w})$ from the functional space $\mathcal{F}$ (see Definition 6.13) such that the matrix in (8.2) is nonsingular.[34]

2. Compute the two-index tensor μ (whose Latin entries are those of the matrix in (8.2), and the remaining components are defined in (8.13) and (8.14)) and its inverse ν. Then, compute the $m_w + 1$ vector fields: $\hat{g}^\alpha$ defined in (8.18) ($\alpha = 0, 1, \ldots, m_w$).

3. Compute the codistribution Ω_0 and Ω_1 (at x_0) by using Algorithm 8.3.

4. Compute the vector fields ${}^i\phi_m^{\alpha_1,\ldots,\alpha_m}$ ($i = 1, \ldots, m_u$ $\forall \alpha_1, \ldots, \alpha_m$) by using Algorithm 8.1, starting from $m = 0$, to check whether the considered system is in the special case contemplated by Lemma 8.11. In this trivial case, set $m' = 0$, use the recursive step of Algorithm 8.3 to build the codistribution Ω_m for $m \geq 2$, and skip to step 7.

5. Compute the three-index tensor $\mathcal{T}$ defined in (8.26) and the gradients of all its components.

6. Use the recursive step of Algorithm 8.3 to build the codistribution Ω_m for $m \geq 2$, and, for each m, check whether $\nabla \mathcal{T}_k^{\alpha,\beta} \in \Omega_m$, $\alpha, \beta = 0, 1, \ldots, m_w$ and $k = 1, \ldots, m_w$. Denote by m' the smallest m such that $\nabla \mathcal{T}_k^{\alpha,\beta} \in \Omega_m$, $\alpha, \beta = 0, 1, \ldots, m_w$ and $k = 1, \ldots, m_w$.

---

[34] If this is not possible, apply the procedure provided in Appendix C in order to set the system in canonic form. In the case where the system cannot be set in canonic form, it means that some of the unknown inputs can be ignored to obtain the observability properties (note that the procedure detects these unknown inputs automatically).

7. For each $m \geq m'$ check whether $\Omega_{m+1} = \Omega_m$ and denote $\Omega = \Omega_m$, where m is the smallest integer such that $m \geq m'$ and $\Omega_{m+1} = \Omega_m$ (note that $m \leq n+2$).

8. If the gradient of a given state component (x_j, $j = 1, \ldots, n$) belongs to Ω (namely if $\nabla x^j \in \Omega$) on a given neighborhood of x_0, then x^j is weakly observable at x_0. If this holds for all the state components, the state x is weakly observable at x_0. Finally, if the dimension of Ω is smaller than n on a given neighborhood of x_0, then the state is not weakly observable at x_0.

8.2 • Applications

In this section we use the method presented in the previous section to study the observability properties of systems that are characterized by the equations in (8.1). Specifically, we go back to the problem of visual inertial sensor fusion, whose observability properties were already derived in Chapter 5, in several conditions (calibrated and uncalibrated sensors, single and multiple agents) but in absence of unknown inputs. In this section, we study the very challenging case when some of the three degrees of freedom of the acceleration and some of the three degrees of freedom of the angular speed are not measured by the inertial sensors. In other words, some of the components of the acceleration and/or of the angular speed are unknown and act on the dynamics of the system as unknown inputs.

We study separately the case of calibrated and uncalibrated sensors. In addition, we study both the planar case (sections 8.2.1 and 8.2.2) and the 3D case (sections 8.2.3 and 8.2.4). The last case (i.e., 3D with uncalibrated sensors) is very important not only in technological sciences but also in neuroscience. As it will be seen, our analysis will allow us to obtain interesting results about the problem of visual-vestibular integration for self-motion perception in mammals.

Note that the method is very powerful and can be used to automatically obtain the observability properties of any system that satisfies (8.1), i.e., independently of the state dimension (intuitive reasoning becomes often prohibitive for systems characterized by high-dimensional states), independently of the type of nonlinearity, and in general independently of the system complexity. In this regard, note that the method can be implemented automatically by using a simple symbolic computation tool (in our examples, we simply used the symbolic computation toolbox of MATLAB).

8.2.1 • Visual-inertial sensor fusion: The planar case with calibrated sensors

We consider a rigid body ($\mathcal{B}$) equipped with inertial sensors (IMU) and a monocular camera. The body moves on a plane. The inertial sensors measure the body acceleration and the angular speed. In 2D, the acceleration is a two-dimensional vector and the angular speed is a scalar. The camera provides the bearing angle of the features in its own local frame.

In this section, we derive the observability properties in the case when the IMU only provides the acceleration along a single axis (instead of two). In other words, the other component of the acceleration and the angular speed are unknown. In addition, we investigate the minimal case of a single point feature.

In general, the camera frame does not coincide with the IMU frame. Additionally, the measurements provided by the IMU are in general biased. In this section, we study the case when the camera frame coincides with the IMU frame. We call this frame the *body frame*. In addition, the inertial measurements are unbiased. Then, in section 8.2.2, we relax both these assumptions

8.2. Applications

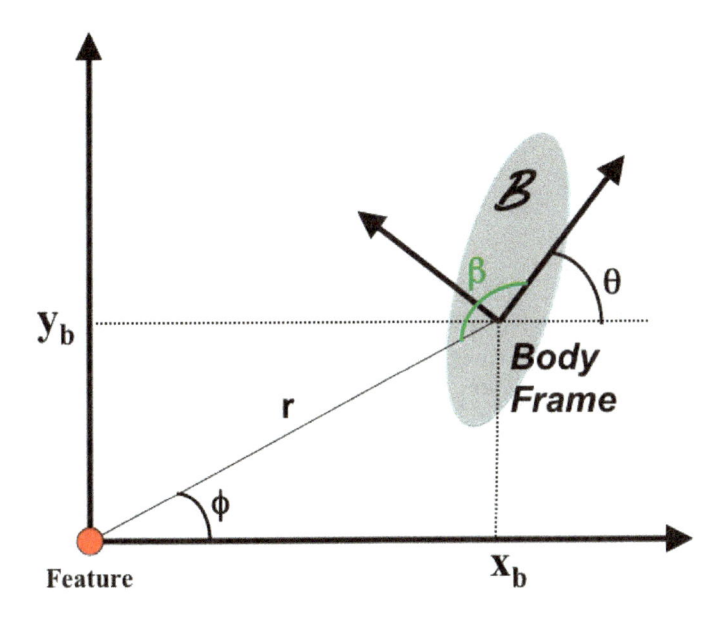

Figure 8.1. *Visual-inertial sensor fusion in 2D with calibrated sensors and in the case of a single point feature. The global frame, the body frame, and the camera observation (β).*

by also assuming that the camera is extrinsically uncalibrated (i.e., the transformation between the IMU frame and the camera frame is unknown). Figure 8.1 depicts our system.

The state that characterizes our system is

$$[x_b, y_b, v_x, v_y, \theta]^T,$$

where the first two components are the body position, the second two components the body speed, and the last component the body orientation. All these quantities are expressed in a common global frame (see Figure 8.1 for an illustration). The dynamics are

$$\begin{bmatrix} \dot{x}_b &= v_x, \\ \dot{y}_b &= v_y, \\ \dot{v}_x &= \cos\theta A_x - \sin\theta A_y, \\ \dot{v}_y &= \sin\theta A_x + \cos\theta A_y, \\ \dot{\theta} &= \omega, \end{bmatrix} \quad (8.27)$$

where $[A_x, A_y]^T$ is the body acceleration expressed in the body frame and ω the angular speed. In order to have a simpler expression for the output, it is better to work in polar coordinates. In other words, we set the following:

- $r \triangleq \sqrt{x_b^2 + y_b^2}$;
- $\phi \triangleq \arctan\left(\frac{y_b}{x_b}\right)$;
- $v \triangleq \sqrt{v_x^2 + v_y^2}$;
- $\alpha \triangleq \arctan\left(\frac{v_y}{v_x}\right)$.

Hence, we define the state
$$X = [r,\ \phi,\ v,\ \alpha,\ \theta]^T. \tag{8.28}$$

Its dynamics are
$$\begin{bmatrix} \dot{r} &= v\cos(\alpha - \phi), \\ \dot{\phi} &= \frac{v}{r}\sin(\alpha - \phi), \\ \dot{v} &= A_x \cos(\alpha - \theta) + A_y \sin(\alpha - \theta), \\ \dot{\alpha} &= -\frac{A_x}{v}\sin(\alpha - \theta) + \frac{A_y}{v}\cos(\alpha - \theta), \\ \dot{\theta} &= \omega. \end{bmatrix} \tag{8.29}$$

Without loss of generality, we assume that the feature is positioned at the origin of the global frame. The camera provides the angle $\beta = \pi - \theta + \phi$. Hence, we can perform the observability analysis by using the output (we ignore π):

$$y = h(X) = \phi - \theta. \tag{8.30}$$

We consider the system characterized by the state in (8.28), the dynamics in (8.29), and the output in (8.30) under the assumptions that the IMU only provides the acceleration along a single axis. Without loss of generality, we assume that it provides A_x. By comparing (8.29) with (8.1) we obtain $m_u = 1$, $m_w = 2$, $p = 1$, $u_1 \triangleq u = A_x$, $w_1 = \omega$, $w_2 = A_y$,

$$g^0 = \begin{bmatrix} v\cos(\alpha - \phi) \\ \frac{v}{r}\sin(\alpha - \phi) \\ 0 \\ 0 \\ 0 \end{bmatrix},\ f \triangleq f^1 = \begin{bmatrix} 0 \\ 0 \\ \cos(\alpha - \theta) \\ -\frac{1}{v}\sin(\alpha - \theta) \\ 0 \end{bmatrix},\ g^1 = \begin{bmatrix} 0 \\ 0 \\ 0 \\ 0 \\ 1 \end{bmatrix},\ g^2 = \begin{bmatrix} 0 \\ 0 \\ \sin(\alpha - \theta) \\ \frac{1}{v}\cos(\alpha - \theta) \\ 0 \end{bmatrix}.$$

Before proceeding with the steps summarized in section 8.1.2, we need to check that the system is in canonic form. If it is not, we need to perform the system canonization, as explained in Appendix C.2.

System canonization

We follow the procedure described in appendix C.2. We start by building the space of functions $\mathcal{F}^0$ and the codistribution $\mathcal{DF}^0$. Since $\mathcal{L}_f h = 0$, we obtain that $\mathcal{F}^0$ only contains the function $h(X)$ in (8.30) and $\mathcal{DF}^0 = \text{span}\{[0, 1, 0, 0, -1]\}$. Additionally, the space $\mathcal{L}_G \mathcal{F}^0$ only contains $\mathcal{L}_G h = \frac{v\sin(\alpha - \phi)}{r} - w_1$. We easily obtain that $\mathcal{D}_w \mathcal{L}_G \mathcal{F}^0 = \text{span}\{[-1, 0]\}$. Hence, its dimension is $1 < m_w = 2$ and the system is not in canonic form. Since the covectors in $\mathcal{D}_w \mathcal{L}_G \mathcal{F}^0$ have the second entry equal to zero, we do not need to change the coordinates according to (C.4). In accordance with (C.6), we include w_1 in the state:

$$X \to [X^T\ w_1]^T.$$

We proceed by computing $\mathcal{F}^1$. Since $\mathcal{L}_f^2 \mathcal{L}_G h = 0$, $\mathcal{F}^1$ only contains the functions h, $\mathcal{L}_G h$, and $\mathcal{L}_f \mathcal{L}_G h$. In particular, they are independent; i.e., the codistribution $\mathcal{DF}^1$ has dimension equal to 3. We compute the space $\mathcal{L}_G \mathcal{F}^1$. Then we define the new unknown input vector in accordance with (C.7):

$$^1w = [w_1^{(1)}\ w_2].$$

We obtain $\mathcal{D}_w \mathcal{L}_G \mathcal{F}^1$ by computing the gradients (with respect to ^1w) of the functions in $\mathcal{L}_G \mathcal{F}^1$. By a direct computation we obtain that the dimension of $\mathcal{D}_w \mathcal{L}_G \mathcal{F}^1$ is $1 < m_w = 2$ and the

8.2. Applications

system is not in canonic form. This time, the second entry of the covectors in $\mathcal{D}_w \mathcal{L}_G \mathcal{F}^1$ is not automatically 0. Hence, we need to change the coordinates in accordance with (C.4). We obtain

$$w_1^{(1)} \to w_1^{(1)} - \frac{\cos(\theta-\phi)}{r} w_2,$$
$$w_2 \to w_2.$$

In accordance with (C.6), we include $w_1^{(1)}$ in the state:

$$X \to [X^T \; w_1^{(1)}]^T.$$

We proceed by computing $\mathcal{F}^2$. In this case we obtain that the independent functions in $\mathcal{F}^2$, i.e., the ones whose gradients with respect to X generate $\mathcal{D}\mathcal{F}^2$, are h, $\mathcal{L}_G h$, $\mathcal{L}_G^2 h$, $\mathcal{L}_f \mathcal{L}_G^2 h$, and $\mathcal{L}_f^2 \mathcal{L}_G^2 h$. We compute the space $\mathcal{L}_G \mathcal{F}^2$. Then we define the new unknown input vector in accordance with (C.7):

$$^2 w = [w_1^{(2)} \; w_2].$$

We obtain $\mathcal{D}_w \mathcal{L}_G \mathcal{F}^2$ by computing the gradients (with respect to ^2w) of the functions in $\mathcal{L}_G \mathcal{F}^2$. By a direct computation we obtain that the dimension of $\mathcal{D}_w \mathcal{L}_G \mathcal{F}^2$ is $2 = m_w$ and the system is in canonic form. In particular, $\mathcal{D}\mathcal{F}^2$ is generated starting from the following two functions in $\mathcal{F}^2$: $\mathcal{L}_G^2 h$, $\mathcal{L}_f \mathcal{L}_G^2 h$.

Observability properties

We study the observability properties of our system by using the procedure provided in section 8.1.2. After the system canonization, we need to characterize our system with the following state:

$$X = [r, \; \phi, \; v, \; \alpha, \; \theta, \; \omega, \; \bar{\omega}]^T, \tag{8.31}$$

where $\bar{\omega}$ is the angular acceleration. The known input and the second unknown input remain the same ($u = A_x$, $w_2 = A_y$). The first unknown input becomes

$$w_1 = \dot{\bar{\omega}}.$$

Regarding the dynamics, the first five components of the state satisfy (8.29). The last two components satisfy the following equations: $\dot{\omega} = \bar{\omega}$ and $\dot{\bar{\omega}} = w_1$.

By comparing the new dynamics with (8.1) we obtain

$$g^0 = \begin{bmatrix} v\cos(\alpha-\phi) \\ \frac{v}{r}\sin(\alpha-\phi) \\ 0 \\ 0 \\ \omega \\ \bar{\omega} \\ 0 \end{bmatrix}, \; f \triangleq f^1 = \begin{bmatrix} 0 \\ 0 \\ \cos(\alpha-\theta) \\ -\frac{1}{v}\sin(\alpha-\theta) \\ 0 \\ 0 \\ 0 \end{bmatrix}, \; g^1 = \begin{bmatrix} 0 \\ 0 \\ 0 \\ 0 \\ 0 \\ 0 \\ 1 \end{bmatrix}, \; g^2 = \begin{bmatrix} 0 \\ 0 \\ \sin(\alpha-\theta) \\ \frac{1}{v}\cos(\alpha-\theta) \\ 0 \\ 0 \\ 0 \end{bmatrix}.$$

We are now ready to apply the method in section 8.1.2.

First step

In accordance with the results of the above system canonization, we can select $\tilde{h}_1$ and $\tilde{h}_2$ as follows:

$$\tilde{h}_1 = \mathcal{L}_G^2 h = -\bar{\omega} - \frac{v^2 \sin(2\alpha - 2\phi)}{r^2},$$
$$\tilde{h}_2 = \mathcal{L}_f \mathcal{L}_G^2 h = -2\frac{v \sin(\alpha - 2\phi + \theta)}{r^2}.$$

From (8.12) we obtain

$$\begin{aligned}
\mu_1^1 &= \mathcal{L}_{g^1}\tilde{h}_1 = -1, \\
\mu_1^2 &= \mathcal{L}_{g^2}\tilde{h}_1 = -2v\frac{\cos(\alpha-2\phi+\theta)}{r^2}, \\
\mu_2^1 &= \mathcal{L}_{g^1}\tilde{h}_2 = 0, \\
\mu_2^2 &= \mathcal{L}_{g^2}\tilde{h}_2 = \frac{2-4\cos(\phi-\theta)^2}{r^2},
\end{aligned}$$

which is nonsingular.

Second step

We compute the inverse of the previous tensor. We easily obtain

$$\begin{aligned}
\nu_1^1 &= -1, \\
\nu_1^2 &= v\frac{\cos(\alpha-2\phi+\theta)}{\cos(2\phi-2\theta)}, \\
\nu_2^1 &= 0, \\
\nu_2^2 &= -\frac{r^2}{2\cos(2\phi-2\theta)}.
\end{aligned}$$

Additionally, we obtain from (8.14)

$$\begin{aligned}
\mu_1^0 &= \mathcal{L}_{g^0}\tilde{h}_1 = \frac{2v^3 \sin(3\alpha-3\phi)}{r^3}, \\
\mu_2^0 &= \mathcal{L}_{g^0}\tilde{h}_2 = 2v\frac{2v\sin(2\alpha-3\phi+\theta)-r\omega\cos(\alpha-2\phi+\theta)}{r^3}.
\end{aligned}$$

Finally, from (8.18) we obtain

$$\hat{g}^1 = \begin{bmatrix} 0 \\ 0 \\ 0 \\ 0 \\ 0 \\ 0 \\ -1 \end{bmatrix}, \quad \hat{g}^2 = \frac{1}{2v\cos(2\phi-2\theta)} \begin{bmatrix} 0 \\ 0 \\ -vr^2\sin(\alpha-\theta) \\ -r^2\cos(\alpha-\theta) \\ 0 \\ -vr\cos(\theta-\phi) \\ 2v^2\cos(\alpha-2\phi+\theta) \end{bmatrix}.$$

We do not provide the expression of $\hat{g}^0$ for the sake of brevity (its expression is more complex).

Third step

We compute the gradient of $\tilde{h}_1$ and $\tilde{h}_2$ (obtained at the first step) with respect to the state in (8.31). By a direct computation we obtain that the dimension of Ω_0 is 2. We compute Ω_1 by using the recursive step of Algorithm 8.3 (we can ignore its last two terms, since the outputs are among the functions $\tilde{h}_1$ and $\tilde{h}_2$). The dimension of Ω_1 is 3.

Fourth step

We have $\mathcal{L}_{\phi_0}\mathcal{L}_{g^2}\tilde{h}_1 = \mathcal{L}_f\mu_1^2 = \frac{4\sin^2(\phi-\theta)-2}{r^2}$, which does not vanish, in general. This suffices to conclude that the considered system is not in the special case examined by Lemma 8.11, and we need to continue with step 5.

8.2. Applications

Fifth step

We compute the three-index tensor in (8.26). We remind the reader that we can consider the lower index as a Latin index since the components of the tensor when this index is zero vanish (this because we are working with a synchronous frame). We obtain $\mathcal{T}_1^{01} = \mathcal{T}_2^{01} = \mathcal{T}_1^{10} = \mathcal{T}_2^{10} = \mathcal{T}_1^{11} = \mathcal{T}_2^{11} = \mathcal{T}_1^{12} = \mathcal{T}_2^{12} = \mathcal{T}_1^{21} = \mathcal{T}_2^{21} = \mathcal{T}_2^{22} = 0$,

$$\mathcal{T}_2^{02} = \frac{2v\sin(2\alpha - 5\phi + 3\theta) + 2v\sin(\phi - \theta) - r^3\omega\sin(2\theta - 2\phi) + r^3\omega\sin(2\phi - 2\theta)}{r^3(1 - 2\sin(\phi - \theta)^2)},$$

$$\mathcal{T}_1^{20} = \frac{v\sin(2\phi - 2\theta)(2v\sin(2\alpha - 3\phi + \theta) - r\omega\cos(\alpha - 2\phi + \theta))}{r\cos(2\phi - 2\theta)^2}$$

$$- 3\frac{v^2\cos(3\phi - 2\alpha - \theta) + v^2\cos(2\alpha - 3\phi + \theta)}{2r\cos(2\phi - 2\theta)},$$

$$\mathcal{T}_2^{20} = \frac{2r\omega\sin(2\phi - 2\theta) - 3v\cos(\alpha - 3\phi + 2\theta) + v\cos(\phi - \alpha) - 4v\cos(3\phi - \alpha - 2\theta)}{2r\cos(2\phi - 2\theta)},$$

$\mathcal{T}_1^{22} = \frac{r^2 \sin(2\phi - 2\theta)}{2\sin(2\phi - 2\theta)^2 - 2}$. We do not provide the expression of $\mathcal{T}_1^{00}$, $\mathcal{T}_2^{00}$, and $\mathcal{T}_1^{02}$ for the sake of brevity (their expression is more complex).

Sixth step

We need to compute Ω_2 and, in order to do this, we need to compute ϕ_1^0, ϕ_1^1, and ϕ_1^2 by using Algorithm 8.1. We obtain that only the first one differs from the null vector:

$$\phi_1^0 = \begin{bmatrix} -\cos(\theta - \phi) \\ -\frac{\sin(\theta - \phi)}{r} \\ -\omega\sin(\theta - \alpha) \\ \frac{\omega\cos(\theta - \alpha)}{v} \\ 0 \\ 0 \\ 0 \end{bmatrix}.$$

Hence, by using Algorithm 8.3 we can compute Ω_2. Its dimension is 6. In addition, we obtain that the gradients of all the components of the tensor $\mathcal{T}_j^{\alpha i}$ belong to Ω_2, meaning that $m' = 2$.

Seventh step

By an explicit computation (by using the subsequent steps of Algorithm 8.3), we obtain $\Omega_3 = \Omega_2$. Hence, Algorithm 8.3 has converged and $\Omega = \Omega_2$.

Eighth step

By computing the distribution orthogonal to the codistribution Ω_2 we can find the continuous symmetry that characterizes the unobservable space. By an explicit computation we obtain the vector

$$[0, 1, 0, 1, 1, 0, 0]^T.$$

This symmetry corresponds to an invariance with respect to a rotation around the vertical axis.[35] Note that the absolute scale is invariant with respect to a rotation around the vertical axis, meaning that it can be estimated. This also holds for the body speed in the local frame.

We remark that the invariance that corresponds to the continuous symmetry detected by our observability analysis also exists by having a complete IMU, i.e., when the IMU provides the acceleration along the two axes and the angular speed. This means that the information obtained by fusing the measurements from a camera with the measurements from a complete IMU is redundant.

8.2.2 ▪ Visual-inertial sensor fusion: The planar case with uncalibrated sensors

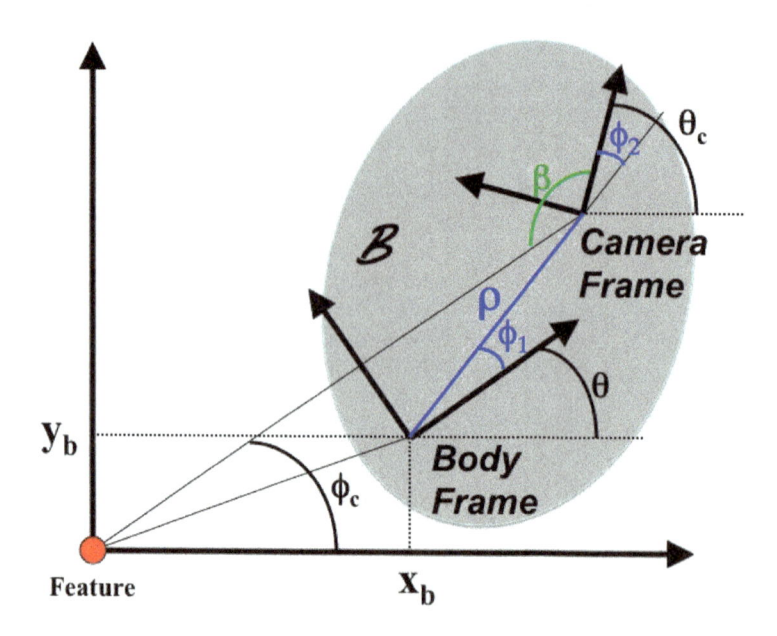

Figure 8.2. *Visual-inertial sensor fusion in 2D with uncalibrated sensors and in the case of a single point feature. The global frame, the body frame, and the camera frame are displayed together with the parameters that characterize the camera extrinsic calibration.*

We consider now the case when the measurements from the accelerometer are biased and the camera is extrinsically uncalibrated, i.e., the camera position and orientation in the IMU frame are unknown. Without loss of generality, we assume that the body frame coincides with the IMU frame. We characterize the camera configuration in the body frame with the three parameters ρ, ϕ_1, and ϕ_2 (see Figure 8.2 for an illustration). As in the previous case, we are assuming that the IMU only provides the acceleration along the x-axis of the local frame (A_x). Hence, we

---

[35] If instead of working in polar coordinates we adopted Cartesian coordinates, we would obtain the following symmetry: $[-y_b, x_b, -v_y, v_x, 1, 0, 0]^T$.

8.2. Applications

consider again the angular speed (ω) and the other component of the acceleration (A_y) as two independent unknown inputs. We characterize our system by the following state (in this case polar coordinates are useless):

$$X = [x_b,\ y_b,\ v_x,\ v_y,\ \theta,\ B,\ \rho,\ \phi_1,\ \phi_2]^T, \tag{8.32}$$

where B is the accelerometer bias. The dynamics of the state are

$$\begin{bmatrix}
\dot{x}_b &= v_x, \\
\dot{y}_b &= v_y, \\
\dot{v}_x &= \cos\theta(A_x + B) - \sin\theta A_y, \\
\dot{v}_y &= \sin\theta(A_x + B) + \cos\theta A_y, \\
\dot{\theta} &= \omega, \\
\dot{B} &= 0, \\
\dot{\rho} &= 0, \\
\dot{\phi}_1 &= 0, \\
\dot{\phi}_2 &= 0.
\end{bmatrix} \tag{8.33}$$

The analytic expression of the output is obtained starting from the expression $\beta = \pi + \phi_c - \theta_c$, where ϕ_c and θ_c are the bearing of the camera and its orientation in the global frame (see Figure 8.2). We compute the tangent of this expression, obtaining

$$y \triangleq \tan\beta = \frac{y_c \cos\theta_c - x_c \sin\theta_c}{x_c \cos\theta_c + y_c \sin\theta_c}, \tag{8.34}$$

where $(x_c,\ y_c)$ is the camera position in the global frame. We can express the previous output in terms of the state components by using the following equations: $x_c = x_b + \rho\cos(\theta + \phi_1)$, $y_c = y_b + \rho\sin(\theta + \phi_1)$, and $\theta_c = \theta + \phi_1 + \phi_2$.

System canonization

We need to check whether the system is in canonic form. If it is not, we need to proceed as explained in Appendix C.2. This is exactly what we have done in the previous case. In particular, as in the case of calibrated sensors, we need to compute the space of functions $\mathcal{F}^2$ and we can check that the dimension of the codistribution $\mathcal{D}_w\mathcal{L}_G\mathcal{F}^2$ is 2. On the other hand, the functions $\widetilde{h}_1$ and $\widetilde{h}_2$ that we automatically select from $\mathcal{F}^2$ have now an analytic expression much more complex than in the case of calibrated sensors. Even if this fact does not prevent the application of the method in section 8.1.2, for educational purposes, we prefer to proceed in a different manner. In particular, we apply the result stated in Remark C.3 (in Appendix C). Specifically, it is possible to check, by running Algorithm 6.1, that the gradient of the scalar $v_x^2 + v_y^2$ belongs to the observable codistribution. Hence, we can assume that the quantity $v_x^2 + v_y^2$ is a further system output. It is possible to check that the system defined by the state in (8.32), the dynamics in (8.33), the outputs $v_x^2 + v_y^2$, and the one in (8.34) is directly in canonic form.

Observability properties

We apply the method in section 8.1.2. By comparing the dynamics in (8.33) with (8.1) we obtain

$$g^0 = \begin{bmatrix} v_x \\ v_y \\ B\cos\theta \\ B\sin\theta \\ 0 \\ 0 \\ 0 \\ 0 \\ 0 \end{bmatrix}, \quad f \triangleq f^1 = \begin{bmatrix} 0 \\ 0 \\ \cos\theta \\ \sin\theta \\ 0 \\ 0 \\ 0 \\ 0 \\ 0 \end{bmatrix}, \quad g^1 = \begin{bmatrix} 0 \\ 0 \\ 0 \\ 0 \\ 1 \\ 0 \\ 0 \\ 0 \\ 0 \end{bmatrix}, \quad g^2 = \begin{bmatrix} 0 \\ 0 \\ -\sin\theta \\ \cos\theta \\ 0 \\ 0 \\ 0 \\ 0 \\ 0 \end{bmatrix}.$$

First step

In accordance with what we mentioned above, we can select $\tilde{h}_1$ and $\tilde{h}_2$ as follows:

$$\tilde{h}_1 = \frac{y_c \cos\theta_c - x_c \sin\theta_c}{x_c \cos\theta_c + y_c \sin\theta_c},$$
$$\tilde{h}_2 = v_x^2 + v_y^2.$$

From (8.12) we obtain $\mu_1^2 = \mu_2^1 = 0$, $\mu_1^1 = \frac{-\rho(x_b \cos(\phi_1+\theta)+y_b \sin(\phi_1+\theta))-x_b^2-y_b^2}{(\rho\cos(\phi_2)+x_b\cos(\phi_1+\phi_2+\theta)+y_b\sin(\phi_1+\phi_2+\theta))^2}$ and $\mu_2^2 = 2v_y\cos\theta - 2v_x\sin\theta$, which is nonsingular (and diagonal).

Second step

We compute the inverse of the previous tensor. Because of the diagonal structure, we easily obtain $\nu_1^2 = \nu_2^1 = 0$, $\nu_1^1 = \frac{1}{\mu_1^1}$ and $\nu_2^2 = \frac{1}{\mu_2^2}$. Additionally, we obtain from (8.14)

$$\mu_1^0 = \frac{\rho(v_y\cos(\phi_1+\theta)-v_x\sin(\phi_1+\theta))+v_y x_b - v_x y_b}{(\rho\cos\phi_2+x_b\cos(\phi_1+\phi_2+\theta)+y_b\sin(\phi_1+\phi_2+\theta))^2},$$
$$\mu_2^0 = 2B(v_x\cos\theta + v_y\sin\theta).$$

Finally, from (8.18) we obtain

$$\widehat{g}^0 = \begin{bmatrix} v_x \\ v_y \\ \frac{Bv_y}{v_y\cos\theta - v_x\sin\theta} \\ \frac{-Bv_x}{v_y\cos\theta - v_x\sin\theta} \\ \frac{\rho(v_y\cos(\phi_1+\theta)-v_x\sin(\phi_1+\theta))+v_y x_b - v_x y_b}{x_b^2+\rho\cos(\phi_1+\theta)x_b+y_b^2+\rho\sin(\phi_1+\theta)y_b} \\ 0 \\ 0 \\ 0 \\ 0 \end{bmatrix}.$$

Regarding the vector fields $\widehat{g}^1$ and $\widehat{g}^2$, we obtain the following expressions. All the entries of the vector field $\widehat{g}^1$ vanish, with the exception of the fifth entry, which is

$$\frac{-(\rho\cos\phi_2 + x_b\cos(\phi_1+\phi_2+\theta) + y_b\sin(\phi_1+\phi_2+\theta))^2}{x_b^2 + \rho\cos(\phi_1+\theta)x_b + y_b^2 + \rho\sin(\phi_1+\theta)y_b}.$$

All the entries of the vector field $\widehat{g}^2$ vanish, with the exception of the third and the fourth, which are $\frac{-\sin\theta}{2v_y\cos\theta - 2v_x\sin\theta}$ and $\frac{\cos\theta}{2v_y\cos\theta - 2v_x\sin\theta}$, respectively.

8.2. Applications

Third step

We compute the gradient of $\widetilde{h}_1$ and $\widetilde{h}_2$ (obtained at the first step) with respect to the state in (8.32). By a direct computation we obtain that the dimension of Ω_0 is 2. Additionally, we compute Ω_1. Its dimension is 3.

Fourth step

We have $\mathcal{L}_{\phi_0}\mathcal{L}_{g^j}\widetilde{h}_i = \mathcal{L}_f \mu_i^j = 0 \; \forall i,j$. In order to check whether we are in the special case examined by Lemma 8.11, we need to compute ϕ_1^0, ϕ_1^1, and ϕ_1^2 by using Algorithm 8.1. We obtain that only the first two differ from the null vector:

$$\phi_1^0 = \begin{bmatrix} -\cos\theta \\ -\sin\theta \\ -\frac{\sin\theta(\rho(v_y\cos(\phi_1+\theta)-v_x\sin(\phi_1+\theta))+v_y x_b - v_x y_b)}{x_b^2+\rho\cos(\phi_1+\theta)x_b+y_b^2+\rho\sin(\phi_1+\theta)y_b} \\ \frac{\cos\theta(\rho(v_y\cos(\phi_1+\theta)-v_x\sin(\phi_1+\theta))+v_y x_b - v_x y_b)}{x_b^2+\rho\cos(\phi_1+\theta)x_b+y_b^2+\rho\sin(\phi_1+\theta)y_b} \\ 0 \\ 0 \\ 0 \\ 0 \\ 0 \end{bmatrix},$$

$$\phi_1^1 = \begin{bmatrix} 0 \\ 0 \\ \frac{\sin\theta(\rho\cos\phi_2 + x\cos(\phi_1+\phi_2+\theta)+y_b\sin(\phi_1+\phi_2+\theta))^2}{x_b^2+\rho\cos(\phi_1+\theta)x_b+y_b^2+\rho\sin(\phi_1+\theta)y_b} \\ -\frac{\cos\theta(\rho\cos\phi_2 + x_b\cos(\phi_1+\phi_2+\theta)+y_b\sin(\phi_1+\phi_2+\theta))^2}{x_b^2+\rho\cos(\phi_1+\theta)x_b+y_b^2+\rho\sin(\phi_1+\theta)y_b} \\ 0 \\ 0 \\ 0 \\ 0 \\ 0 \end{bmatrix}.$$

We have $\mathcal{L}_{\phi_1^1}\mathcal{L}_{g^2}\widetilde{h}_2 = \mathcal{L}_{\phi_1^1}\mu_2^2 = -2\frac{(r\cos\phi_2+x_b\cos(\phi_1+\phi_2+\theta)+y_b\sin(\phi_1+\phi_2+\theta))^2}{x_b^2+r\cos(\phi_1+\theta)x_b+y_b^2+r\sin(\phi_1+\theta)y_b}$, which does not vanish, in general. This suffices to conclude that the considered system is not in the special case examined by Lemma 8.11, and we need to continue with step 5.

Fifth step

We compute the three-index tensor in (8.26). We do not provide here the expression of its components, for brevity's sake. We only mention that the nonvanishing components are the following eleven: $\mathcal{T}_1^{00}$, $\mathcal{T}_2^{00}$, $\mathcal{T}_1^{01}$, $\mathcal{T}_2^{02}$, $\mathcal{T}_1^{10}$, $\mathcal{T}_2^{10}$, $\mathcal{T}_1^{20}$, $\mathcal{T}_2^{20}$, $\mathcal{T}_1^{11}$, $\mathcal{T}_2^{12}$, and $\mathcal{T}_2^{22}$.

Sixth step

By using Algorithm 8.3 and the vector fields ϕ_1^0, ϕ_1^1, and ϕ_1^2 previously computed, we can compute Ω_2. Its dimension is 7. We obtain that the gradients of all the components of $\mathcal{T}$ belong to Ω_2, with the exception of $\mathcal{T}_1^{00}$, $\mathcal{T}_1^{01}$, $\mathcal{T}_1^{10}$, $\mathcal{T}_1^{20}$, and $\mathcal{T}_1^{11}$. Hence, we need to compute Ω_3 by using Algorithm 8.3. We do not provide all the steps. It is possible to check that also the gradients of the remaining components of $\mathcal{T}$ belong to Ω_3, whose dimension is 8. Therefore, we have $m' = 3$.

Seventh step

By an explicit computation (by using the subsequent steps of Algorithm 8.3), we obtain $\Omega_4 = \Omega_3$. Hence, Algorithm 8.3 has coverged and $\Omega = \Omega_3$.

Eighth step

By computing the distribution orthogonal to the codistribution Ω_3, we can find the continuous symmetry that characterizes the unobservable space. By an explicit computation, we obtain the vector

$$[-y_b,\ x_b,\ -v_y,\ v_x,\ 1,\ 0,\ 0,\ 0,\ 0]^T,$$

which corresponds to a rotation around the vertical axis. Note that the absolute scale is invariant with respect to a rotation around the vertical axis, meaning that it can be estimated. This also holds for the body speed in the local frame.

We remark that we obtain the same result that holds in the case of calibrated sensors. This means that the sensors, even if not calibrated, provide enough information to perform their self-calibration. Additionally, as in the case of calibrated sensors, we remark that the invariance that corresponds to the continuous symmetry detected by our observability analysis would be present also by having a complete IMU, i.e., when the IMU provides the acceleration along the two axes and the angular speed. This means that the information obtained by fusing the measurements from a camera with the measurements from a complete IMU is redundant.

8.2.3 ▪ Visual-inertial sensor fusion: The 3D case with calibrated sensors

We consider again a rigid body ($\mathcal{B}$) equipped with inertial sensors (IMU) and a monocular camera. However, we now assume that $\mathcal{B}$ moves on a 3D environment. Without loss of generality, we define the body local frame as the IMU frame. The inertial sensors measure the body acceleration and the angular speed. In 3D, the acceleration and the angular speed are three-dimensional vectors. The camera provides the bearing angles of the features in its own local frame.

We derive the observability properties in the case when the IMU only provides the acceleration along a single axis (instead of three). In other words, we will consider the other components of the acceleration and the angular speed as unknown inputs. Therefore, the system is characterized by five unknown inputs. Finally, we consider the minimal case of a single point feature and we fix the global frame on it.

As we saw in the planar case, in general the camera frame does not coincide with the IMU frame. Additionally, the measurements provided by the IMU are in general biased. In this section we study the simple case when the camera frame coincides with the IMU frame and the inertial measurements are unbiased. Then, in section 8.2.4, we relax both these assumptions by also assuming that the camera is extrinsically uncalibrated (i.e., the transformation between the IMU frame and the camera frame is unknown).

Our system can be characterized by the following state (see Figure 8.3 for an illustration):

$$X \triangleq [F_x,\ F_y,\ F_z,\ V_x,\ V_y,\ V_z,\ q_t,\ q_x,\ q_y,\ q_z,\ |g|]^T, \tag{8.35}$$

8.2. Applications

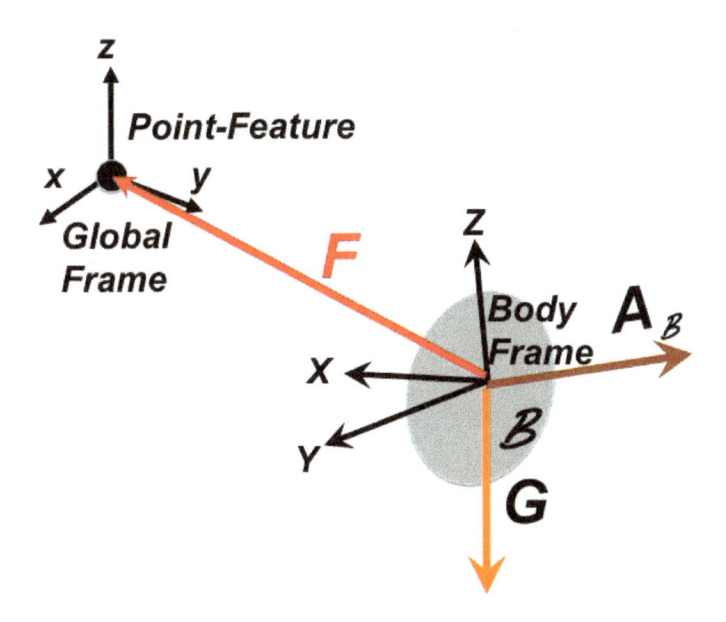

Figure 8.3. *Visual-inertial sensor fusion in 3D with calibrated sensors and in the case of a single point feature.*

where $F = [F_x, F_y, F_z]$ is the position of the point feature in the local frame, $V = [V_x, V_y, V_z]$ is the body speed in the local frame, $q = q_t + q_x i + q_y j + q_z k$ is the unit quaternion that describes the transformation change between the global and the local frame, and $|g|$ is the magnitude of the gravity (that is assumed unknown). The dynamics are

$$\begin{bmatrix} \dot{F} & = -\Omega \wedge F - V, \\ \dot{V} & = -\Omega \wedge V + A + G, \\ \dot{q} & = \frac{1}{2} q \Omega_q, \\ \dot{|g|} & = 0, \end{bmatrix} \quad (8.36)$$

where $\Omega \triangleq [\Omega_x \; \Omega_y \; \Omega_z]$ is the unknown angular speed of the camera, Ω_q is the imaginary quaternion associated with Ω, i.e., $\Omega_q \triangleq \Omega_x i + \Omega_y j + \Omega_z k$, G is the gravity in the local frame, and A is the acceleration that would be perceived by a noiseless tri-axis accelerometer. In other words, since the accelerometer also perceives the gravity, $A^{\text{inertial}} \triangleq A + G$ is the inertial acceleration expressed in the local frame. Without loss of generality, we assume that the accelerometer provides the third component of A. In other words, the first two components of A act as unknown inputs.

The monocular camera provides the position of the feature in the local frame (F) up to a scale. Hence, it provides the ratios of the components of F:

$$h_1(X) = \frac{F_x}{F_z}, \qquad h_2(X) = \frac{F_y}{F_z}. \quad (8.37)$$

We have also to consider the constraint $q^*q = 1$. This provides the further output

$$h_3(X) = q^*q = q_t^2 + q_x^2 + q_y^2 + q_z^2. \quad (8.38)$$

We consider the system characterized by the state in (8.35), the dynamics in (8.36), and the three outputs h_1, h_2, and h_3 in (8.37) and (8.38) under the assumptions that the IMU only provides the acceleration (inertial and gravitational) along the z-axis of the local frame. Hence, we have a single known input ($m_u = 1$) that is $u = u_1 = A_z$. Additionally, we have five unknown inputs ($m_w = 5$) that are $w_1 = \Omega_x$, $w_2 = \Omega_y$, $w_3 = \Omega_z$, $w_4 = A_x$, and $w_5 = A_y$. By comparing (8.36) with (8.1), we obtain

$$g^0 = \begin{bmatrix} -V_x \\ -V_y \\ -V_z \\ -2|g|(q_t q_y - q_x q_z) \\ 2|g|(q_t q_x + q_y q_z) \\ |g|(q_t^2 - q_x^2 - q_y^2 + q_z^2) \\ 0 \\ 0 \\ 0 \\ 0 \\ 0 \end{bmatrix}, \quad f \triangleq f^1 = \begin{bmatrix} 0 \\ 0 \\ 0 \\ 0 \\ 0 \\ 0 \\ 1 \\ 0 \\ 0 \\ 0 \\ 0 \end{bmatrix}, \quad g^1 = \begin{bmatrix} 0 \\ F_z \\ -F_y \\ 0 \\ V_z \\ -V_y \\ -q_x/2 \\ q_t/2 \\ q_z/2 \\ -q_y/2 \\ 0 \end{bmatrix}, \quad g^2 = \begin{bmatrix} -F_z \\ 0 \\ F_x \\ -V_z \\ 0 \\ V_x \\ -q_y/2 \\ -q_z/2 \\ q_t/2 \\ q_x/2 \\ 0 \end{bmatrix},$$

$$g^3 = \begin{bmatrix} F_y \\ -F_x \\ 0 \\ V_y \\ -V_x \\ 0 \\ -q_z/2 \\ q_y/2 \\ -q_x/2 \\ q_t/2 \\ 0 \end{bmatrix}, \quad g^4 = \begin{bmatrix} 0 \\ 0 \\ 0 \\ 1 \\ 0 \\ 0 \\ 0 \\ 0 \\ 0 \\ 0 \\ 0 \end{bmatrix}, \quad g^5 = \begin{bmatrix} 0 \\ 0 \\ 0 \\ 0 \\ 1 \\ 0 \\ 0 \\ 0 \\ 0 \\ 0 \\ 0 \end{bmatrix}.$$

System canonization

We need to check whether the system is in canonic form. If it is not, we need to proceed as explained in Appendix C.2. This is exactly what we have done in the previous cases. In particular, we need to compute the space of functions $\mathcal{F}^3$ and we can check that the dimension of the codistribution $\mathcal{D}_w \mathcal{L}_G \mathcal{F}^3$ is 5. On the other hand, the functions $\tilde{h}_1$, $\tilde{h}_2$, $\tilde{h}_3$, $\tilde{h}_4$, and $\tilde{h}_5$ that we automatically select from $\mathcal{F}^3$ have a complex analytic expression. Even if this fact does not prevent the application of the method in section 8.1.2, for educational purposes, we prefer to proceed in a different manner. In particular, we apply the result stated in Remark C.2 (in Appendix C). Specifically, it is possible to check that the functions $\tilde{h}_1 = F_x$, $\tilde{h}_2 = F_y$, $\tilde{h}_3 = V_x$, $\tilde{h}_4 = V_y$, and $\tilde{h}_5 = V_z$ belong to $\mathcal{F}^3$ (the first two functions also belong to $\mathcal{F}^2$). Additionally, it is possible to check that the system defined by the state in (8.35), the dynamics in (8.36), and the outputs $\tilde{h}_1$, $\tilde{h}_2$, $\tilde{h}_3$, $\tilde{h}_4$, and $\tilde{h}_5$ is directly in canonic form. Note that we applied the result stated by Remark C.2 (in Appendix C), with $k = 0$, because the five functions only depend on the original state.

Observability properties

We apply the method in section 8.1.2 to obtain the observability properties of the system characterized by the state in (8.35), the dynamics in (8.36), and the outputs $\tilde{h}_1 = F_x$, $\tilde{h}_2 = F_y$, $\tilde{h}_3 = V_x$, $\tilde{h}_4 = V_y$, $\tilde{h}_5 = V_z$, and $h_1 = q_t^2 + q_x^2 + q_y^2 + q_z^2$.

First step

In accordance with what we mentioned above, the system is in canonic form. From (8.12) we obtain

$$\mu_j^i = \begin{bmatrix} 0 & -F_z & F_y & 0 & 0 \\ F_z & 0 & -F_x & 0 & 0 \\ 0 & -V_z & V_y & 1 & 0 \\ V_z & 0 & -V_x & 0 & 1 \\ -V_y & V_x & 0 & 0 & 0 \end{bmatrix},$$

where the upper index corresponds to the column and the lower index to the line. This tensor is nonsingular.

Second step

We compute the inverse of the previous tensor. We easily obtain

$$\nu_j^i = \frac{1}{F_z(F_x V_y - F_y V_x)}$$

$$\times \begin{bmatrix} -F_x V_x & -F_y V_x & 0 & 0 & -F_x F_z \\ -F_x V_y & -F_y V_y & 0 & 0 & -F_y F_z \\ -F_z V_x & -F_z V_y & 0 & 0 & -F_z^2 \\ -V_y(F_x V_z - F_z V_x) & F_z V_y^2 - F_y V_y V_z & F_z(F_x V_y - F_y V_x) & 0 & -F_z(F_y V_z - F_z V_y) \\ F_x V_x V_z - F_z V_x^2 & V_x(F_y V_z - F_z V_y) & 0 & F_z(F_x V_y - F_y V_x) & F_z(F_x V_z - F_z V_x) \end{bmatrix}.$$

Additionally, we obtain from (8.14)

$$\begin{aligned}
\mu_1^0 &= -V_x, \\
\mu_2^0 &= -V_y, \\
\mu_3^0 &= -2|g|(q_t q_y - q_x q_z), \\
\mu_4^0 &= 2|g|(q_t q_x + q_y q_z), \\
\mu_5^0 &= |g|(q_t^2 - q_x^2 - q_y^2 + q_z^2).
\end{aligned}$$

Finally, from (8.18) we obtain

$$\hat{g}^1 = \frac{1}{2\eta} \begin{bmatrix} 2F_z(F_x V_y - F_y V_x) \\ 0 \\ 2F_x F_y V_x - 2F_x^2 V_y \\ 0 \\ 0 \\ F_x V_x q_x + F_x V_y q_y + F_z V_x q_z \\ F_x V_y q_z - F_x V_x q_t - F_z V_x q_y \\ F_z V_x q_x - F_x V_x q_z - F_x V_y q_t \\ F_x V_x q_y - F_z V_x q_t - F_x V_y q_x \\ 0 \end{bmatrix}, \hat{g}^2 = \frac{1}{2\eta} \begin{bmatrix} 0 \\ 2F_z(F_x V_y - F_y V_x) \\ 2F_y^2 V_x - 2F_x F_y V_y \\ 0 \\ 0 \\ F_y V_x q_x + F_y V_y q_y + F_z V_y q_z \\ F_y V_y q_z - F_y V_x q_t - F_z V_y q_y \\ F_z V_y q_x - F_y V_x q_z - F_y V_y q_t \\ F_y V_x q_y - F_z V_y q_t - F_y V_y q_x \\ 0 \end{bmatrix}, \hat{g}^3 = \begin{bmatrix} 0 \\ 0 \\ 0 \\ 1 \\ 0 \\ 0 \\ 0 \\ 0 \\ 0 \\ 0 \end{bmatrix},$$

$$\widehat{g}^4 = \begin{bmatrix} 0 \\ 0 \\ 0 \\ 0 \\ 1 \\ 0 \\ 0 \\ 0 \\ 0 \\ 0 \end{bmatrix}, \quad \widehat{g}^5 = \frac{F_z}{2\eta} \begin{bmatrix} 0 \\ 0 \\ 0 \\ 0 \\ 0 \\ 2F_xV_y - 2F_yV_x \\ F_xq_x + F_yq_y + F_zq_z \\ F_yq_z - F_xq_t - F_zq_y \\ F_zq_x - F_xq_z - F_yq_t \\ F_xq_y - F_zq_t - F_yq_x \\ 0 \end{bmatrix},$$

where $\eta = F_z(F_xV_y - F_yV_x)$. We do not provide the expression of g^0 for brevity's sake.

Third step

We compute the gradient of $\widetilde{h}_1, \widetilde{h}_2, \widetilde{h}_3, \widetilde{h}_4, \widetilde{h}_5$, and h_1 with respect to the state in (8.35). By a direct computation we obtain that the dimension of Ω_0 is 6. Additionally, we obtain $\Omega_0 = \Omega_1$.

Fourth step

We have $\mathcal{L}_{\phi_0}\mathcal{L}_{g^1}\widetilde{h}_4 = \mathcal{L}_f \mu_4^1 = 1 \neq 0$. This suffices to conclude that the considered system is not in the special case examined by Lemma 8.11, and we need to continue with step 5.

Fifth step

We compute the three-index tensor in (8.26). We remind the reader that we can consider the lower index as a Latin index since the components of the tensor when this index is zero vanish (this is because we are working with a synchronous frame). Since the Latin index takes the values $1,\ldots,5$, and the Greek indexes $0,\ldots,5$, this tensor has 180 components that can be different from zero. To display these components, we provide separately $\mathcal{T}_i^{1j}, \mathcal{T}_i^{2j}, \mathcal{T}_i^{3j}, \mathcal{T}_i^{4j}, \mathcal{T}_i^{5j}$. We do not provide $\mathcal{T}_i^{00}, \mathcal{T}_i^{0j}, \mathcal{T}_i^{j0}$ for brevity's sake. We have

$$\mathcal{T}_i^{1j} = \frac{1}{\eta}\begin{bmatrix} -(F_x^2 V_y)/F_z & -(F_xF_yV_y)/F_z & 0 & 0 & -F_xF_y \\ (V_x(F_x^2+F_z^2))/F_z & (V_yF_z^2+F_xF_yV_x)/F_z & 0 & 0 & F_x^2+F_z^2 \\ 0 & 0 & 0 & 0 & 0 \\ 0 & 0 & 0 & 0 & 0 \\ 0 & 0 & 0 & 0 & 0 \end{bmatrix},$$

$$\mathcal{T}_i^{2j} = \frac{1}{\eta}\begin{bmatrix} -(V_xF_z^2+F_xF_yV_y)/F_z & -(V_y(F_y^2+F_z^2))/F_z & 0 & 0 & -F_y^2-F_z^2 \\ (F_xF_yV_x)/F_z & (F_y^2V_x)/F_z & 0 & 0 & F_xF_y \\ 0 & 0 & 0 & 0 & 0 \\ 0 & 0 & 0 & 0 & 0 \\ 0 & 0 & 0 & 0 & 0 \end{bmatrix},$$

$$\mathcal{T}_i^{3j} = \frac{1}{\eta}\begin{bmatrix} 0 & 0 & 0 & 0 & 0 \\ 0 & 0 & 0 & 0 & 0 \\ 0 & 0 & 0 & 0 & 0 \\ F_zV_x & F_zV_y & 0 & 0 & F_z^2 \\ -F_xV_y & -F_yV_y & 0 & 0 & -F_yF_z \end{bmatrix}, \quad \mathcal{T}_i^{4j} = \frac{1}{\eta}\begin{bmatrix} 0 & 0 & 0 & 0 & 0 \\ 0 & 0 & 0 & 0 & 0 \\ -F_zV_x & -F_zV_y & 0 & 0 & -F_z^2 \\ 0 & 0 & 0 & 0 & 0 \\ F_xV_x & F_yV_x & 0 & 0 & F_xF_z \end{bmatrix},$$

8.2. Applications

$$\mathcal{T}_i^{5j} = \frac{1}{\eta} \begin{bmatrix} 0 & 0 & 0 & 0 & 0 \\ 0 & 0 & 0 & 0 & 0 \\ F_x V_y & F_y V_y & 0 & 0 & F_y F_z \\ -F_x V_x & -F_y V_x & 0 & 0 & -F_x F_z \\ 0 & 0 & 0 & 0 & 0 \end{bmatrix}.$$

Sixth step

We need to compute Ω_2. In order to do this, we need to compute ϕ_1^0, ϕ_1^1, ϕ_1^2, ϕ_1^3, ϕ_1^4, and ϕ_1^5 by using Algorithm 8.1. We obtain that ϕ_1^3 and ϕ_1^4 are null. The remaining four are as follows:

$$\phi_1^0 = \frac{1}{\eta} \begin{bmatrix} 0 \\ 0 \\ F_z(F_x V_y - F_y V_x) \\ g(F_y q_t^2 - F_y q_x^2 - F_y q_y^2 + F_y q_z^2) F_z - (F_y V_y^2 + F_x V_x V_y) \\ (F_x V_x^2 + F_y V_y V_x) - F_z g(F_x q_t^2 - F_x q_x^2 - F_x q_y^2 + F_x q_z^2) \\ 0 \\ 0 \\ 0 \\ 0 \\ 0 \end{bmatrix},$$

$$\phi_1^1 = \frac{1}{\eta} \begin{bmatrix} 0 \\ 0 \\ 0 \\ -F_x V_y \\ F_x V_x \\ 0 \\ 0 \\ 0 \\ 0 \\ 0 \end{bmatrix}, \quad \phi_1^2 = \frac{1}{\eta} \begin{bmatrix} 0 \\ 0 \\ 0 \\ -F_y V_y \\ F_y V_x \\ 0 \\ 0 \\ 0 \\ 0 \\ 0 \end{bmatrix}, \quad \phi_1^5 = \frac{1}{\eta} \begin{bmatrix} 0 \\ 0 \\ 0 \\ -F_y F_z \\ F_x F_z \\ 0 \\ 0 \\ 0 \\ 0 \\ 0 \end{bmatrix}.$$

Hence, by using Algorithm 8.3 we obtain Ω_2. Its dimension is 8. We obtain that the gradients of all the components of $\mathcal{T}$ belong to Ω_2, with the exception of $\mathcal{T}_3^{00}$, $\mathcal{T}_4^{00}$, $\mathcal{T}_5^{00}$, $\mathcal{T}_3^{10}$, $\mathcal{T}_4^{10}$, $\mathcal{T}_5^{10}$, $\mathcal{T}_3^{20}$, $\mathcal{T}_4^{20}$, $\mathcal{T}_5^{20}$, $\mathcal{T}_3^{50}$, $\mathcal{T}_4^{50}$, and $\mathcal{T}_5^{50}$.

We need to compute Ω_3 by using Algorithm 8.3. We do not provide all the steps. We obtain that the dimension of Ω_3 is 10. Additionally, we obtain that the gradients of all the components of $\mathcal{T}$ belong to Ω_3. Hence, we have $m' = 3$.

Seventh step

We compute Ω_4 by using Algorithm 8.3. We obtain $\Omega_4 = \Omega_3$. Hence, Algorithm 8.3 has converged to the codistribution $\Omega = \Omega_3$.

Eighth step

By computing the distribution orthogonal to the codistribution Ω we can find the continuous symmetry that characterizes the unobservable space. By an explicit computation we obtain the vector

$$[0,\ 0,\ 0,\ 0,\ 0,\ 0,\ -q_z,\ -q_y,\ q_x,\ q_t,\ 0]^T,$$

which corresponds to a rotation around the vertical axis (the axis aligned with the gravity).

We remark that the invariance that corresponds to the continuous symmetry detected by our observability analysis would be present also by having a complete IMU, i.e., when the IMU provides the acceleration along the three axes and the three components of the angular speed. This means that the information obtained by fusing the measurements from a camera with the measurements from a complete IMU is redundant.

8.2.4 ▪ Visual-inertial sensor fusion: The 3D case with uncalibrated sensors

We now investigate the same sensor fusion problem studied in the previous section, but we assume that the camera is not extrinsically calibrated and the inertial measurements are biased. We need to include in the state the inertial bias and the parameters that describe the transformation change between the camera and the IMU. We remind the reader that the IMU only consists of a single-axis accelerometer.

We assume that the body local frame coincides with the frame attached to the single-axis accelerometer and, without loss of generality, we assume that this local frame has its z-axis coincident with the axis of the accelerometer. The position of the camera optical center in the local body frame will be denoted by $P^c = [X_c,\ Y_c,\ Z_c]^T$, and the camera orientation will be characterized through the three Euler angles $\alpha,\ \beta,\ \gamma$. Specifically, a vector with orientation $\hat{r}$ in the body frame will have the orientation $R\hat{r}$ in the camera frame, where $R = R(\alpha, \beta, \gamma) = R_\alpha^z R_\beta^x R_\gamma^z$ and R_η^z and R_η^x rotates the unit vector $\hat{r}$ clockwise through the angle η about the z-axis and the x-axis, respectively. The vector P^c and the three angles $\alpha,\ \beta,\ \gamma$ characterize the extrinsic camera calibration and are assumed to be unknown. As in the previous section, we consider the minimal case when a single point feature is available and we denote its position in the camera frame with $F \triangleq [F_x\ F_y\ F_z]^T$. Figure 8.4 depicts the system with the three frames (the global frame, the body frame, which coincides with the IMU frame, and the camera frame).

We characterize our system through the following state:

$$X \triangleq [F_x,\ F_y,\ F_z,\ V_x,\ V_y,\ V_z,\ q_t,\ q_x,\ q_y,\ q_z,\ |g|,\ P_x^c,\ P_y^c,\ P_z^c,\ \alpha,\ \beta,\ \gamma,\ B]^T, \quad (8.39)$$

where B is the accelerometer bias. Note that the position of the point feature is expressed in the camera frame, while the speed, $V = [V_x,\ V_y,\ V_z]^T$, is in the body (IMU) frame. Additionally, $q = q_t + q_x i + q_y j + q_z k$ is the unit quaternion that describes the transformation change between the global and the body frames. The dynamics are

$$\begin{bmatrix} \dot{F} &= -{}^c\Omega \wedge F - R(V + \Omega \wedge P^c), \\ \dot{V} &= -\Omega \wedge V + A + G, \\ \dot{q} &= \frac{1}{2}q\Omega_q, \\ \dot{P_c} &= [0,0,0]^T\ |\dot{g}| = \dot{B} = \dot{\alpha} = \dot{\beta} = \dot{\gamma} = 0, \end{bmatrix} \quad (8.40)$$

where ${}^c\Omega$ is the angular speed expressed in the camera frame, which is related to the one expressed in the body frame through the rotation matrix R, ${}^c\Omega = R\Omega$. As in the previous section,

8.2. Applications

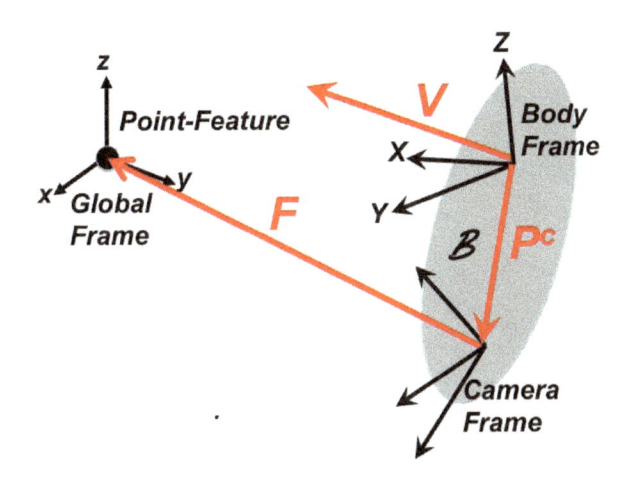

Figure 8.4. *The camera frame does not coincide with the body frame, defined as the IMU frame. F is the position of the point feature in the camera frame. V is the speed of the body in the local (body) frame.*

the accelerometer provides the third component of the acceleration perceived in the body frame, which includes the gravity. We assume that the measurements are affected by the bias B.

The monocular camera provides the position of the feature in the camera frame (F) up to a scale. Hence, it provides the ratios of the components of F, i.e., the two outputs in (8.37). In addition, as in the previous case, we need to include the constraint $q^*q = 1$. This provides the further output given in (8.38).

We consider the system characterized by the state in (8.39), the dynamics in (8.40), and the three outputs h_1, h_2, and h_3 in (8.37) and (8.38) under the assumptions that the IMU only provides the acceleration (inertial and gravitational) along the z-axis of the body frame. Hence, we have a single known input ($m_u = 1$) that is $u = u_1 = A_z$. Additionally, we have five unknown inputs ($m_w = 5$) that are $w_1 = \Omega_x$, $w_2 = \Omega_y$, $w_3 = \Omega_z$, $w_4 = A_x$, and $w_5 = A_y$.

By comparing (8.40) with (8.1) we obtain the expressions of the vector fields: g^0, g^1, g^2, g^3, g^4, g^5, and f. In particular the following hold:

- g^0 is obtained by setting to zero all the components of both A and Ω in (8.40), with the exception of the third component of A, which is set equal to $-B$;

- f is obtained by removing g^0 from (8.40) and then by setting $A_x = A_y = \Omega_x = \Omega_y = \Omega_z = 0$ and $A_z = 1$;

- g^j (for $j = 1, \ldots, 5$) are obtained by removing g^0 from (8.40) and then by setting to zero all the components of both A and Ω, with the exception of one of them depending on j (e.g., to obtain g^1 we set $A_x = A_y = A_z = \Omega_y = \Omega_z = 0$ and $\Omega_x = 1$).

System canonization

We need to check whether the system is in canonic form. If it is not, we need to proceed as explained in Appendix C. This is exactly what we have done in all the previous cases. By proceeding with the procedure described in Appendix C it is possible to set the system in canonic form. On the other hand, the analytic expressions of the quantities computed by the application of this procedure are complex. For educational purposes, we prefer to proceed differently. In particular, by using Algorithm 6.1, it is possible to prove that the gradient of the following scalar functions, which only depend on the components of the state in (8.39), belong to the observable codistribution: F_x, F_y, F_z, $V_x^2 + V_y^2$, V_z, $\frac{V_y \cos\gamma - V_x \sin\gamma}{V_x \cos\gamma + V_y \sin\gamma}$, α, β, and B. Hence, we can use these functions as system outputs. In particular, we find that the system is directly in canonic form by using the following outputs:

$$\tilde{h}_1 = F_x, \ \tilde{h}_2 = F_y, \ \tilde{h}_3 = \frac{V_y \cos\gamma - V_x \sin\gamma}{V_x \cos\gamma + V_y \sin\gamma}, \ \tilde{h}_4 = V_x^2 + V_y^2, \ \tilde{h}_5 = V_z.$$

Therefore, we use these five scalar functions for the m_w scalar functions that we have to select, in accordance with the first step of the method in section 8.1.2. Before proceeding, in order to further simplify the analytic computation, we can remove α, β, and B from the state. In particular, we can set their values to zero. Hence we refer to the following state:

$$X \triangleq \left[F_x, \ F_y, \ F_z, \ V_x, \ V_y, \ V_z, \ q_t, \ q_x, \ q_y, \ q_z, \ |g|, \ P_x^c, \ P_y^c, \ P_z^c, \ \gamma \right]^T, \qquad (8.41)$$

whose dynamics are easily obtained from (8.40). Note that now the matrix R only describes a rotation of γ about the accelerometer axis. By comparing these dynamics with (8.1) we obtain

$$g^0 = \begin{bmatrix} -V_x \cos\gamma - V_y \sin\gamma \\ V_x \sin\gamma - V_y \cos\gamma \\ -V_z \\ -2|g|(q_t q_y - q_x q_z) \\ 2|g|(q_t q_x + q_y q_z) \\ |g|(q_t^2 - q_x^2 - q_y^2 + q_z^2) \\ 0 \\ 0 \\ 0 \\ 0 \\ 0 \\ 0 \\ 0 \\ 0 \end{bmatrix}, \ f \triangleq f^1 = \begin{bmatrix} 0 \\ 0 \\ 0 \\ 0 \\ 0 \\ 1 \\ 0 \\ 0 \\ 0 \\ 0 \\ 0 \\ 0 \\ 0 \\ 0 \end{bmatrix}, \ g^1 = \begin{bmatrix} \sin\gamma(F_z + Z_c) \\ \cos\gamma(F_z + Z_c) \\ -Y_c - F_y \cos\gamma - F_x \sin\gamma \\ 0 \\ V_z \\ -V_y \\ -q_x/2 \\ q_t/2 \\ q_z/2 \\ -q_y/2 \\ 0 \\ 0 \\ 0 \\ 0 \end{bmatrix},$$

$$g^2 = \begin{bmatrix} -\cos\gamma(F_z + Z_c) \\ \sin\gamma(F_z + Z_c) \\ X_c + F_x \cos\gamma - F_y \sin\gamma \\ -V_z \\ 0 \\ V_x \\ -q_y/2 \\ -q_z/2 \\ q_t/2 \\ q_x/2 \\ 0 \\ 0 \\ 0 \\ 0 \end{bmatrix}, \ g^3 = \begin{bmatrix} 0 \\ 0 \\ 0 \\ 1 \\ 0 \\ 0 \\ 0 \\ 0 \\ 0 \\ 0 \\ 0 \\ 0 \\ 0 \\ 0 \end{bmatrix}, \ g^4 = \begin{bmatrix} 0 \\ 0 \\ 0 \\ 0 \\ 1 \\ 0 \\ 0 \\ 0 \\ 0 \\ 0 \\ 0 \\ 0 \\ 0 \\ 0 \end{bmatrix}, \ g^5 = \begin{bmatrix} F_y + Y_c \cos\gamma - X_c \sin\gamma \\ -F_x - X_c \cos\gamma - Y_c \sin\gamma \\ 0 \\ V_y \\ -V_x \\ 0 \\ -q_z/2 \\ q_y/2 \\ -q_x/2 \\ q_t/2 \\ 0 \\ 0 \\ 0 \\ 0 \end{bmatrix}.$$

Observability properties

We apply the method in section 8.1.2 to obtain the observability properties of the system characterized by the state in (8.41), the dynamics in (8.40) (without α, β, and B), and the outputs:

$$\begin{aligned}
\widetilde{h}_1 &= F_x, \\
\widetilde{h}_2 &= F_y, \\
\widetilde{h}_3 &= \frac{V_y \cos\gamma - V_x \sin\gamma}{V_x \cos\gamma + V_y \sin\gamma}, \\
\widetilde{h}_4 &= V_x^2 + V_y^2, \\
\widetilde{h}_5 &= V_z,
\end{aligned}$$

and $h_1 = F_z$ and $h_2 = q_t^2 + q_x^2 + q_y^2 + q_z^2$.

First step

In accordance with what we mentioned in section 8.2.4, the system is directly in canonic form. From (8.12) we obtain $\mu_j^i =$

$$\begin{bmatrix}
\sin\gamma(F_z + Z_c) & -\cos\gamma(F_z + Z_c) & 0 & 0 & F_y + Y_c\cos\gamma - X_c\sin\gamma \\
\cos\gamma(F_z + Z_c) & \sin\gamma(F_z + Z_c) & 0 & 0 & F_y + Y_c\cos\gamma - X_c\sin\gamma \\
\frac{V_x V_z}{(V_x\cos\gamma + V_y\sin\gamma)^2} & \frac{V_y V_z}{(V_x\cos\gamma + V_y\sin\gamma)^2} & \frac{-V_y}{(V_x\cos\gamma + V_y\sin\gamma)^2} & \frac{V_x}{(V_x\cos\gamma + V_y\sin\gamma)^2} & \frac{-(V_x^2+V_y^2)}{(V_x\cos\gamma + V_y\sin\gamma)^2} \\
2V_y V_z & -2V_x V_z & 2V_x & 2V_y & 0 \\
-V_y & V_x & 0 & 0 & 0
\end{bmatrix}$$

where the upper index corresponds to the column and the lower index to the line. This tensor is nonsingular.

Second step

We compute the inverse of the previous tensor. We do not provide the expression for brevity's sake. Additionally, we obtain from (8.14)

$$\begin{aligned}
\mu_1^0 &= -V_x \cos\gamma - V_y \sin\gamma, \\
\mu_2^0 &= V_x \sin\gamma - V_y \cos\gamma, \\
\mu_3^0 &= \frac{V_x(2gq_tq_x + 2gq_yq_z) + V_y(2gq_tq_y - 2gq_xq_z)}{(V_x\cos\gamma + V_y\sin\gamma)^2}, \\
\mu_4^0 &= 4V_y g(q_t q_x + q_y q_z) - 4V_x g(q_t q_y - q_x q_z), \\
\mu_5^0 &= g(q_t^2 - q_x^2 - q_y^2 + q_z^2).
\end{aligned}$$

Finally, from (8.18) we obtain the expressions of $\widehat{g}^0$, $\widehat{g}^1$, $\widehat{g}^2$, $\widehat{g}^3$, $\widehat{g}^4$, and $\widehat{g}^5$. For brevity's sake, we only provide the expressions of $\widehat{g}^3$ and $\widehat{g}^4$, which are simple:

$$\widehat{g}^3 = \frac{(V_x\cos\gamma + V_y\sin\gamma)^2}{V_x^2 + V_y^2}[0, 0, 0, -V_y, V_x, 0, 0, 0, 0, 0, 0, 0, 0, 0, 0]^T,$$

$$\widehat{g}^4 = \frac{1}{2(V_x^2 + V_y^2)}[0, 0, 0, V_x, V_y, 0, 0, 0, 0, 0, 0, 0, 0, 0, 0]^T.$$

Third step

We compute the gradient of $\widetilde{h}_1$, $\widetilde{h}_2$, $\widetilde{h}_3$, $\widetilde{h}_4$, $\widetilde{h}_5$, h_1, and h_2 with respect to the state in (8.41). By a direct computation we obtain that the dimension of Ω_0 is 7. Additionally, we compute Ω_1 and we obtain $\Omega_1 = \Omega_0$.

Fourth step

We have $\mathcal{L}_{\phi_0}\mathcal{L}_{g^1}\widetilde{h}_4 = \mathcal{L}_f \mu_4^1 = 2V_y \neq 0$, in general. This suffices to conclude that the considered system is not in the special case examined by Lemma 8.11, and we need to continue with step 5.

Fifth step

We compute the three-index tensor in (8.26). We remind the reader that we can consider the lower index as a Latin index since the components of the tensor when this index is zero vanish (this is because we are working with a synchronous frame). Since the Latin index takes the values $1, \ldots, 5$ and the Greek indexes $0, \ldots, 5$, this tensor has 180 components that can differ from zero. For brevity's sake, we do not provide here the analytic expressions of its components. We only notify the components that are not vanishing. They are the following 63: $\mathcal{T}_1^{00}$, $\mathcal{T}_2^{00}$, $\mathcal{T}_3^{00}$, $\mathcal{T}_4^{00}$, $\mathcal{T}_5^{00}$, $\mathcal{T}_1^{10}$, $\mathcal{T}_2^{10}$, $\mathcal{T}_3^{10}$, $\mathcal{T}_4^{10}$, $\mathcal{T}_5^{10}$, $\mathcal{T}_1^{20}$, $\mathcal{T}_2^{20}$, $\mathcal{T}_3^{20}$, $\mathcal{T}_4^{20}$, $\mathcal{T}_5^{20}$, $\mathcal{T}_1^{30}$, $\mathcal{T}_2^{30}$, $\mathcal{T}_4^{30}$, $\mathcal{T}_5^{30}$, $\mathcal{T}_1^{40}$, $\mathcal{T}_2^{40}$, $\mathcal{T}_3^{40}$, $\mathcal{T}_5^{40}$, $\mathcal{T}_3^{50}$, $\mathcal{T}_4^{50}$, $\mathcal{T}_5^{50}$, $\mathcal{T}_1^{01}$, $\mathcal{T}_1^{02}$, $\mathcal{T}_1^{05}$, $\mathcal{T}_2^{01}$, $\mathcal{T}_2^{02}$, $\mathcal{T}_2^{05}$, $\mathcal{T}_1^{11}$, $\mathcal{T}_1^{12}$, $\mathcal{T}_1^{15}$, $\mathcal{T}_2^{11}$, $\mathcal{T}_2^{12}$, $\mathcal{T}_2^{15}$, $\mathcal{T}_1^{21}$, $\mathcal{T}_1^{22}$, $\mathcal{T}_1^{25}$, $\mathcal{T}_2^{21}$, $\mathcal{T}_2^{22}$, $\mathcal{T}_2^{25}$, $\mathcal{T}_3^{33}$, $\mathcal{T}_3^{34}$, $\mathcal{T}_4^{33}$, $\mathcal{T}_4^{31}$, $\mathcal{T}_4^{32}$, $\mathcal{T}_4^{35}$, $\mathcal{T}_5^{31}$, $\mathcal{T}_5^{32}$, $\mathcal{T}_5^{35}$, $\mathcal{T}_3^{41}$, $\mathcal{T}_3^{42}$, $\mathcal{T}_3^{43}$, $\mathcal{T}_3^{45}$, $\mathcal{T}_4^{44}$, $\mathcal{T}_5^{45}$, $\mathcal{T}_3^{51}$, $\mathcal{T}_3^{52}$, $\mathcal{T}_3^{55}$, and $\mathcal{T}_4^{55}$.

Sixth step

We need to compute Ω_2, and, in order to do this, we need to compute ϕ_1^0, ϕ_1^1, ϕ_1^2, ϕ_1^3, ϕ_1^4, and ϕ_1^5 by using Algorithm 8.1. We obtain that ϕ_1^3 and ϕ_1^4 are null. We do not provide the expression of the remaining four. By using Algorithm 8.3 we obtain Ω_2. Its dimension is 10. We obtain that the gradients of the following 19 components of $\mathcal{T}$ belong to Ω_2: $\mathcal{T}_1^{03}$, $\mathcal{T}_2^{03}$, $\mathcal{T}_5^{03}$, $\mathcal{T}_1^{04}$, $\mathcal{T}_2^{04}$, $\mathcal{T}_5^{04}$, $\mathcal{T}_3^{33}$, $\mathcal{T}_3^{34}$, $\mathcal{T}_4^{33}$, $\mathcal{T}_3^{43}$, $\mathcal{T}_4^{44}$, $\mathcal{T}_5^{45}$, $\mathcal{T}_4^{55}$, $\mathcal{T}_5^{31}$, $\mathcal{T}_5^{32}$, $\mathcal{T}_5^{35}$, $\mathcal{T}_3^{51}$, $\mathcal{T}_3^{52}$, and $\mathcal{T}_3^{55}$. Hence, the gradients of $63 - 19 = 44$ components of $\mathcal{T}$ do not belong to Ω_2.

We compute Ω_3 by using the subsequent steps of Algorithm 8.3. Its dimension is 13. Additionally, the gradients of all the components of $\mathcal{T}$ belong to Ω_3. Hence, $m' = 3$;

Seventh step

We need to compute Ω_4 by using Algorithm 8.3. We do not provide all the steps. We obtain that $\Omega_4 = \Omega_3$. Hence, Algorithm 8.3 has converged to the codistribution $\Omega = \Omega_3$.

Eighth step

By computing the distribution orthogonal to the codistribution Ω we can find the continuous symmetries that characterize the unobservable space. By an explicit computation we obtain the

8.2. Applications

following two vectors:

$$\begin{bmatrix} 0 \\ 0 \\ 0 \\ 0 \\ 0 \\ 0 \\ -q_z/2 \\ -q_y/2 \\ q_x/2 \\ q_t/2 \\ 0 \\ 0 \\ 0 \\ 0 \\ 0 \end{bmatrix}, \begin{bmatrix} 0 \\ 0 \\ 0 \\ -V_y \\ V_x \\ 0 \\ q_z/2 \\ -q_y/2 \\ q_x/2 \\ -q_t/2 \\ 0 \\ -Y_c \\ X_c \\ 0 \\ 1 \end{bmatrix}.$$

The former corresponds to a rotation around the vertical axis (the axis aligned with the gravity). This was the only continuous symmetry that characterized the case with calibrated sensors, discussed in the previous section. It is not surprising that this symmetry remains. The latter corresponds to a rotation around the accelerometer axis.

We conclude this section by remarking that if the camera is not extrinsically calibrated, an internal symmetry arises. As a result, it is not possible to distinguish all the physical quantities rotated around the accelerometer axis, independently of the accomplished trajectory. This means that, in this setting, it is not possible to fully perceive self-motion. If an additional inertial sensor is introduced, the latter symmetry is broken, provided that this additional sensor is not aligned with the accelerometer.

We summarize the results of the last two sections as follows.

In the visual-inertial sensor fusion problem with only two inertial sensors, the observability properties are the same as in the standard case, provided that the two inertial sensors are along two distinct axes and with at least one of them that is an accelerometer. In other words, when $\mathcal{B}$ is equipped with a monocular camera and two single-axis inertial sensors, it is able to perceive its self-motion exactly as in the case when $\mathcal{B}$ is equipped with a monocular camera and a complete inertial measurement unit. This holds even in the most challenging scenario, i.e., in the case of unknown camera-inertial sensor transformation, unknown magnitude of the gravity, unknown biases, and a single point feature available. In the case when the inertial sensors only consist of a single accelerometer, a new internal symmetry arises. As a result, the initial speed and orientation and the camera-inertial sensor transformation are not fully observable: all these quantities cannot be distinguished from the same quantities rotated around the accelerometer axis. All the remaining states are observable, as in the standard visual-inertial sensor fusion problem.

It is very interesting to remark that these results provide a new insight about the problem of visual-vestibular integration for self-motion perception in neuroscience. Most vertebrates are equipped with two distinct organs in the inner ear that are able to sense acceleration (both inertial acceleration and gravity). These are called otoliths: the saccule and the utricle. The interesting point is that each of them is able to sense acceleration along two distinct axes. In accordance with our results, this makes it possible to autocalibrate these sensors separately. Specifically, our results clearly prove that there is enough information to estimate the position and orientation of a two-axis accelerometer with respect to the visual sensor, by only using the data provided by the two-axis accelerometer and the visual sensor. The autocalibration is a fundamental step to be accomplished in order to have the possibility of properly using the measurements provided by the sensor. In other words, if each otolith was constituted by a single-axis accelerometer, the calibration was not possible. As a result, the measurements provided by this sensor were not properly used and the self-perception of motion was not possible.

8.3 ▪ Analytic derivations

In this section, we provide all the analytic derivations to obtain the analytic solution presented in section 8.1. In other words, we prove that the entire observable codistribution, for the systems characterized by (8.1), is generated by Algorithm 8.3. This analytic derivation starts from the results obtained in Chapter 6, in particular from Algorithm 6.1 and Theorem 6.12 introduced in section 6.3, which together provide a partial tool to determine the observability properties of systems characterized by (6.1).

For simplicity's sake, we provide all the proofs by working in the space of the states and not in the chronospace. In this regard, we wish to emphasize the following two aspects:

- The chronospace was in any case fundamental since it allowed us to detect the group of invariance denoted by $\mathcal{SUIO}$, which is defined on the chronospace. This allowed us to detect all the tensorial fields, namely the tensors introduced in section 8.1.1, which play a key role in our derivations. In addition, it allowed us to detect the simplified system, obtained by requiring that $\mathcal{SUIO}$ is Abelian (section 3.3.3).

- Working in the space of the states (and not in the chronospace) prevents us from dealing with time-variant systems. This case will be discussed separately in section 8.4, where we derive the observability rank condition for systems characterized by (8.68), which generalizes (8.1) by accounting for an explicit time dependence. To deal with this last more general case, we need to work in the chronospace.

In section 6.3, we provided an analytic procedure to obtain the observability properties in the presence of unknown inputs. This procedure is based on Algorithm 6.1 and Theorem 6.12. In particular, it consists of the implementation of a certain number of steps of Algorithm 6.1 together with the test to check whether the components of the original state are observable or not (Theorem 6.12). In section 6.3.1 we applied this procedure to two specific systems, and in section 6.3.2 we discussed its limits of applicability. These are basically the following two:

- Algorithm 6.1 does not converge automatically. This means that if for a given step k the gradients of the components of the original state belong to $\overline{\Omega}_k$, we can conclude that the original state is weakly observable. On the other hand, if this is not true, we cannot exclude that it is true for a larger k.

- The codistribution returned by Algorithm 6.1 is defined in the extended space; i.e., its covectors are the gradients of scalar functions with respect to the augmented state. As a result, the computation dramatically increases with the state augmentation and it becomes prohibitive after few steps.

The goal of the next two subsections is precisely to deal with the above fundamental issues. All these results are first obtained in the case with a single known input ($m_u = 1$), and then they will be extended to the case when $m_u > 1$ (section 8.3.3).

8.3. Analytic derivations

We remind the reader that, to reduce notational complexity, we adopt the Einstein notation. According to this notation, a repeated dummy index is summed over. When the dummy index is Latin, the sum is from 1 to m_w. When it is Greek, the sum is from 0 to m_w. In addition, to refer to the components of a tensor, e.g., Γ_k^α, $\alpha = 0, 1, \ldots, m_w$ and $k = 1, \ldots, m_w$, we simply write $\Gamma_k^\alpha \; \forall \alpha, k$.

We also remind the reader about our notation about differential operators:

$$\nabla = \frac{\partial}{\partial x},$$

$$\underline{\nabla} = \left(\frac{\partial}{\partial t}, \frac{\partial}{\partial x}\right),$$

$$\nabla_k = \frac{\partial}{\partial^k x} = \left(\frac{\partial}{\partial x}, \frac{\partial}{\partial E^k}\right)$$
$$= \left(\frac{\partial}{\partial x_1}, \ldots, \frac{\partial}{\partial x_n}, \frac{\partial}{\partial w_1^{(0)}}, \ldots, \frac{\partial}{\partial w_{m_w}^{(0)}}, \ldots, \frac{\partial}{\partial w_1^{(k-1)}}, \ldots, \frac{\partial}{\partial w_{m_w}^{(k-1)}}\right).$$

We also denote by ∇_w the gradient with respect to w, i.e.,

$$\nabla_w = \frac{\partial}{\partial w}.$$

Finally, we remind the reader that the vector field G is defined in (6.6), i.e.,

$$G \triangleq \begin{bmatrix} g^0 + \sum_{j=1}^{m_w} g^j w_j \\ w^{(1)} \\ w^{(2)} \\ \ldots \\ w^{(k-1)} \\ 0_{m_w} \end{bmatrix}.$$

8.3.1 ▪ Observable codistribution

In the general case, we do not have the same result of separation stated by Theorem 7.7. This fact, however, does not prevent us from deriving a codistribution that only depends on the original state and that fully characterizes the observability properties. This is the codistribution generated by Algorithm 8.3, which is convergent, as is proven in section 8.3.2.

In Appendix C we prove that any system that satisfies (8.1) either can be set in canonic form or some of the unknown inputs can be spurious; i.e., they can be eliminated in order to derive the system observability properties. In the latter case, the system can be set in canonic form with respect to the remaining unknown inputs. Therefore, without loss of generality, we assume that our system is in canonic form. We select the scalar functions

$$\widetilde{h}_1, \ldots, \widetilde{h}_{m_w}$$

such that the codistribution $\mathcal{D}_w \mathcal{L}_G \mathcal{F}$, which is defined in Appendix C, is generated by

$$\nabla_w \mathcal{L}_G \widetilde{h}_1, \ldots, \nabla_w \mathcal{L}_G \widetilde{h}_{m_w}.$$

Note that, as mentioned at the beginning of Appendix C, the codistribution

$$\text{span}\left\{\nabla_k \widetilde{h}_1, \ldots, \nabla_k \widetilde{h}_{m_w}\right\} \subseteq \overline{\Omega}_m$$

for a given integer m.

For each integer m, we generate the codistribution Ω_m by using Algorithm 8.3 (note that we are always assuming $m \leq k$; in addition, here we are considering $m_u = 1$). By construction, the generators of Ω_m are the gradients of scalar functions that only depend on the original state (x) and not on its extension. In what follows, we need to embed this codistribution in the extended space. We will denote by $[\Omega_m, 0_{km_w}]$ the codistribution made by covectors whose first n components are covectors in Ω_m and the last components are all zero.

Additionally, we will denote by $L^m_{m_w}$ the codistribution that is the span of the Lie derivatives of $\nabla_k \widetilde{h}_1, \ldots, \nabla_k \widetilde{h}_{m_w}$ up to the order m along the vector G, i.e.,

$$L^m_{m_w} \triangleq \text{span}\left\{\mathcal{L}^1_G \nabla_k \widetilde{h}_1, \ldots, \mathcal{L}^1_G \nabla_k \widetilde{h}_{m_w}, \ldots, \mathcal{L}^m_G \nabla_k \widetilde{h}_1, \ldots, \mathcal{L}^m_G \nabla_k \widetilde{h}_{m_w}\right\}.$$

We finally introduce the following codistribution.

Definition 8.2 ($\widetilde{\Omega}$ Codistribution). *This codistribution is defined as follows:* $\widetilde{\Omega}_m \triangleq [\Omega_m, 0_{km_w}] \oplus L^m_{m_w}$.

The codistribution $\widetilde{\Omega}_m$ consists of two parts. Specifically, we can select a basis that consists of exact differentials, which are the gradients of functions that only depend on the original state (x) and not on its extension (these are the generators of $[\Omega_m, 0_{m_w k}]$) and the gradients

$$\mathcal{L}^1_G \nabla_k \widetilde{h}_1, \ldots, \mathcal{L}^1_G \nabla_k \widetilde{h}_{m_w}, \ldots, \mathcal{L}^m_G \nabla_k \widetilde{h}_1, \ldots, \mathcal{L}^m_G \nabla_k \widetilde{h}_{m_w}.$$

The second set of generators, i.e., the gradients $\mathcal{L}^1_G \nabla_k \widetilde{h}_1, \ldots, \mathcal{L}^1_G \nabla_k \widetilde{h}_{m_w}, \ldots, \mathcal{L}^m_G \nabla_k \widetilde{h}_1, \ldots, \mathcal{L}^m_G \nabla_k \widetilde{h}_{m_w}$, are $m m_w$ and, with respect to the first set, they are gradients of functions that also depend on the state extension $[w, w^{(1)}, \ldots, w^{(m-1)}]^T$. We have the following result, which is exactly the extension of Lemma 7.2.

Lemma 8.3. $\nabla x_j \in \Omega_m$ *if and only if* $\nabla_k x_j \in \widetilde{\Omega}_m$, *where x_j is the jth component of the state* ($j = 1, \ldots, n$).

Proof. As for Lemma 7.2, the fact that $\nabla x_j \in \Omega_m$ implies that $\nabla_k x_j \in \widetilde{\Omega}_m$ is obvious since $[\Omega_m, 0_{m_w k}] \subseteq \widetilde{\Omega}_m$ by definition. Let us prove that also the contrary holds, i.e., that if $\nabla_k x_j \in \widetilde{\Omega}_m$, then $\nabla x_j \in \Omega_m$. Since $\nabla_k x_j \in \widetilde{\Omega}_m$, we have $\nabla_k x_j = \sum_{i=1}^{N_1} c_i^1 \omega_i^1 + \sum_{i=1}^{N_2} c_i^2 \omega_i^2$, where $\omega_1^1, \omega_2^1, \ldots, \omega_{N_1}^1$ are N_1 generators of $[\Omega_m, 0_{m_w k}]$ and $\omega_1^2, \omega_2^2, \ldots, \omega_{N_2}^2$ are N_2 generators of $L^m_{m_w}$. We want to prove that $N_2 = 0$.

We proceed by contradiction. Let us suppose that $N_2 \geq 1$. We remark that the first set of generators have the last $m_w k$ entries equal to zero, as for $\nabla_k x_j$. The second set of generators consists of the Lie derivatives of $\nabla_k \widetilde{h}_1$ and $\nabla_k \widetilde{h}_{m_w}$ along G up to the m order. Let us select the following m_w generators among this second set. They are the highest-order Lie derivatives along G of $\nabla_k \widetilde{h}_1$ and $\nabla_k \widetilde{h}_{m_w}$, respectively. Let us denote by $j'_1, \ldots, j'_{m_w}$ these highest orders, and let us denote by j' the largest value among $j'_1, \ldots, j'_{m_w}$. This means that at least one of the integers among $j'_1, \ldots, j'_{m_w}$ is equal to j'. We denote by $j'_{m_1}, \ldots, j'_{m_r}$ all the $r(\geq 1)$ highest

8.3. Analytic derivations

orders that are equal to j'. It is immediate to realize that $\mathcal{L}_G^{j'}\nabla_k \tilde{h}_{m_1}, \ldots, \mathcal{L}_G^{j'}\nabla_k \tilde{h}_{m_r}$ are the only generators among the N_2 generators above whose entries between the $(n+(j'-1)m_w+1)$th and $(n+j'm_w)$th can be different from zero. This is because the functions $\mathcal{L}_G^{j'}\tilde{h}_{m_1}, \ldots, \mathcal{L}_G^{j'}\tilde{h}_{m_r}$ are the only ones that depend on $w^{(j'-1)}$. By a direct computation, we can derive this dependency. We easily obtain that $\mathcal{L}_G^{j'}\tilde{h}_{m_j}$ ($j = 1, \ldots, r$) depends on $w^{(j'-1)}$, by the following linear expression: $\mu_{m_j}^l w_l^{(j'-1)}$. Let us consider any linear combination of $\mathcal{L}_G^{j'}\nabla_k \tilde{h}_{m_1}, \ldots, \mathcal{L}_G^{j'}\nabla_k \tilde{h}_{m_r}$, i.e., $\alpha^1 \mathcal{L}_G^{j'}\nabla_k \tilde{h}_{m_1} + \cdots + \alpha^r \mathcal{L}_G^{j'}\nabla_k \tilde{h}_{m_r}$, with α non-null. We remark that the function $\alpha^1 \mathcal{L}_G^{j'}\tilde{h}_{m_1} + \cdots + \alpha^r \mathcal{L}_G^{j'}\tilde{h}_{m_r}$ depends on $w^{(j'-1)}$ as follows: $\sum_{j=1}^{r} \alpha^j \mu_{m_j}^l w_l^{(j'-1)}$. We also remark that there must exist at least one value of l such that $\sum_{j=1}^{r} \alpha^j \mu_{m_j}^l \neq 0$ (if this is not true, it means that the tensor μ is singular). Hence, we obtain that $\nabla_k x_j$ has at least one entry, among the last $m_w k$ entries, different from zero, and this is not possible. ◀

In the general case, we do not have the same result stated by Theorem 7.7. We prove that the codistribution generated by Algorithm 8.3, which is convergent (see section 8.3.1), fully characterizes the observability properties of the original state. This is proven by using the following three results, which hold for a scalar function $\theta(x)$ of the original state:

1. If $\theta(x)$ is weakly observable, $\exists m$ such that $\nabla_k \theta(x) \in \overline{\Omega}_m$.

2. For any integer m we have $\overline{\Omega}_m \subseteq \tilde{\Omega}_m$.

3. If for a given integer m, $\nabla \theta(x) \in \Omega_m$, then $\theta(x)$ is weakly observable.

The first result is proven by Theorem 6.12. In the rest of this section we prove the last two results (Propositions 8.7 and 8.9).

We start by proving the second result (i.e., the one stated by Proposition 8.7). This proof follows several steps, which are similar to the ones operated to prove Theorem 7.7.

As in that case, for a given $m \leq k$ we define the vector Φ_m by the following algorithm:

1. $\Phi_0 = F$;

2. $\Phi_m = [\Phi_{m-1}, G]$,

where now the field G is the one given in (6.6) and the Lie brackets $[\cdot, \cdot]$ are computed with respect to the extended state, whose dimension is now $n + km_w$.

As in the case of Theorem 7.7, it is immediate to realize that Φ_m has the last components (in this case km_w) identically null. In what follows, we will denote by $\check{\Phi}_m$ the vector that contains the first n components of Φ_m. In other words, $\Phi_m \triangleq [\check{\Phi}_m^T, 0_{km_w}^T]^T$. Additionally, we set $\widehat{\phi}_m \triangleq \begin{bmatrix} \phi_m \\ 0_{km_w} \end{bmatrix}$, where, for brevity's sake, here we denote by ϕ_m the vector $\phi_m^{\alpha_1, \ldots, \alpha_m}$ defined by Algorithm 8.1 (we adopt this simplified notation when there is no ambiguity).

The result stated by Lemma 7.3 still holds, and the proof is identical. Also, the result stated by Lemma 7.4 still holds. However, the proof is more complicated.

Lemma 8.4. $\check{\Phi}_m = \sum_{j=1}^{m} c_{\alpha_1, \ldots, \alpha_j}^j (\mathcal{L}_G \tilde{h}_1, \ldots, \mathcal{L}_G \tilde{h}_{m_w}, \ldots, \mathcal{L}_G^m \tilde{h}_1, \ldots, \mathcal{L}_G^m \tilde{h}_{m_w}) \phi_j^{\alpha_1, \ldots, \alpha_j}$, i.e., the vector $\check{\Phi}_m$ is a linear combination of the vectors $\phi_j^{\alpha_1, \ldots, \alpha_j}$ ($j = 1, \ldots, m$), where the coefficients ($c_{\alpha_1, \ldots, \alpha_j}^j$) depend on the state only through the functions that generate the codistribution $L_{m_w}^m$.

Proof. We proceed by induction. By definition, $\check{\Phi}_0 = \phi_0$.
Inductive step: Let us assume that

$$\check{\Phi}_{m-1} = \sum_{j=1}^{m-1} c^j_{\alpha_1,\ldots,\alpha_j} (\mathcal{L}_G \tilde{h}_1, \ldots, \mathcal{L}_G \tilde{h}_{m_w}, \ldots, \mathcal{L}_G^m \tilde{h}_1, \ldots, \mathcal{L}_G^m \tilde{h}_{m_w}) \phi_j^{\alpha_1,\ldots,\alpha_j}.$$

We have

$$\Phi_m = [\Phi_{m-1}, G] = \sum_{j=1}^{m-1} \left[c^j_{\alpha_1,\ldots,\alpha_j} \begin{bmatrix} \phi_j^{\alpha_1,\ldots,\alpha_j} \\ 0_{km_w} \end{bmatrix}, G \right]$$

$$= \sum_{j=1}^{m-1} c^j_{\alpha_1,\ldots,\alpha_j} \left[\begin{bmatrix} \phi_j^{\alpha_1,\ldots,\alpha_j} \\ 0_{km_w} \end{bmatrix}, G \right] - \mathcal{L}_G c^j_{\alpha_1,\ldots,\alpha_j} \begin{bmatrix} \phi_j^{\alpha_1,\ldots,\alpha_j} \\ 0_{km_w} \end{bmatrix}. \tag{8.42}$$

We directly compute the Lie bracket in the sum (note that $\phi_j^{\alpha_1,\ldots,\alpha_j}$ is independent of the unknown input vector w and its time derivatives):

$$\left[\begin{bmatrix} \phi_j^{\alpha_1,\ldots,\alpha_j} \\ 0_{km_w} \end{bmatrix}, G \right] = \begin{bmatrix} [\phi_j^{\alpha_1,\ldots,\alpha_j}, g^0] + [\phi_j^{\alpha_1,\ldots,\alpha_j}, g^i] w_i \\ 0_{km_w} \end{bmatrix}. \tag{8.43}$$

On the other hand, we have

$$\mathcal{L}_G \tilde{h}_k = \mathcal{L}_{g^0} \tilde{h}_k + \mathcal{L}_{g^j} \tilde{h}_k \, w_j = \mu_k^0 + \mu_k^j w_j,$$

where we used (8.12) and (8.14), and we remind the reader that we are restricting the analysis to systems that do not explicitly depend on time (this will be studied in section 8.4).

By multiplying both members by ν_i^k and summing up over $k = 1, \ldots, m_w$ we obtain

$$w_i = \nu_i^k \mathcal{L}_G \tilde{h}_k - \nu_i^k \mu_k^0 = \nu_i^k \mathcal{L}_G \tilde{h}_k + \nu_i^0 \quad \forall i, \tag{8.44}$$

where we used (8.16). By substituting (8.44) in (8.43) and by using (8.22) and (8.16) we obtain

$$[\phi_j^{\alpha_1,\ldots,\alpha_j}, g^0] + [\phi_j^{\alpha_1,\ldots,\alpha_j}, g^i] w_i = [\phi_j^{\alpha_1,\ldots,\alpha_j}]^0 + [\phi_j^{\alpha_1,\ldots,\alpha_j}]^k \mathcal{L}_G \tilde{h}_k$$

$$= \phi_{j+1}^{\alpha_1,\ldots,\alpha_j,0} + \phi_{j+1}^{\alpha_1,\ldots,\alpha_j,k} \mathcal{L}_G \tilde{h}_k \quad \forall \alpha_1, \ldots, \alpha_j. \tag{8.45}$$

The last equality is due to the definition of the fields $\phi_{j+1}^{\alpha_1,\ldots,\alpha_j,\alpha_{j+1}}$ provided by Algorithm 8.1. Regarding the second term in (8.42), we remark that $\mathcal{L}_G c^j_{\alpha_1,\ldots,\alpha_j} = \sum_{k=1}^{m-1} \frac{\partial c^j_{\alpha_1,\ldots,\alpha_j}}{\partial (\mathcal{L}_G^k \tilde{h}_i)} \mathcal{L}_G^{k+1} \tilde{h}_i$. By substituting this equality and the one in (8.45) in (8.42) and by proceeding exactly as in the last part of the proof of Lemma 7.4, we easily obtain the proof of the statement. ◂

In the general case, the result stated by Lemma 7.5 does not hold, and this is one of the reasons why the separation property (Theorem 7.7) does not hold. In particular, by only using Lemma 8.4 and not the analogue of Lemma 7.5, we cannot prove the analogue of Proposition 7.6. On the other hand, by only using Lemma 8.4 it is immediate to obtain the following weaker result.

Proposition 8.5. *For any scalar function $h(x)$ we have*

$$\mathrm{span}\{\mathcal{L}_{\Phi_0} \nabla_k h, \mathcal{L}_{\Phi_1} \nabla_k h, \ldots, \mathcal{L}_{\Phi_m} \nabla_k h\} \oplus L_{m_w}^m \subseteq \mathrm{span}\{\mathcal{L}_{\widehat{\phi}_0} \nabla_k h, \mathcal{L}_{\widehat{\phi}_1^{\alpha_1}} \nabla_k h, \ldots,$$

$$\mathcal{L}_{\widehat{\phi}_m^{\alpha_1,\ldots,\alpha_m}} \nabla_k h\} \oplus L_{m_w}^m.$$

8.3. Analytic derivations

We also have the following result (note that also this result was an equality in the proof of Theorem 7.7).

Proposition 8.6.

$$L^1_{m_w} \oplus \mathcal{L}_G[\Omega_{m-1}, 0_{km_w}] \subseteq L^1_{m_w} \oplus \sum_{\alpha=0}^{m_w}[\mathcal{L}_{\widehat{g}^\alpha}\Omega_{m-1}, 0_{km_w}].$$

Proof. Let us consider a scalar function $\theta(x)$ such that $\nabla\theta \in \Omega_{m-1}$. We have

$$\mathcal{L}_G\theta = \mathcal{L}_{g^0}\theta + \mathcal{L}_{g^i}\theta\, w_i,$$

by using (8.44), we have

$$\mathcal{L}_G\theta = \mathcal{L}_{g^0}\theta + \mathcal{L}_{g^i}\theta(\nu_i^k \mathcal{L}_G\widetilde{h}_k + \nu_i^0),$$

and by using (8.18), we obtain

$$\mathcal{L}_G\theta = \mathcal{L}_{\widehat{g}^0}\theta + \mathcal{L}_{\widehat{g}^k}\theta \mathcal{L}_G\widetilde{h}_k,$$

from which the proof follows. ◄

We are now ready to prove Proposition 8.7.

Proposition 8.7. *For any integer* m, $\overline{\Omega}_m \subseteq \widetilde{\Omega}_m \triangleq [\Omega_m, 0_{km_w}] \oplus L^m_{m_w}$.

Proof. This proof follows the same steps as the proof of Theorem 7.7. We proceed by induction. By definition, $\overline{\Omega}_0 = \widetilde{\Omega}_0$ since they are both the span of the output (or the outputs in case of multiple outputs).
Inductive step: Let us assume that $\overline{\Omega}_{m-1} \subseteq \widetilde{\Omega}_{m-1}$. We have $\overline{\Omega}_m = \overline{\Omega}_{m-1} \oplus \mathcal{L}_F\overline{\Omega}_{m-1} \oplus \mathcal{L}_G\overline{\Omega}_{m-1} = \overline{\Omega}_{m-1}\oplus\mathcal{L}_F\widetilde{\Omega}_{m-1}\oplus\mathcal{L}_G\overline{\Omega}_{m-1} = \overline{\Omega}_{m-1}\oplus[\mathcal{L}_f\Omega_{m-1},0_{km_w}]\oplus\mathcal{L}_F L^{m-1}_{m_w}\oplus\mathcal{L}_G\overline{\Omega}_{m-1}$.
On the other hand,

$$\mathcal{L}_F L^{m-1}_{m_w} = \mathrm{span}\{\mathcal{L}_F\mathcal{L}^1_G\nabla_k\widetilde{h}_i\} \oplus \cdots \oplus \mathrm{span}\{\mathcal{L}_F\mathcal{L}^{m-2}_G\nabla_k\widetilde{h}_i\} \oplus \mathrm{span}\{\mathcal{L}_F\mathcal{L}^{m-1}_G\nabla_k\widetilde{h}_i\}.$$

The only terms which are not in $\overline{\Omega}_{m-1}$ are $\mathrm{span}\{\mathcal{L}_F\mathcal{L}^{m-1}_G\nabla_k\widetilde{h}_i\}$.
 Hence we have

$$\overline{\Omega}_m = \overline{\Omega}_{m-1} \oplus [\mathcal{L}_f\Omega_{m-1}, 0_{km_w}] \oplus \mathrm{span}\{\mathcal{L}_F\mathcal{L}^{m-1}_G\nabla_k\widetilde{h}_i\} \oplus \mathcal{L}_G\overline{\Omega}_{m-1}.$$

By using Lemma 7.3 we obtain

$$\overline{\Omega}_m = \overline{\Omega}_{m-1} \oplus [\mathcal{L}_f\Omega_{m-1}, 0_{km_w}] \oplus \mathrm{span}\{\mathcal{L}_{\Phi_{m-1}}\nabla_k\widetilde{h}_i\} \oplus \mathcal{L}_G\overline{\Omega}_{m-1}.$$

By using again the induction assumption we obtain

$$\overline{\Omega}_m = [\Omega_{m-1},0_{km_w}]\oplus L^{m-1}_{m_w}\oplus[\mathcal{L}_f\Omega_{m-1},0_{km_w}]\oplus\mathrm{span}\{\mathcal{L}_{\Phi_{m-1}}\nabla_k\widetilde{h}_i\}\oplus\mathcal{L}_G[\Omega_{m-1},0_{km_w}]\oplus\mathcal{L}_G L^{m-1}_{m_w}$$

$$= [\Omega_{m-1},0_{km_w}] \oplus L^m_{m_w} \oplus [\mathcal{L}_f\Omega_{m-1},0_{km_w}] \oplus \mathrm{span}\{\mathcal{L}_{\Phi_{m-1}}\nabla_k\widetilde{h}_i\} \oplus \mathcal{L}_G[\Omega_{m-1},0_{km_w}],$$

and by using Propositions 8.5 and 8.6 we obtain $\overline{\Omega}_m \subseteq [\Omega_{m-1}, 0_{km_w}] \oplus L_{m_w}^m \oplus [\mathcal{L}_f \Omega_{m-1}, 0_{km_w}] \oplus$ span$\{\mathcal{L}_{\hat{\phi}_{m-1}^{\alpha_1, \ldots, \alpha_{m-1}}} \nabla_k \widetilde{h}_i\} \oplus [\mathcal{L}_{\hat{g}^\alpha} \Omega_{m-1}, 0_{km_w}] = \widetilde{\Omega}_m$. ◄

Before proceeding with the proof of Proposition 8.9, we need to prove the following lemma.

Lemma 8.8. *Let us consider the scalar function* $\lambda(x, w) = \theta(x) + \theta^i(x) \mathcal{L}_G \widetilde{h}_i$, *and let us assume that it is weakly observable (in $\Sigma^{(1)}$). Then, all the $m_w + 1$ functions $\theta(x), \theta^1(x), \ldots, \theta^{m_w}(x)$ are weakly observable (in $\Sigma^{(0)}$).*

Proof. This proof will exploit the concept of indistinguishability (see Definition 6.15).

Let us consider two points x_a and x_b, indistinguishable in $\Sigma^{(0)}$. We have to prove that $\theta(x_a) = \theta(x_b)$ and $\theta^i(x_a) = \theta^i(x_b)$ ($\forall i$). From Definition 6.15, we know that there exist $m_w + 1$ distinct pairs of vectors (w_a^0, w_b^0), (w_a^1, w_b^1), ..., $(w_a^{m_w}, w_b^{m_w})$, such that the two points $[x_a, w_a^\alpha]$ and $[x_b, w_b^\alpha]$ $\forall \alpha$ are indistinguishable in $\Sigma^{(1)}$, and the vector space generated by the vectors $t^j \triangleq w_a^j - w_a^0$ ($\forall j$) has dimension m_w.

From the fact that the function λ is observable, and by using the previous $m_w + 1$ indistinguishable points, we obtain the following $m_w + 1$ equations:

$$\lambda(x_a, w_a^\alpha) = \lambda(x_b, w_b^\alpha) \quad \forall \alpha.$$

By using the expression $\lambda(x, w) = \theta(x) + \theta^i(x) \mathcal{L}_G \widetilde{h}_i$, and the fact that the functions $\mathcal{L}_G \widetilde{h}_i$ are observable, we obtain

$$\theta(x_a) + \theta^i(x_a) \mathcal{L}_G \widetilde{h}_i(x_a, w_a^\alpha) = \theta(x_b) + \theta^i(x_b) \mathcal{L}_G \widetilde{h}_i(x_a, w_a^\alpha) \quad \forall \alpha.$$

By using the expression $\mathcal{L}_G \widetilde{h}_i = \mu_i^0 + \mu_i^k w_k$ we obtain

$$\theta(x_a) + \theta^i(x_a)[\mu_i^0(x_a) + \mu_i^k(x_a)(w_a^\alpha)_k] = \theta(x_b) + \theta^i(x_b)[\mu_i^0(x_a) + \mu_i^k(x_a)(w_a^\alpha)_k] \quad \forall \alpha.$$

By subtracting from the last m_w equations the first equation we obtain

$$\theta^i(x_a) \mu_i^k(x_a)(w_a^j - w_a^0)_k = \theta^i(x_b) \mu_i^k(x_a)(w_a^j - w_a^0)_k \quad \forall j,$$

namely

$$\mu_i^k t_k^j (\theta^i(x_a) - \theta^i(x_b)) = 0 \quad \forall j.$$

Since both μ_i^k and t_k^j are nonsingular, $\theta^i(x_a) = \theta^i(x_b)$. By using the first equation ($\alpha = 0$) we also obtain $\theta(x_a) = \theta(x_b)$. ◄

We are now ready to prove the following result.

Proposition 8.9. *For any integer m, the codistribution $\widetilde{\Omega}_m$ is observable.*

Proof. We proceed by induction. By definition, $\widetilde{\Omega}_0$ is the span of the gradients of the observable functions $\widetilde{h}_1, \ldots, \widetilde{h}_{m_w}$ (and, in the case of multiple outputs that do not appear in the selection of these further outputs).

8.3. Analytic derivations

Inductive step: Let us assume that the codistribution $\widetilde{\Omega}_{m-1}$ is observable. We want to prove that also $\widetilde{\Omega}_m$ is observable. From Algorithm 8.3, we need to prove that the following codistributions are observable:

$$\mathcal{L}_f \Omega_{m-1}, \quad \mathcal{L}_{\widehat{g}^\alpha} \Omega_{m-1}, \quad \text{span}\{\mathcal{L}_{\phi_{m-1}^{\alpha_1,\ldots,\alpha_{m-1}}} \nabla_k \widetilde{h}_i\}$$

$\forall \alpha, \alpha_1, \ldots, \alpha_{m-1}, i$.

Since $\widetilde{\Omega}_{m-1}$ is observable, so is $\mathcal{L}_F \widetilde{\Omega}_{m-1}$. Hence, $\mathcal{L}_F [\Omega_{m-1}, 0_{km_w}]$ is observable, and consequently, we obtain that $\mathcal{L}_f \Omega_{m-1}$ is observable.

We also have that $\mathcal{L}_G \widetilde{\Omega}_{m-1}$ is observable, and consequently, so is $\mathcal{L}_G [\Omega_{m-1}, 0_{km_w}]$. Let us consider a function $\theta(x)$ such that $\nabla_k \theta \in \Omega_{m-1}$. We obtain that $\nabla_k \mathcal{L}_G \theta$ belongs to the observable codistribution. On the other hand we have

$$\mathcal{L}_G \theta = \mathcal{L}_{g^0} \theta + \mathcal{L}_{g^i} \theta w_i = \mathcal{L}_{g^0} \theta + \mathcal{L}_{g^i} \theta (\nu_i^k \mathcal{L}_G \widetilde{h}_k + \nu_i^0) = \mathcal{L}_{\widehat{g}^0} \theta + \mathcal{L}_{\widehat{g}^i} \theta \mathcal{L}_G \widetilde{h}_i,$$

where we used (8.44). From lemma 8.8 we immediately obtain that the codistribution $\mathcal{L}_{\widehat{g}^\alpha} \Omega_{m-1}$ is observable $\forall \alpha$.

It remains to show that also the codistribution $\text{span}\{\mathcal{L}_{\phi_{m-1}^{\alpha_1,\ldots,\alpha_{m-1}}} \nabla_k \widetilde{h}_i\}$ is observable $\forall \alpha_1, \ldots, \alpha_{m-1}, i$.

To prove this, we start by remarking that, by applying $\mathcal{L}_F$ and $\mathcal{L}_G$ repetitively on the observable codistribution, starting from $[\Omega_{m-1}, 0_{km_w}]$ and by proceeding as before, we finally obtain an observable codistribution in the space of the original state, which is invariant under $\mathcal{L}_{\widehat{g}^\alpha}$ ($\forall \alpha$) and $\mathcal{L}_f$. As in the driftless case with a single unknown input (see Proposition 7.11), it is possible to show that this codistribution is also invariant under $\mathcal{L}_{\phi_{m-2}^{\alpha_1,\ldots,\alpha_{m-2}}}$ ($\forall \alpha_1, \ldots, \alpha_{m-2}$). This means that the function $\mathcal{L}_{[\widehat{\phi}_{m-2}, G]} \widetilde{h}_i$ is observable $\forall i$. Let us compute the Lie bracket of $\widehat{\phi}_{m-2} = [\phi_{m-2}^{\alpha_1,\ldots,\alpha_{m-2}}, 0_{km_w}]^T$ with G (for brevity's sake we omit the first $m-2$ indexes (α)). We have

$$[\widehat{\phi}_{m-2}, G] = \begin{bmatrix} [\phi_{m-2}, g^0] \\ 0_{km_w} \end{bmatrix} + \begin{bmatrix} [\phi_{m-2}, g^i] w_i \\ 0_{km_w} \end{bmatrix}.$$

By using (8.44) we obtain

$$[\widehat{\phi}_{m-2}, G] = \begin{bmatrix} [\phi_{m-2}, g^0] + [\phi_{m-2}, g^i](\nu_i^k \mathcal{L}_G \widetilde{h}_k + \nu_i^0) \\ 0_{km_w} \end{bmatrix} = \begin{bmatrix} [\phi_{m-2}]^0 + [\phi_{m-2}]^k \mathcal{L}_G \widetilde{h}_k \\ 0_{km_w} \end{bmatrix}$$

$$= \begin{bmatrix} \phi_{m-1}^{\ldots,0} + \phi_{m-1}^{\ldots,k} \mathcal{L}_G \widetilde{h}_k \\ 0_{km_w} \end{bmatrix}.$$

Hence we have

$$\mathcal{L}_{[\widehat{\phi}_{m-2}, G]} \widetilde{h}_i = \mathcal{L}_{\phi_{m-1}^{\ldots,0}} \widetilde{h}_i + \mathcal{L}_{\phi_{m-1}^{\ldots,k}} \widetilde{h}_i \mathcal{L}_G \widetilde{h}_k$$

and the observability of $\mathcal{L}_{\phi_{m-1}^{\alpha_1,\ldots,\alpha_{m-1}}} \nabla_k \widetilde{h}_i$ $\forall \alpha_1, \ldots, \alpha_{m-1}, i$ follows from Lemma 8.8. ◀

The results of this section can be summarized by the following theorem.

Theorem 8.10 (Observable Codistribution). $\theta(x)$ is weakly observable iff $\exists m$ such that $\nabla \theta(x) \in \Omega_m$.

Proof. If $\theta(x)$ is weakly observable, from Theorem 6.12 we have that $\exists m$ such that $\nabla_k \theta(x) \in \overline{\Omega}_m$. From Proposition 8.7 and Lemma 8.3 we obtain that $\nabla \theta(x) \in \Omega_m$. Conversely, if for a given integer m, $\nabla \theta(x) \in \Omega_m$, then, from Proposition 8.9 and Lemma 8.3, $\theta(x)$ is weakly observable. ◂

8.3.2 ▪ Convergence

The goal of this section is to investigate the convergence properties of Algorithm 8.3. We will show that the algorithm converges in a finite number of steps, and we will also provide the criterion to establish that the algorithm has converged (Theorem 8.17). This theorem will be proved at the end of this section since we need to introduce several important new quantities and properties.

When investigating the convergence properties of Algorithm 8.3, we remark that the main difference between Algorithms 4.2 and 8.3 is the presence of the last term in the recursive step of the latter. Without this term, the convergence criterion would simply consist of the inspection of the equality $\Omega_{m+1} = \Omega_m$, as for Algorithm 4.2.

The following result provides the convergence criterion in a very special case that basically occurs when the contribution due to the last term in the recursive step of Algorithm 8.3 is included in the other terms. In this case, we obviously obtain that the convergence criterion consists of the inspection of the equality $\Omega_{m+1} = \Omega_m$, as for Algorithm 4.2.

We have the following result (which extends the result given by Lemma 7.8 that holds for the simplified system).

Lemma 8.11. *Let us denote by Λ_j the distribution generated by $\phi_0, \phi_1^{\alpha_1}, \ldots, \phi_j^{\alpha_1,\ldots,\alpha_j} \, \forall \alpha_1, \ldots, \alpha_j$, and let us denote by $m(\leq n-1)$ the smallest integer for which $\Lambda_{m+1} = \Lambda_m$ (n is the dimension of the state x). In the very special case when $\mathcal{L}_{\phi_j^{\alpha_1,\ldots,\alpha_j}} \mathcal{L}_{g^\alpha} \widetilde{h}_k = 0 \, \forall j = 0, \ldots, m$, $\forall \alpha, \alpha_1, \ldots, \alpha_j, k$, Algorithm 8.3 converges at the integer j such that $\Omega_{j+1} = \Omega_j$, and this occurs in at most $n - 1$ steps.*

Proof. First of all, we remark that the existence of an integer $m(\leq n-1)$ such that $\Lambda_{m+1} = \Lambda_m$ is a simple extension of the result proved in [24], when the bracket defined in Definition 8.1 is adopted instead of the Lie bracket. In particular, it is possible to prove that the distribution converges to Λ^* and that the convergence is achieved at the smallest integer for which we have $\Lambda_{m+1} = \Lambda_m$. Additionally, we have $\Lambda_{m+1} = \Lambda_m = \Lambda^*$ and m cannot exceed $n - 1$ (see Lemmas 1.8.2 and 1.8.3 in [24]).

In the very special case when $\mathcal{L}_{\phi_j^{\alpha_1,\ldots,\alpha_j}} \mathcal{L}_{g^\alpha} \widetilde{h}_k = 0 \, \forall j = 0, \ldots, m, \, \forall \alpha, \alpha_1, \ldots, \alpha_j, k$, thanks to the aforementioned convergence of the distribution Λ_j, we easily obtain that $\mathcal{L}_{\phi_j^{\alpha_1,\ldots,\alpha_j}} \mathcal{L}_{g^\alpha} \widetilde{h}_k = 0 \, \forall j \geq 1$.

We have (by avoiding writing all the upper indexes) $\phi_j = [\phi_{j-1}]^{\alpha_j} = \nu_\beta^{\alpha_j} [\phi_{j-1}, g^\beta]$. Hence,

$$\mathcal{L}_{\phi_j} \widetilde{h}_m = \nu_\beta^{\alpha_j} \left(\mathcal{L}_{\phi_{j-1}} \mathcal{L}_{g^\beta} \widetilde{h}_m - \mathcal{L}_{g^\beta} \mathcal{L}_{\phi_{j-1}} \widetilde{h}_m \right) = \nu_\beta^{\alpha_j} \mathcal{L}_{\phi_{j-1}} \mathcal{L}_{g^\beta} \widetilde{h}_m - \mathcal{L}_{\widetilde{g}^{\alpha_j}} \mathcal{L}_{\phi_{j-1}} \widetilde{h}_m.$$

Since $\mathcal{L}_{\phi_{j-1}} \mathcal{L}_{g^\beta} \widetilde{h}_m = 0 \, \forall j \geq 1$, we have $\mathcal{L}_{\phi_j} \widetilde{h}_m = -\mathcal{L}_{\widetilde{g}^{\alpha_j}} \mathcal{L}_{\phi_{j-1}} \widetilde{h}_m$ for any $j \geq 1$. Therefore, in this case the last term in the recursive step of Algorithm 8.3 is included in the second to last term. Therefore, we conclude that Algorithm 8.3 converges when $\Omega_{m+1} = \Omega_m$. This occurs in at most $n - 1$ steps, as for Algorithm 4.2. ◂

8.3. Analytic derivations

Let us consider now the general case. To proceed we need to introduce several important new quantities and properties. As in the driftless case with a single unknown input we introduce the analogue of the field ψ. We start by defining the following new operation. Given a vector field in the chronospace $\underline{a}$ we define the following $m_w + 1$ vector fields as follows:

$$\{\underline{a}\}^\alpha \triangleq [\underline{a},\, \widehat{\underline{g}}^\alpha]. \tag{8.46}$$

By using (8.17) we obtain

$$\{\underline{a}\}^\alpha = [\underline{a},\, \nu_\beta^\alpha \underline{g}^\beta] = \nu_\beta^\alpha [\underline{a},\, \underline{g}^\beta] + \left(\mathcal{L}_{\underline{a}} \nu_\beta^\alpha\right) \mu_\gamma^\beta \widehat{\underline{g}}^\gamma = [\underline{a}]^\alpha + \left(\mathcal{L}_{\underline{a}} \nu_\beta^\alpha\right) \mu_\gamma^\beta \widehat{\underline{g}}^\gamma.$$

On the other hand, for synchronous frames (where (8.16) holds), we have

$$\mathcal{L}_{\underline{a}} \nu_0^\alpha = 0.$$

Hence,

$$\{\underline{a}\}^\alpha = [\underline{a}]^\alpha + \left(\mathcal{L}_{\underline{a}} \nu_j^\alpha\right) \mu_\gamma^j \widehat{\underline{g}}^\gamma = [\underline{a}]^\alpha + \left(\mathcal{L}_{\underline{a}} \nu_j^\alpha\right) \mu_i^j \widehat{\underline{g}}^i,$$

where we also used (8.13).

Let us consider the case (very common when we work in a synchronous frame) when the operand has the following structure:

$$\underline{a} = \begin{bmatrix} 0 \\ a \end{bmatrix}. \tag{8.47}$$

By using (8.24) we obtain

$$\{\underline{a}\}^\alpha = \begin{bmatrix} 0 \\ [a]^\alpha + \left(\mathcal{L}_a \nu_j^\alpha\right) \mu_i^j \widehat{g}^i \end{bmatrix} \triangleq \begin{bmatrix} 0 \\ \{a\}^\alpha \end{bmatrix}. \tag{8.48}$$

For a given integer $0 \leq m \leq k$ we define $(m_w + 1)^m$ vectors $\underline{\psi}_m^{\alpha_1, \ldots, \alpha_m}$ ($\forall \alpha_j$) by the following recursive algorithm:

1. $\underline{\psi}_0 = \underline{f}$;
2. $\underline{\psi}_m^{\alpha_1, \ldots, \alpha_m} = \{\underline{\psi}_m^{\alpha_1, \ldots, \alpha_{m-1}}\}^{\alpha_m}$.

Based on (8.48), we define the analogue in the space of the states, namely:

1. $\psi_0 = f$;
2. $\psi_m^{\alpha_1, \ldots, \alpha_m} = \{\psi_m^{\alpha_1, \ldots, \alpha_{m-1}}\}^{\alpha_m}$.

It is possible to find a useful expression that relates these vectors to the vectors ϕ_j, previously defined. This is the analogue of the relation given in Lemma 7.9. On the other hand, we provide here a simplified version of this relation. For what follows, we do not need the complete relation.

Lemma 8.12. *For $m \geq 1$, the following equation holds:*

$$\psi_m^{\alpha_1, \ldots, \alpha_m} = \phi_m^{\alpha_1, \ldots, \alpha_m} + \sum_{k=1}^{k_m} \gamma_k \widehat{j}^k, \tag{8.49}$$

where the following hold:

- k_m is a strictly positive integer that depends on m;

- the vectors $\widehat{j}^k$ are among the vectors $\widehat{g}^\alpha$ ($\forall \alpha$) and their Lie brackets up to $m-1$ times;

- γ_k are scalar functions of the state x that satisfy $\nabla \gamma_k \in \Omega_{m+1}$.

Before providing the proof of this lemma, we remark that (7.18), which holds in the special case $m_w = 1$, $g^0 = 0$, agrees with (8.49). In particular, in (7.18), the scalar functions γ_k are explicitly computed (and, because of Lemma 7.10, their gradients belong to Ω_{m+1}). Regarding the vectors $\widehat{j}^k$, in the case $m_w = 1$, $g^0 = 0$, we only have the vector $\widehat{g}^1 \triangleq \frac{g^1}{L_g^1}$, since all the Lie brackets of this vector with itself vanish.

Proof. We prove the analogue of (8.49) in the chronospace, i.e.,

$$\underline{\psi}_m^{\alpha_1, \ldots, \alpha_m} = \underline{\phi}_m^{\alpha_1, \ldots, \alpha_m} + \sum_{k=1}^{k_m} \gamma_k \underline{\widehat{j}}^k. \tag{8.50}$$

We proceed by induction. We first need to consider $m = 1$, and we compute the equation that relates ψ_1^α to ϕ_1^α $\forall \alpha$.

We have

$$\underline{\psi}_1^\alpha = \{\underline{f}\}^\alpha = [\underline{f}]^\alpha + (\mathcal{L}_f \nu_j^\alpha) \mu_i^j \underline{\widehat{g}}^i = \underline{\phi}_1^\alpha + (\mathcal{L}_f \nu_j^\alpha) \mu_i^j \underline{\widehat{g}}^i.$$

This equation agrees with (8.50), provided that we are able to prove that the gradients of all the components of the two-index tensor $(\mathcal{L}_f \nu_j^\alpha) \mu_i^j$ belong to $\Omega_2 (= \Omega_{m+1})$. To prove this, we consider the operator $\mathcal{L}_{(\mathcal{L}_f \nu_j^\alpha) \mu_i^j \widehat{g}^i}$. From the previous equation, we easily obtain

$$\mathcal{L}_{(\mathcal{L}_f \nu_j^\alpha) \mu_i^j \widehat{g}^i} = \mathcal{L}_{\underline{\psi}_1^\alpha} - \mathcal{L}_{\underline{\phi}_1^\alpha}.$$

We apply this operator on $\widetilde{h}_k$ ($\forall k$). The result is a function whose gradient belongs to $\Omega_2 (= \Omega_{m+1})$, by construction. We obtain

$$\mathcal{L}_{(\mathcal{L}_f \nu_j^\alpha) \mu_i^j \widehat{g}^i} \widetilde{h}_k = (\mathcal{L}_f \nu_j^\alpha) \mu_i^j \mathcal{L}_{\widehat{g}^i} \widetilde{h}_k = (\mathcal{L}_f \nu_j^\alpha) \mu_i^j \nu_\beta^i \mathcal{L}_{g^\beta} \widetilde{h}_k$$
$$= (\mathcal{L}_f \nu_j^\alpha) \mu_i^j \nu_\beta^i \mu_k^\beta = (\mathcal{L}_f \nu_j^\alpha) \mu_i^j \delta_k^i = (\mathcal{L}_f \nu_j^\alpha) \mu_k^j.$$

Hence, the gradients of all the components of the two-index tensor $(\mathcal{L}_f \nu_j^i) \mu_k^j$ belong to $\Omega_2 (= \Omega_{m+1})$.

Inductive step: Let us assume that (8.50) holds for $m-1$. We need to prove that

$$\underline{\psi}_m^{\alpha_1, \ldots, \alpha_{m-1}, \alpha_m} = \underline{\phi}_m^{\alpha_1, \ldots, \alpha_{m-1}, \alpha_m} + \sum_{k=1}^{k_m} \gamma_k \underline{\widehat{j}}^k, \tag{8.51}$$

with $\nabla \gamma_k \in \Omega_{m+1}$.

8.3. Analytic derivations

We have

$$\underline{\psi}_m^{\alpha_1,\ldots,\alpha_{m-1},\alpha_m} = \{\underline{\psi}_{m-1}^{\alpha_1,\ldots,\alpha_{m-1}}\}^{\alpha_m} = \left\{\underline{\phi}_{m-1}^{\alpha_1,\ldots,\alpha_{m-1}} + \sum_{k=1}^{k_{m-1}} \gamma_k \widehat{\underline{j}}^k\right\}^{\alpha_m}$$

$$= \left\{\underline{\phi}_{m-1}^{\alpha_1,\ldots,\alpha_{m-1}}\right\}^{\alpha_m} + \left\{\sum_{k=1}^{k_{m-1}} \gamma_k \widehat{\underline{j}}^k\right\}^{\alpha_m}. \tag{8.52}$$

Let us consider the second term:

$$\left\{\sum_{k=1}^{k_{m-1}} \gamma_k \widehat{\underline{j}}^k\right\}^{\alpha_m} = \left[\sum_{k=1}^{k_{m-1}} \gamma_k \widehat{\underline{j}}^k, \widehat{\underline{g}}^{\alpha_m}\right]$$

$$= \sum_{k=1}^{k_{m-1}} \gamma_k \left[\widehat{\underline{j}}^k, \widehat{\underline{g}}^{\alpha_m}\right] - \sum_{k=1}^{k_{m-1}} (\mathcal{L}_{\widehat{\underline{g}}^{\alpha_m}} \gamma_k)\widehat{\underline{j}}^k = \sum_{k=1}^{k_m} \gamma_k \widehat{\underline{j}}^k,$$

with $\nabla \gamma_k \in \Omega_{m+1}$, and the vectors $\widehat{\underline{j}}^k$ are among the vectors $\widehat{\underline{g}}^\alpha$ ($\forall \alpha$) and their Lie brackets up to $m-1$ times. We also have

$$\left\{\underline{\phi}_{m-1}^{\alpha_1,\ldots,\alpha_{m-1}}\right\}^{\alpha_m} = \underline{\phi}_m^{\alpha_1,\ldots,\alpha_m} + (\mathcal{L}_{\underline{\phi}_{m-1}^{\alpha_1,\ldots,\alpha_{m-1}}} \nu_j^{\alpha_m})\mu_l^j \widetilde{\underline{g}}^l.$$

Substituting this in (8.52) we obtain

$$\underline{\psi}_m^{\alpha_1,\ldots,\alpha_m} = \underline{\phi}_m^{\alpha_1,\ldots,\alpha_m} + (\mathcal{L}_{\underline{\phi}_{m-1}^{\alpha_1,\ldots,\alpha_{m-1}}} \nu_j^{\alpha_m})\mu_l^j \widetilde{\underline{g}}^l + \sum_{k=1}^{k_m} \gamma_k \widehat{\underline{j}}^k.$$

This equation agrees with (8.50), provided that we are able to prove that the gradients of all the components of the $(m+1)$-index tensor $(\mathcal{L}_{\underline{\phi}_{m-1}^{\alpha_1,\ldots,\alpha_{m-1}}} \nu_j^{\alpha_m})\mu_l^j$ belong to Ω_{m+1}. To prove this, we consider the following operator:

$$\mathcal{L}_{(\mathcal{L}_{\underline{\phi}_{m-1}^{\alpha_1,\ldots,\alpha_{m-1}}} \nu_j^{\alpha_m})\mu_l^j \widehat{\underline{g}}^l}.$$

We apply this operator on $\widetilde{h}_k$ ($\forall k$). The result is a function whose gradient belongs to Ω_{m+1}, by construction. We obtain

$$\mathcal{L}_{(\mathcal{L}_{\underline{\phi}_{m-1}^{\alpha_1,\ldots,\alpha_{m-1}}} \nu_j^{\alpha_m})\mu_l^j \widehat{\underline{g}}^l} \widetilde{h}_k = (\mathcal{L}_{\underline{\phi}_{m-1}^{\alpha_1,\ldots,\alpha_{m-1}}} \nu_j^{\alpha_m})\mu_l^j \mathcal{L}_{\widehat{\underline{g}}^l} \widetilde{h}_k (\mathcal{L}_{\underline{\phi}_{m-1}^{\alpha_1,\ldots,\alpha_{m-1}}} \nu_j^{\alpha_m})\mu_l^j \delta_k^l$$

$$= (\mathcal{L}_{\underline{\phi}_{m-1}^{\alpha_1,\ldots,\alpha_{m-1}}} \nu_j^{\alpha_m})\mu_k^j.$$

Hence, the gradients of all the components of the $(m+1)$-index tensor $(\mathcal{L}_{\underline{\phi}_{m-1}^{\alpha_1,\ldots,\alpha_{m-1}}} \nu_j^{\alpha_m})\mu_l^j$ belong to Ω_{m+1} and this proves the inductive step. ◀

For any integer $m \geq 1$, we introduce two fundamental $(m+1)$-index tensors. They are both characterized by m upper indexes and 1 lower index. They are

$$\mathcal{O}_\gamma^{\alpha_1,\ldots,\alpha_{m-1},\alpha_m} \triangleq (\mathcal{L}_{\underline{\phi}_{m-1}^{\alpha_1,\ldots,\alpha_{m-1}}} \nu_\beta^{\alpha_m})\mu_\gamma^\beta = (\mathcal{L}_{\underline{\phi}_{m-1}^{\alpha_1,\ldots,\alpha_{m-1}}} \nu_\beta^{\alpha_m})\mu_\gamma^\beta,$$

$$\mathcal{P}_\gamma^{\alpha_1,\ldots,\alpha_{m-1},\alpha_m} \triangleq (\mathcal{L}_{\underline{\phi}_{m-1}^{\alpha_1,\ldots,\alpha_{m-1}}} \mu_\gamma^\beta)\nu_\beta^{\alpha_m} = (\mathcal{L}_{\underline{\phi}_{m-1}^{\alpha_1,\ldots,\alpha_{m-1}}} \mu_\gamma^\beta)\nu_\beta^{\alpha_m}.$$

For brevity's sake, we avoid writing all the indexes $\alpha_1, \ldots, \alpha_{m-1}$. The previous tensors can be written as follows:

$$\mathcal{O}_\gamma^{\ldots,\alpha_m} \triangleq (\mathcal{L}_{\phi_{m-1}} \nu_\beta^{\alpha_m})\mu_\gamma^\beta, \tag{8.53}$$

$$\mathcal{P}_\gamma^{\ldots,\alpha_m} \triangleq (\mathcal{L}_{\phi_{m-1}} \mu_\gamma^\beta)\nu_\beta^{\alpha_m}. \tag{8.54}$$

We remark that $(\mathcal{L}_{\phi_{m-1}} \nu_\beta^\alpha \mu_\gamma^\beta) = (\mathcal{L}_{\phi_{m-1}} \delta_\gamma^\alpha) = 0$. Hence,

$$\mathcal{O}_\gamma^{\ldots,\alpha} = -\mathcal{P}_\gamma^{\ldots,\alpha}. \tag{8.55}$$

The following result extends the one given in Lemma 7.10.

Lemma 8.13. *For any integer $m \geq 1$, the gradients of all the components of the two $(m+1)$-index tensors given in (8.53) and (8.54) belong to Ω_{m+1}.*

Proof. By using (8.55), we only need to consider one of these tensors. Let us refer to $\mathcal{O}$. The proof that the gradients of these components belong to Ω_{m+1} is available in the proof of Lemma 8.12. ◄

The result stated by Proposition 7.11 also holds in the general case.

Proposition 8.14. *If Ω_m is invariant with respect to $\mathcal{L}_f$ and $\mathcal{L}_{\hat{g}^\alpha}$ $\forall \alpha$, then it is also invariant with respect to $\mathcal{L}_{\phi_j^{\alpha_1,\ldots,\alpha_j}}$ for $j = 1, \ldots, m-1$.*

Proof. From (8.49) we obtain the following operator equality:

$$\mathcal{L}_{\phi_j^{\alpha_1,\ldots,\alpha_j}} = \mathcal{L}_{\psi_j^{\alpha_1,\ldots,\alpha_j}} - \sum_{k=1}^{k_j} \mathcal{L}_{\gamma_k \hat{j}^k},$$

with $\nabla \gamma_k \in \Omega_{j+1}$. Let us apply the previous equality on a given covector in Ω_m. Because of the invariance with respect to $\mathcal{L}_f$ and $\mathcal{L}_{\hat{g}^\alpha}$ $\forall \alpha$, we also have the invariance with respect to $\mathcal{L}_{\hat{j}^k}$ and $\mathcal{L}_{\psi_j^{\alpha_1,\ldots,\alpha_j}}$. If $j \leq m-1$, $\nabla \gamma_k \in \Omega_m$ and we obtain the invariance with respect to $\mathcal{L}_{\phi_j^{\alpha_1,\ldots,\alpha_j}}$. ◄

The following result extends the one given in Lemma 7.12. As for (7.20), we need to substitute the expression of ϕ_j in terms of ϕ_{j-2} in the term $\mathcal{L}_{\phi_j} h$. This will allow us to detect the key quantity that governs the convergence of Algorithm 8.3, in particular regarding the contribution due to the last term in the recursive step of Algorithm 8.3. In the case $m_w = 1$, $g^0 = 0$, this quantity was a scalar and it was the one provided in (7.3). In the general case, the derivation was very troublesome since we did not know a priori that, instead of a scalar, the key quantity becomes a three-index tensor. Specifically, it is the tensor defined by (8.26), of type $(2, 1)$. For the sake of clarity, we provide (8.26) below:

$$\mathcal{T}_\gamma^{\alpha\beta} \triangleq \nu_\eta^\beta (\mathcal{L}_{\hat{g}^\alpha} \mu_\gamma^\eta) \quad \forall \alpha, \beta, \gamma.$$

8.3. Analytic derivations

In general, this tensor has $(m_w + 1) \times (m_w + 1) \times (m_w + 1)$ components. On the other hand, for synchronous frames it is immediate to obtain that $\mathcal{T}_0^{\alpha,\beta} = 0 \, \forall \alpha, \beta$. In other words, for these frames we can consider the lower index as a Latin index and the components of this tensor are $(m_w + 1) \times (m_w + 1) \times m_w$.

We have the following fundamental analytic result.

Lemma 8.15. *For $j \geq 2$, we have the following key equality (for brevity's sake, we denote by three dots $(\cdots)$ the first $j - 2$ Greek indexes, $\alpha_1, \ldots, \alpha_{j-2}$):*

$$\mathcal{L}_{\phi_j}{}^{\cdots \alpha_{j-1} \alpha_j} \tilde{h}_m \tag{8.56}$$

$$= \mathcal{L}_{\phi_{j-2}} \mathcal{T}_m^{\alpha_{j-1} \alpha_j} - \mathcal{O}_k^{\cdots \alpha_j} \mathcal{T}_m^{\alpha_{j-1} k} - \mathcal{O}_k^{\cdots \alpha_{j-1}} \mathcal{T}_m^{k \alpha_j} - \mathcal{L}_{\widehat{g}^{\alpha_{j-1}}} \mathcal{P}_m^{\cdots \alpha_j} - \mathcal{T}_k^{\alpha_{j-1} \alpha_j} \mathcal{P}_m^{\cdots k} - \mathcal{L}_{\widehat{g}^{\alpha_j}} \mathcal{L}_{\phi_{j-1}} \tilde{h}_m.$$

Proof. We will prove this equality by an explicit computation.

We have $\phi_j = [\phi_{j-1}]^{\alpha_j} = \nu_\beta^{\alpha_j} [\phi_{j-1}, g^\beta]$. Hence,

$$\mathcal{L}_{\phi_j} \tilde{h}_m = \nu_\beta^{\alpha_j} \left(\mathcal{L}_{\phi_{j-1}} \mathcal{L}_{g^\beta} \tilde{h}_m - \mathcal{L}_{g^\beta} \mathcal{L}_{\phi_{j-1}} \tilde{h}_m \right) = \nu_\beta^{\alpha_j} \mathcal{L}_{\phi_{j-1}} \mathcal{L}_{g^\beta} \tilde{h}_m - \mathcal{L}_{\widehat{g}^{\alpha_j}} \mathcal{L}_{\phi_{j-1}} \tilde{h}_m. \tag{8.57}$$

The second term coincides with the last term in (8.56). Hence, we need to prove that

$$\nu_\beta^{\alpha_j} \mathcal{L}_{\phi_{j-1}} \mathcal{L}_{g^\beta} \tilde{h}_m \tag{8.58}$$

$$= \mathcal{L}_{\phi_{j-2}} \mathcal{T}_m^{\alpha_{j-1} \alpha_j} - \mathcal{O}_k^{\cdots \alpha_j} \mathcal{T}_m^{\alpha_{j-1} k} - \mathcal{O}_k^{\cdots \alpha_{j-1}} \mathcal{T}_m^{k \alpha_j} - \mathcal{L}_{\widehat{g}^{\alpha_{j-1}}} \mathcal{P}_m^{\cdots \alpha_j} - \mathcal{T}_k^{\alpha_{j-1} \alpha_j} \mathcal{P}_m^{\cdots k}.$$

We have

$$\nu_\beta^{\alpha_j} \mathcal{L}_{\phi_{j-1}} \mathcal{L}_{g^\beta} \tilde{h}_m = \nu_\beta^{\alpha_j} \nu_\eta^{\alpha_{j-1}} \mathcal{L}_{\phi_{j-2}} \mathcal{L}_{g^\eta} \mathcal{L}_{g^\beta} \tilde{h}_m - \nu_\beta^{\alpha_j} \nu_\eta^{\alpha_{j-1}} \mathcal{L}_{g^\eta} \mathcal{L}_{\phi_{j-2}} \mathcal{L}_{g^\beta} \tilde{h}_m. \tag{8.59}$$

Let us compute these two terms separately. For the first we obtain

$$\nu_\beta^{\alpha_j} \nu_\eta^{\alpha_{j-1}} \mathcal{L}_{\phi_{j-2}} \mathcal{L}_{g^\eta} \mathcal{L}_{g^\beta} \tilde{h}_m = \nu_\beta^{\alpha_j} \nu_\eta^{\alpha_{j-1}} \mathcal{L}_{\phi_{j-2}} \mathcal{L}_{g^\eta} \mu_m^\beta$$

$$= \mathcal{L}_{\phi_{j-2}} (\nu_\beta^{\alpha_j} \nu_\eta^{\alpha_{j-1}} \mathcal{L}_{g^\eta} \mu_m^\beta) - \mathcal{L}_{\phi_{j-2}} (\nu_\beta^{\alpha_j} \nu_\eta^{\alpha_{j-1}}) \mathcal{L}_{g^\eta} \mu_m^\beta$$

$$= \mathcal{L}_{\phi_{j-2}} (\nu_\beta^{\alpha_j} \mathcal{L}_{\widehat{g}^{\alpha_{j-1}}} \mu_m^\beta) - \mathcal{L}_{\phi_{j-2}} (\nu_\beta^{\alpha_j} \nu_{\eta'}^{\alpha_{j-1}}) \delta_\eta^{\eta'} \mathcal{L}_{g^\eta} \mu_m^\beta$$

$$= \mathcal{L}_{\phi_{j-2}} \mathcal{T}_m^{\alpha_{j-1} \alpha_j} - \mathcal{L}_{\phi_{j-2}} (\nu_\beta^{\alpha_j} \nu_{\eta'}^{\alpha_{j-1}}) \mu_\gamma^{\eta'} \nu_\eta^\gamma \mathcal{L}_{g^\eta} \mu_m^\beta$$

$$= \mathcal{L}_{\phi_{j-2}} \mathcal{T}_m^{\alpha_{j-1} \alpha_j} - \mathcal{L}_{\phi_{j-2}} (\nu_\beta^{\alpha_j} \nu_{\eta'}^{\alpha_{j-1}}) \mu_\gamma^{\eta'} \mathcal{L}_{\widehat{g}^\gamma} \mu_m^\beta$$

$$= \mathcal{L}_{\phi_{j-2}} \mathcal{T}_m^{\alpha_{j-1} \alpha_j} - \mathcal{L}_{\phi_{j-2}} (\nu_\beta^{\alpha_j}) \nu_{\eta'}^{\alpha_{j-1}} \mu_\gamma^{\eta'} \mathcal{L}_{\widehat{g}^\gamma} \mu_m^\beta - \mathcal{L}_{\phi_{j-2}} (\nu_{\eta'}^{\alpha_{j-1}}) \nu_\beta^{\alpha_j} \mu_\gamma^{\eta'} \mathcal{L}_{\widehat{g}^\gamma} \mu_m^\beta$$

$$= \mathcal{L}_{\phi_{j-2}} \mathcal{T}_m^{\alpha_{j-1} \alpha_j} - \mathcal{L}_{\phi_{j-2}} (\nu_\beta^{\alpha_j}) \mathcal{L}_{\widehat{g}^{\alpha_{j-1}}} \mu_m^\beta - \mathcal{O}_\gamma^{\alpha_{j-1}} \mathcal{T}_m^{\gamma \alpha_j}$$

$$= \mathcal{L}_{\phi_{j-2}} \mathcal{T}_m^{\alpha_{j-1} \alpha_j} - \mathcal{L}_{\phi_{j-2}} (\nu_{\beta'}^{\alpha_j}) \mu_\gamma^{\beta'} \nu_\beta^\gamma \mathcal{L}_{\widehat{g}^{\alpha_{j-1}}} \mu_m^\beta - \mathcal{O}_\gamma^{\alpha_{j-1}} \mathcal{T}_m^{\gamma \alpha_j}$$

$$= \mathcal{L}_{\phi_{j-2}} \mathcal{T}_m^{\alpha_{j-1} \alpha_j} - \mathcal{O}_\gamma^{\alpha_j} \mathcal{T}_m^{\alpha_{j-1} \gamma} - \mathcal{O}_\gamma^{\alpha_{j-1}} \mathcal{T}_m^{\gamma \alpha_j}.$$

Hence, for this first term on the left-hand side of (8.59) we obtain

$$\nu_\beta^{\alpha_j} \nu_\eta^{\alpha_{j-1}} \mathcal{L}_{\phi_{j-2}} \mathcal{L}_{g^\eta} \mathcal{L}_{g^\beta} \tilde{h}_m = \mathcal{L}_{\phi_{j-2}} \mathcal{T}_m^{\alpha_{j-1} \alpha_j} - \mathcal{O}_\gamma^{\alpha_j} \mathcal{T}_m^{\alpha_{j-1} \gamma} - \mathcal{O}_\gamma^{\alpha_{j-1}} \mathcal{T}_m^{\gamma \alpha_j}. \tag{8.60}$$

Regarding the second term on the left-hand side of (8.59) we have

$$-\nu_\beta^{\alpha_j}\nu_\eta^{\alpha_{j-1}}\mathcal{L}_{g^\eta}\mathcal{L}_{\phi_{j-2}}\mathcal{L}_{g^\beta}\tilde{h}_m = -\nu_\beta^{\alpha_j}\nu_\eta^{\alpha_{j-1}}\mathcal{L}_{g^\eta}\mathcal{L}_{\phi_{j-2}}\mu_m^\beta = -\nu_\beta^{\alpha_j}\mathcal{L}_{\hat{g}^{\alpha_{j-1}}}\mathcal{L}_{\phi_{j-2}}\mu_m^\beta$$

$$= -\mathcal{L}_{\hat{g}^{\alpha_{j-1}}}(\nu_\beta^{\alpha_j}\mathcal{L}_{\phi_{j-2}}\mu_m^\beta) + \mathcal{L}_{\hat{g}^{\alpha_{j-1}}}(\nu_\beta^{\alpha_j})\mathcal{L}_{\phi_{j-2}}\mu_m^\beta$$

$$= -\mathcal{L}_{\hat{g}^{\alpha_{j-1}}}\mathcal{P}_m^{\alpha_j} + \mathcal{L}_{\hat{g}^{\alpha_{j-1}}}(\nu_\beta^{\alpha_j})\mathcal{L}_{\phi_{j-2}}\mu_m^\beta$$

$$= -\mathcal{L}_{\hat{g}^{\alpha_{j-1}}}\mathcal{P}_m^{\alpha_j} + \mathcal{L}_{\hat{g}^{\alpha_{j-1}}}(\nu_{\beta'}^{\alpha_j})\mu_\gamma^{\beta'}\nu_\beta^\gamma\mathcal{L}_{\phi_{j-2}}\mu_m^\beta$$

$$= -\mathcal{L}_{\hat{g}^{\alpha_{j-1}}}\mathcal{P}_m^{\alpha_j} + \mathcal{L}_{\hat{g}^{\alpha_{j-1}}}(\nu_{\beta'}^{\alpha_j}\mu_\gamma^{\beta'})\nu_\beta^\gamma\mathcal{L}_{\phi_{j-2}}\mu_m^\beta - \mathcal{L}_{\hat{g}^{\alpha_{j-1}}}(\nu_{\beta'}^{\alpha_j}\mu_\gamma^{\beta'})\nu_\beta^\gamma\mathcal{L}_{\phi_{j-2}}\mu_m^\beta$$

$$= -\mathcal{L}_{\hat{g}^{\alpha_{j-1}}}\mathcal{P}_m^{\alpha_j} - \nu_{\beta'}^{\alpha_j}\mathcal{L}_{\hat{g}^{\alpha_{j-1}}}(\mu_\gamma^{\beta'})\nu_\beta^\gamma\mathcal{L}_{\phi_{j-2}}\mu_m^\beta$$

$$= -\mathcal{L}_{\hat{g}^{\alpha_{j-1}}}\mathcal{P}_m^{\alpha_j} - \mathcal{T}_\gamma^{\alpha_{j-1}\alpha_j}\mathcal{P}_m^\gamma.$$

Hence, for this second term in (8.59) we obtain

$$-\nu_\beta^{\alpha_j}\nu_\eta^{\alpha_{j-1}}\mathcal{L}_{g^\eta}\mathcal{L}_{\phi_{j-2}}\mathcal{L}_{g^\beta}\tilde{h}_m = -\mathcal{L}_{\hat{g}^{\alpha_{j-1}}}\mathcal{P}_m^{\alpha_j} - \mathcal{T}_\gamma^{\alpha_{j-1}\alpha_j}\mathcal{P}_m^\gamma. \tag{8.61}$$

By substituting (8.60) and (8.61) in (8.59) and by reminding the reader that for synchronous frames the components of the tensor $\mathcal{T}$ that have the lower index equal to zero vanish, we immediately obtain (8.58). ◄

Lemma 8.16. *In general, there exists a finite* $m \leq n+2$ *such that* $\nabla \mathcal{T}_k^{\alpha,\beta} \in \Omega_m \; \forall \alpha,\beta,k$.

Proof. From (8.56) we have

$$\mathcal{L}_{\phi_{j+2}^{\ldots,\alpha_{j+1},\alpha_{j+2}}}\tilde{h}_m \tag{8.62}$$

$$= \mathcal{L}_{\phi_j^{\ldots}}\mathcal{T}_m^{\alpha_{j+1},\alpha_{j+2}} - \mathcal{O}_k^{\ldots,\alpha_{j+2}}\mathcal{T}_m^{\alpha_{j+1},k} - \mathcal{O}_k^{\ldots,\alpha_{j+1}}\mathcal{T}_m^{k,\alpha_{j+2}} + \mathcal{O}_m^{\ldots,k}\mathcal{T}_k^{\alpha_{j+1},\alpha_{j+2}}$$

$$- \mathcal{L}_{\hat{g}^{\alpha_{j+1}}}\mathcal{P}_m^{\ldots,\alpha_{j+2}} - \mathcal{L}_{\hat{g}^{\alpha_{j+2}}}\mathcal{L}_{\phi_{j+1}^{\ldots,\alpha_{j+1}}}\tilde{h}_m.$$

Let us introduce the following two tensors for a given integer j:

- ${}^j\mathcal{Z}_m^{\ldots,\alpha_{j+1},\alpha_{j+2}} \triangleq \mathcal{L}_{\phi_{j+2}^{\ldots,\alpha_{j+1},\alpha_{j+2}}}\tilde{h}_m$;

- ${}^j\mathcal{B}_m^{\ldots,\alpha_{j+1},\alpha_{j+2}} \triangleq \mathcal{L}_{\phi_j^{\ldots}}\mathcal{T}_m^{\alpha_{j+1},\alpha_{j+2}}$.

Both of them have $j+2$ upper indexes (Greek) and one lower index (Latin). By construction, the gradients of all the $(m_w+1)^{j+2}m_w$ components of the tensor ${}^j\mathcal{Z} \in \Omega_{j+3}$. On the other hand, from (8.62) we immediately obtain

$$\nabla({}^j\mathcal{Z}_m^{\ldots,\alpha_{j+1},\alpha_{j+2}}) = \nabla({}^j\mathcal{B}_m^{\ldots,\alpha_{j+1},\alpha_{j+2}}) - \nabla\mathcal{O}_k^{\ldots,\alpha_{j+2}}\mathcal{T}_m^{\alpha_{j+1},k} - \mathcal{O}_k^{\ldots,\alpha_{j+2}}\nabla\mathcal{T}_m^{\alpha_{j+1},k} \tag{8.63}$$

$$- \nabla\mathcal{O}_k^{\ldots,\alpha_{j+1}}\mathcal{T}_m^{k,\alpha_{j+2}} - \mathcal{O}_k^{\ldots,\alpha_{j+1}}\nabla\mathcal{T}_m^{k,\alpha_{j+2}} + \nabla\mathcal{O}_m^{\ldots,k}\mathcal{T}_k^{\alpha_{j+1},\alpha_{j+2}} + \mathcal{O}_m^{\ldots,k}\nabla\mathcal{T}_k^{\alpha_{j+1},\alpha_{j+2}}$$

$$- \nabla\mathcal{L}_{\hat{g}^{\alpha_{j+1}}}\mathcal{P}_m^{\ldots,\alpha_{j+2}} - \nabla\mathcal{L}_{\hat{g}^{\alpha_{j+2}}}\mathcal{L}_{\phi_{j+1}^{\ldots,\alpha_{j+1}}}\tilde{h}_m.$$

8.3. Analytic derivations

By using Lemma 8.13 we obtain the following results:

- $-\nabla \mathcal{O}_k^{\cdots,\alpha_{j+2}} \mathcal{T}_m^{\alpha_{j+1},k} - \nabla \mathcal{O}_k^{\cdots,\alpha_{j+1}} \mathcal{T}_m^{k,\alpha_{j+2}} + \nabla \mathcal{O}_m^{\cdots,k} \mathcal{T}_k^{\alpha_{j+1},\alpha_{j+2}} \in \Omega_{j+2}$;

- $-\nabla \mathcal{L}_{\hat{g}^{\alpha_{j+1}}} \mathcal{P}_m^{\cdots,\alpha_{j+2}} \in \Omega_{j+3}$.

Additionally, $-\nabla \mathcal{L}_{\hat{g}^{\alpha_{j+2}}} \mathcal{L}_{\phi_{j+1}^{\cdots,\alpha_{j+1}}} \widetilde{h}_m \in \Omega_{j+3}$. Hence, from (8.63), we obtain that the covector

$$^j \mathcal{Z}_m^{\prime\cdots,\alpha_{j+1},\alpha_{j+2}} \tag{8.64}$$

$$= \nabla(^j\mathcal{B}_m^{\cdots,\alpha_{j+1},\alpha_{j+2}}) - \mathcal{O}_k^{\cdots,\alpha_{j+2}}\nabla \mathcal{T}_m^{\alpha_{j+1},k} - \mathcal{O}_k^{\cdots,\alpha_{j+1}}\nabla \mathcal{T}_m^{k,\alpha_{j+2}} + \mathcal{O}_m^{\cdots,k}\nabla \mathcal{T}_k^{\alpha_{j+1},\alpha_{j+2}}$$

belongs to Ω_{j+3}. We proceed as in the case $m_w = 1$, $g^0 = 0$. Let us denote by j^* the smallest integer such that the gradients of all the $(m_w+1)^{j^*+2}m_w$ components of the tensor $^{j^*}\mathcal{B}$ can be expressed as linear combinations of the gradients of the components of all the tensors $^j\mathcal{B}$, $j = 0, 1, \ldots, j^* - 1$,

$$\nabla(^{j^*}\mathcal{B}_m^{\cdots,\alpha_{j^*+1},\alpha_{j^*+2}}) = \sum_{j=0}^{j^*-1} {}^jc^k_{\cdots,\beta_{j+1},\beta_{j+2}} \nabla(^j\mathcal{B}_k^{\cdots,\beta_{j+1},\beta_{j+2}}) + c^k \nabla \widetilde{h}_k, \tag{8.65}$$

where the dummy indexes $k, \beta_1, \ldots, \beta_{j+2}$ are summed up, according to the Einstein notation. Note that j^* is a finite integer and in particular $j^* \leq n-1$. Indeed, if this would not be the case, the dimension of the codistribution generated by ∇h and the gradients of all the components of the tensors $^j\mathcal{B}$, $j = 0, \ldots, n-1$, would be $n+1$, i.e., larger than n. From (8.65) and (8.64) we obtain

$$^{j^*}\mathcal{Z}_m^{\prime\cdots,\alpha_{j^*+1},\alpha_{j^*+2}} = \sum_{j=0}^{j^*-1} {}^jc^k_{\cdots,\beta_{j+1},\beta_{j+2}} \nabla(^j\mathcal{B}_k^{\cdots,\beta_{j+1},\beta_{j+2}}) + c^k \nabla \widetilde{h}_k \tag{8.66}$$

$$-\mathcal{O}_k^{\cdots,\alpha_{j^*+2}}\nabla \mathcal{T}_m^{\alpha_{j^*+1},k} - \mathcal{O}_k^{\cdots,\alpha_{j^*+1}}\nabla \mathcal{T}_m^{k,\alpha_{j^*+2}} + \mathcal{O}_m^{\cdots,k}\nabla \mathcal{T}_k^{\alpha_{j^*+1},\alpha_{j^*+2}}.$$

From (8.64), for $j = 0, \ldots, j^*-1$, we obtain

$$\nabla(^j\mathcal{B}_k^{\cdots,\beta_{j+1},\beta_{j+2}}) = {}^j\mathcal{Z}_k^{\prime\cdots,\beta_{j+1},\beta_{j+2}} + \mathcal{O}_l^{\cdots,\beta_{j+2}}\nabla \mathcal{T}_k^{\beta_{j+1},l} + \mathcal{O}_l^{\cdots,\beta_{j+1}}\nabla \mathcal{T}_k^{l,\beta_{j+2}} - \mathcal{O}_k^{\cdots,l}\nabla \mathcal{T}_l^{\beta_{j+1},\beta_{j+2}}.$$

By substituting in (8.66) we obtain

$$^{j^*}\mathcal{Z}_m^{\prime\cdots,\alpha_{j^*+1},\alpha_{j^*+2}} - c^k\nabla \widetilde{h}_k - \sum_{j=0}^{j^*-1} {}^jc^k_{\cdots,\beta_{j+1},\beta_{j+2}} {}^j\mathcal{Z}_k^{\prime\cdots,\beta_{j+1},\beta_{j+2}} \tag{8.67}$$

$$= \sum_{j=0}^{j^*-1} {}^jc^k_{\cdots,\beta_{j+1},\beta_{j+2}} \{\mathcal{O}_l^{\cdots,\beta_{j+2}}\nabla \mathcal{T}_k^{\beta_{j+1},l} + \mathcal{O}_l^{\cdots,\beta_{j+1}}\nabla \mathcal{T}_k^{l,\beta_{j+2}} - \mathcal{O}_k^{\cdots,l}\nabla \mathcal{T}_l^{\beta_{j+1},\beta_{j+2}}\}$$

$$-\mathcal{O}_k^{\cdots,\alpha_{j^*+2}}\nabla \mathcal{T}_m^{\alpha_{j^*+1},k} - \mathcal{O}_k^{\cdots,\alpha_{j^*+1}}\nabla \mathcal{T}_m^{k,\alpha_{j^*+2}} + \mathcal{O}_m^{\cdots,k}\nabla \mathcal{T}_k^{\alpha_{j^*+1},\alpha_{j^*+2}}.$$

Since this equality holds $\forall m, \alpha_1, \ldots, \alpha_{j^*+2}$, (8.67) consists of $m_w(m_w+1)^{j^*+2}$ equations. We remark that the left-hand side of these equations consists of the sum of covectors that belong to

Ω_{j^*+3}. On the right-hand side we have the gradients of all the $m_w(m_w+1)^2$ components of the tensor $\mathcal{T}$. In general, (8.67) can be used to express these last gradients in terms of covectors that belong to Ω_{j^*+3}. Therefore, by setting $m \triangleq j^* + 3$, we have $m \leq n+2$ and $\nabla \mathcal{T}_k^{\alpha,\beta} \in \Omega_m$ $\forall \alpha, \beta, k$. ◂

The previous lemma ensures that, in general, there exists a finite $m \leq n+2$ such that $\nabla \mathcal{T}_k^{\alpha,\beta} \in \Omega_m$ $\forall \alpha, \beta, k$. Note that the previous proof holds if the matrix that expresses the dependency of (8.67) on the terms $\nabla \mathcal{T}_k^{\alpha,\beta}$ can be inverted. This holds in general, with the exception of the trivial case considered in Lemma 8.11.

The following theorem allows us to obtain the criterion to stop Algorithm 8.3.

Theorem 8.17. *If $\nabla \mathcal{T}_k^{\alpha\beta} \in \Omega_m$ $\forall \alpha, \beta, k$, and $\Omega_{m+1} = \Omega_m$ (namely, Ω_m is invariant under $\mathcal{L}_f$ and $\mathcal{L}_{\hat{g}^\alpha}$ $\forall \alpha$ simultaneously), then $\Omega_{m+p} = \Omega_m$ $\forall p \geq 0$.*

Proof. We proceed by induction. Obviously, the equality holds for $p = 0$.
Inductive step: Let us assume that $\Omega_{m+p} = \Omega_m$, and let us prove that $\Omega_{m+p+1} = \Omega_m$. We have to prove that $\nabla \mathcal{L}_{\phi_{m+p}^{\alpha_1, \ldots, \alpha_{m+p}}} h \in \Omega_m$ $\forall \alpha_1, \ldots, \alpha_{m+p}$. Indeed, from the inductive assumption, we know that $\Omega_{m+p}(= \Omega_m)$ is invariant under $\mathcal{L}_f$ and $\mathcal{L}_{\hat{g}^\alpha}$ $\forall \alpha$. Additionally, because of this invariance, by using Proposition 8.14, we obtain that Ω_m is also invariant under $\mathcal{L}_{\phi_j^{\alpha_1, \ldots, \alpha_j}}$ $\forall \alpha_1, \ldots, \alpha_j$, for $j = 1, 2, \ldots, m+p-1$. Since $\nabla \mathcal{T}_k^{\alpha,\beta} \in \Omega_m$ $\forall \alpha, \beta, k$, by computing the gradient of (8.56) for $j = m+p$, it is immediate to obtain that $\nabla \mathcal{L}_{\phi_{m+p}^{\alpha_1, \ldots, \alpha_{m+p}}} h \in \Omega_m$ $\forall \alpha_1, \ldots, \alpha_{m+p}$. ◂

We conclude this section by providing an upper bound for the number of steps that are in general necessary to achieve the convergence. The dimension of Ω_{j^*+2} is at least the dimension of the span of the covectors ∇h, ${}^0 \mathcal{Z}_k^{\prime \alpha_1, \alpha_2}$, ${}^1 \mathcal{Z}_k^{\prime \alpha_1, \alpha_2, \alpha_3}$, ..., ${}^{j^*-1} \mathcal{Z}_k^{\prime \alpha_1, \ldots, \alpha_{j^*+1}}$. From the definition of j^*, we know that among the vectors ∇h, $\nabla {}^0 \mathcal{B}_k^{\alpha_1, \alpha_2}$, $\nabla {}^1 \mathcal{B}_k^{\alpha_1, \alpha_2, \alpha_3}$, ..., $\nabla {}^{j^*-1} \mathcal{B}_k^{\alpha_1, \ldots, \alpha_{j^*+1}}$ at least $j^* + 1$ are independent meaning that the dimension of their span is at least $j^* + 1$. Hence, from (8.64), it easily follows that the dimension of the span of the vectors ∇h, ${}^0 \mathcal{Z}_k^{\prime \alpha_1, \alpha_2}$, ${}^1 \mathcal{Z}_k^{\prime \alpha_1, \alpha_2, \alpha_3}$, ..., ${}^{j^*-1} \mathcal{Z}_k^{\prime \alpha_1, \ldots, \alpha_{j^*+1}}$, $\nabla \mathcal{T}_k^{\alpha, \beta}$ is at least $j^* + 1$. Since Ω_{j^*+3} contains this span, its dimension is at least $j^* + 1$. Therefore, the condition $\Omega_{m+1} = \Omega_m$ for $m \geq j^* + 3$ is achieved for $m \leq n+2$.

8.3.3 ▪ Extension to the case of multiple known inputs

It is immediate to repeat all the steps carried out in the previous two subsections and extend the validity of Theorem 8.10 to the case of multiple known inputs ($m_u > 1$). Additionally, Theorem 8.17 can be easily extended to cope with the case of multiple known inputs. In this case, requiring that $\Omega_{m+1} = \Omega_m$ means that Ω_m must be invariant with respect to all $\mathcal{L}_{\hat{g}^\alpha}$ and all $\mathcal{L}_{f^i}$ simultaneously.

8.4 ▪ Extension to time-variant systems

In this section we extend the results provided in section 8.1 to a more general class of systems than the ones characterized by (8.1). In particular, the new systems explicitly depend on time; namely, all the functions that characterize their dynamics and/or their outputs explicitly depend

8.4. Extension to time-variant systems

on time. In other words, we provide the same extension given in section 4.7 for the unknown input case. The new systems are characterized by the following equations:

$$\begin{cases} \dot{x} = g^0(x,\,t) + \sum_{i=1}^{m_u} f^i(x,\,t)u_i + \sum_{j=1}^{m_w} g^j(x,\,t)w_j, \\ y = [h_1(x,\,t),\ldots,h_p(x,\,t)]^T. \end{cases} \quad (8.68)$$

As in all the other cases analyzed in this book, the observability properties are obtained by computing the observable codistribution. Therefore, the goal of our study is to find the new algorithm that computes the observable codistribution and that extends Algorithm 8.3.

By working in the chronospace and by repeating the steps carried out in section 8.1 and in section 8.3, we obtain the following algorithm for generating the entire observable codistribution in the chronospace.

ALGORITHM 8.4. Observable codistribution in the chronospace (in the presence of unknown inputs).

Set $k = 0$
Set $\underline{\Omega}_k = \text{span}\left\{\underline{\nabla}\tilde{h}_0, \underline{\nabla}h_1, \ldots, \underline{\nabla}h_p\right\}$
Set $k = k + 1$
Set $\underline{\Omega}_k = \underline{\Omega}_{k-1} \oplus \bigoplus_{i=1}^{m_u} \mathcal{L}_{\underline{f}^i}\underline{\Omega}_{k-1} \oplus \bigoplus_{\alpha=0}^{m_w} \mathcal{L}_{\underline{\hat{g}}^\alpha}\underline{\Omega}_{k-1}$
$\qquad + \bigoplus_{i=1}^{m_u} \bigoplus_{\beta=0}^{m_w} \bigoplus_{\alpha_1=0}^{m_w} \cdots \bigoplus_{\alpha_{k-1}=0}^{m_w} \text{span}\left\{\mathcal{L}_{i\underline{\phi}_{k-1}^{\alpha_1,\ldots,\alpha_{k-1}}}\underline{\nabla}\tilde{h}_\beta\right\}$
while $\dim(\underline{\Omega}_k) > \dim(\underline{\Omega}_{k-1})$ **do**
$\quad$Set $k = k + 1$
$\quad$Set $\underline{\Omega}_k = \underline{\Omega}_{k-1} \oplus \bigoplus_{i=1}^{m_u} \mathcal{L}_{\underline{f}^i}\underline{\Omega}_{k-1} \oplus \bigoplus_{\alpha=0}^{m_w} \mathcal{L}_{\underline{\hat{g}}^\alpha}\underline{\Omega}_{k-1}$
$\qquad + \bigoplus_{i=1}^{m_u} \bigoplus_{\beta=0}^{m_w} \bigoplus_{\alpha_1=0}^{m_w} \cdots \bigoplus_{\alpha_{k-1}=0}^{m_w} \text{span}\left\{\mathcal{L}_{i\underline{\phi}_{k-1}^{\alpha_1,\ldots,\alpha_{k-1}}}\underline{\nabla}\tilde{h}_\beta\right\}$
end while

Our final objective is to obtain the observable codistribution in the space of the states, i.e., the codistribution that includes all the observability properties of the state (and not the chronostate). We proceed as in section 4.7. We know that the time is observable. In particular, the function $\tilde{h}_0$ only depends on t ($\tilde{h}_0 = t$) and, as a result, $\underline{\nabla}\tilde{h}_0 = [1,\,0_n] \in \underline{\Omega}_0$. Hence, for each k, we can split $\underline{\Omega}_k$ into two codistributions as follows:

$$\underline{\Omega}_k = [1,\,0_n] + [0,\,\Omega_k],$$

where $[0,\,\Omega_k]$ includes covectors whose first component is zero. As a result, Ω_k is precisely the observable codistribution we want to compute, i.e., the one that only includes the observability properties of the state. Our objective is to derive the algorithm that directly provides the above Ω_k. We proceed as follows. From the structure of $\hat{g}^\alpha$ given in (8.19) we obtain, for any scalar field $\theta = \theta(x,\,t)$,

$$\mathcal{L}_{\hat{g}^\alpha}\theta = \begin{bmatrix} \frac{\partial\theta}{\partial t} + \mathcal{L}_{\hat{g}^0}\theta, & \alpha = 0, \\ \mathcal{L}_{\hat{g}^i}\theta, & \alpha = i = 1,\ldots,m_w. \end{bmatrix} \quad (8.69)$$

As a result, for any θ such that $\underline{\nabla}\theta \in \underline{\Omega}_k$, the following $m_w + 1$ functions belong to $\underline{\Omega}_{k+1}$:

$$\frac{\partial\theta}{\partial t} + \mathcal{L}_{\hat{g}^0}\theta, \qquad \mathcal{L}_{\hat{g}^i}\theta,\ i = 1,\ldots,m_w.$$

Hence, the observable codistribution in the space of the states is given by the following algorithm.

ALGORITHM 8.5. Observable codistribution for time-variant systems (in the presence of unknown inputs).

Set $m = 0$
Set $\Omega_m = \text{span}\{\nabla h_1, \ldots, \nabla h_p\}$
Set $^i\phi_m = f^i$ for $i = 1, \ldots, m_u$
Set $m = m + 1$
Set $\Omega_m = \Omega_{m-1} \oplus \bigoplus_{i=1}^{m_u} \mathcal{L}_{f^i}\Omega_{m-1} \oplus \widetilde{\mathcal{L}}_{\hat{g}^0}\Omega_{m-1} \oplus \bigoplus_{j=1}^{m_w} \mathcal{L}_{\hat{g}^j}\Omega_{m-1}$
Set $^i\phi_m^{\alpha_1} = [^i\phi_{m-1}]^{\alpha_1}$ for $i = 1, \ldots, m_u$
Set $m = m + 1$
Set $\Omega_m = \Omega_{m-1} \oplus \bigoplus_{i=1}^{m_u} \mathcal{L}_{f^i}\Omega_{m-1} \oplus \widetilde{\mathcal{L}}_{\hat{g}^0}\Omega_{m-1} \oplus \bigoplus_{j=1}^{m_w} \mathcal{L}_{\hat{g}^j}\Omega_{m-1}$
$\qquad + \bigoplus_{i=1}^{m_u} \bigoplus_{l=1}^{m_w} \bigoplus_{\alpha_1=0}^{m_w} \text{span}\left\{\mathcal{L}_{^i\phi_{m-1}^{\alpha_1}}\nabla \widetilde{h}_l\right\}$
Set $^i\phi_m^{\alpha_1, \alpha_2} = [^i\phi_{m-1}^{\alpha_1}]^{\alpha_2}$ for $i = 1, \ldots, m_u$
while $\dim(\Omega_m) > \dim(\Omega_{m-1})$ OR $\nabla \mathcal{T} \notin \Omega_m$ **do**
$\quad$ Set $m = m + 1$
$\quad$ Set $\Omega_m = \Omega_{m-1} \oplus \bigoplus_{i=1}^{m_u} \mathcal{L}_{f^i}\Omega_{m-1} \oplus \widetilde{\mathcal{L}}_{\hat{g}^0}\Omega_{m-1} \oplus \bigoplus_{j=1}^{m_w} \mathcal{L}_{\hat{g}^j}\Omega_{m-1}$
$\qquad + \bigoplus_{i=1}^{m_u} \bigoplus_{l=1}^{m_w} \bigoplus_{\alpha_1=0}^{m_w} \cdots \bigoplus_{\alpha_{m-1}=0}^{m_w} \text{span}\left\{\mathcal{L}_{^i\phi_{m-1}^{\alpha_1, \ldots, \alpha_{m-1}}}\nabla \widetilde{h}_l\right\}$
$\quad$ Set $^i\phi_m^{\alpha_1, \ldots, \alpha_m} = [^i\phi_{m-1}^{\alpha_1, \ldots, \alpha_{m-1}}]^{\alpha_m}$ for $i = 1, \ldots, m_u$
end while
Set $\Omega = \Omega_m$ and $s = \dim(\Omega)$

where we introduced the following operator:

$$\widetilde{\mathcal{L}}_{\hat{g}^0} \triangleq \frac{\partial}{\partial t} + \mathcal{L}_{\hat{g}^0}. \tag{8.70}$$

Appendix A
Proof of Theorem 4.5

By construction, we can select the generators $l_1, \ldots, l_s$ among the Lie derivatives of the system of order smaller than n, where n is the dimension of the state (since Algorithm 4.2 converges in at most $n-1$ steps). We show that, by suitably setting the inputs, we can obtain from the output the values that all these Lie derivatives take at x_0.

By proceeding as in the first part of the proof of Theorem 4.3, we obtain (4.13). We compute this equation at $t = 0$. For each k, the member on the left-hand side is known. We can express all the Lie derivatives that appear in F_k in terms of the values that the generators take at x_0 (i.e., $l_1(x_0), \ldots, l_s(x_0)$). Hence, by considering one equation for each k, we have available infinite equations in $s \leq n$ unknowns (i.e., $l_1(x_0), \ldots, l_s(x_0)$). These equations also depend on the time derivatives of the inputs. These quantities not only are known, but can be suitably set in order to extract from the aforementioned system a set of equations that allows us to retrieve $l_1(x_0), \ldots, l_s(x_0)$. We show that, by suitably setting the time derivatives of the inputs at $t = 0$, we can build a linear system where the unknowns are all the Lie derivatives up to the $(n-1)$-order of the output at x_0. As mentioned above, these unknowns include $l_1(x_0), \ldots, l_s(x_0)$.

The proof is complex. The strategy of the proof is to set the time derivatives of the inputs at $t = 0$ in order to isolate as much as possible the unknown Lie derivatives.

For educational purposes, we provide the strategy for the case when $m = 2$ (i.e., the dynamics is driven by two inputs) and there is no drift (absence of $f^0(x)$ in (4.1)). The general case is more laborious.

Setting the time derivatives of the inputs at $t = 0$ means to set the following values:

$$u_i^{(N)}(0), \quad i = 1, \ldots, m, \ N = 0, 1, 2, \ldots.$$

The expression of the j-order time derivative of the output has the following structure:

$$\frac{d^j h}{dt^j} = \sum_{p=1}^{j} \sum_{i_1,\ldots,i_p=1}^{2} L_{i_1,\ldots,i_p}^p \sum_{k_1,\ldots,k_p|\sum_i k_i = j-p} C_{k_1,\ldots,k_p}^{j,p} u_{i_1}^{(k_1)} \cdots u_{i_p}^{(k_p)}, \tag{A.1}$$

where the coefficients $C_{k_1,\ldots,k_p}^{j,p}$ can be obtained by recursion (i.e., by differentiating the above expression with respect to t). We have

$$\frac{d}{dt}\frac{d^j h}{dt^j} = \sum_{p=1}^{j} \sum_{k_1,\ldots,k_p \mid \sum_i k_i = j-p} C^{j,p}_{k_1,\ldots,k_p}$$

$$\left[L^{p+1}_{i_1,\ldots,i_{p+1}} u^{(k_1)}_{i_1} \cdots u^{(k_p)}_{i_p} u^{(0)}_{i_{p+1}} + L^{p}_{i_1,\ldots,i_p} \left(u^{(k_1+1)}_{i_1} \cdots u^{(k_p)}_{i_p} + \cdots + u^{(k_1)}_{i_1} \cdots u^{(k_p+1)}_{i_p} \right) \right],$$

from which we obtain the following law:

$$C^{j+1,p}_{k_1,\ldots,k_p} = \begin{bmatrix} \sum_i C^{j,p}_{k_1,\ldots,k_i-1,\ldots,k_p} & \text{if } k_p \neq 0, \\ C^{j,p-1}_{k_1,\ldots,k_{p-1}} + \sum_i C^{j,p}_{k_1,\ldots,k_i-1,\ldots,k_p} & \text{if } k_p = 0. \end{bmatrix} \quad (A.2)$$

For instance, we have ($j = 1, 2, 3, 4$)

$$\frac{dh}{dt} = L^1_{i_1} u^{(0)}_{i_1}, \qquad C^{1,1}_0 = 1,$$

$$\frac{d^2 h}{dt^2} = L^2_{i_1,i_2} u^{(0)}_{i_1} u^{(0)}_{i_2} + L^1_{i_1} u^{(1)}_{i_1}, \qquad C^{2,2}_{0,0} = 1, \ C^{2,1}_1 = 1,$$

$$\frac{d^3 h}{dt^3} = L^3_{i_1,i_2,i_3} u^{(0)}_{i_1} u^{(0)}_{i_2} u^{(0)}_{i_3} + 2 L^2_{i_1,i_2} u^{(1)}_{i_1} u^{(0)}_{i_2} + L^2_{i_1,i_2} u^{(0)}_{i_1} u^{(1)}_{i_2} + L^1_{i_1} u^{(2)}_{i_1},$$

$$C^{3,3}_{0,0,0} = 1, \ C^{3,2}_{1,0} = 2, \ C^{3,2}_{0,1} = 1, \ C^{3,1}_2 = 1,$$

$$\frac{d^4 h}{dt^4} = L^4_{i_1,i_2,i_3,i_4} u^{(0)}_{i_1} u^{(0)}_{i_2} u^{(0)}_{i_3} u^{(0)}_{i_4} + 3 L^3_{i_1,i_2,i_3} u^{(1)}_{i_1} u^{(0)}_{i_2} u^{(0)}_{i_3}$$

$$+ 2 L^3_{i_1,i_2,i_3} u^{(0)}_{i_1} u^{(1)}_{i_2} u^{(0)}_{i_3} + L^3_{i_1,i_2,i_3} u^{(0)}_{i_1} u^{(0)}_{i_2} u^{(1)}_{i_3}$$

$$+ 3 L^2_{i_1,i_2} u^{(2)}_{i_1} u^{(0)}_{i_2} + 3 L^2_{i_1,i_2} u^{(1)}_{i_1} u^{(1)}_{i_2} + L^2_{i_1,i_2} u^{(0)}_{i_1} u^{(2)}_{i_2} + L^1_{i_1} u^{(3)}_{i_1},$$

$$C^{4,4}_{0,0,0,0} = 1, \ C^{4,3}_{1,0,0} = 3, \ C^{4,3}_{0,1,0} = 2, \ C^{4,3}_{0,0,1} = 1, \ C^{4,2}_{2,0} = 3, \ C^{4,2}_{1,1} = 3, \ C^{4,2}_{0,2} = 1, \ C^{4,1}_3 = 1.$$

We remark that, by setting all $u^{(N)}_i(0)$ to zero with the exception of a single $N = K$, for which we set

$$u^{(K)}_1(0) = 1, \qquad u^{(K)}_2(0) = 0,$$

we obtain that the only $\frac{d^j h}{dt^j}$ that are not identically null are the ones for which j is a multiple of $K + 1$. In particular, we have for any integer $p = 1, 2, \ldots$

$$\frac{d^{K+1} h}{dt^{K+1}} = L^1_{i_1} C^{K+1,1}_K u^{(K)}_{i_1} = L^1_1,$$

$$\frac{d^{2(K+1)} h}{dt^{2(K+1)}} = L^2_{i_1,i_2} C^{2(K+1),2}_{K,K} u^{(K)}_{i_1} u^{(K)}_{i_2} = C^{2(K+1),2}_{KK} L^2_{1,1},$$

$$\frac{d^{p(K+1)} h}{dt^{p(K+1)}} = L^p_{i_1,\ldots,i_p} C^{p(K+1),p}_{K,\ldots,K} u^{(K)}_{i_1} \cdots u^{(K)}_{i_p} = C^{p(K+1),p}_{K,\ldots,K} L^p_{1,\ldots,1}.$$

These equations allow us to obtain all the Lie derivatives along f^1. Note that, after a given $p \leq n - 1$, these Lie derivatives become dependent. Hence, it suffices to consider at most the first $n - 1$ equations.

Appendix A. Proof of Theorem 4.5

On the other hand, if we simultaneously set

$$u_1^{(J)}(0) = 0, \quad u_2^{(J)}(0) = 1,$$

the j-order time derivatives of the output ($\frac{d^j h}{dt^j}$) that do not vanish are the ones for which $j = p(K+1) + q(J+1)$, for any pair of nonnegative integers p and q. Let us suppose that $K < J$. Our goal is to choose K and J in such a way that the decomposition of the integer $j = p(K+1) + q(J+1)$, in terms of $K+1$ and $J+1$ and limited to the values of j that do not exceed $(n-1)(J+1)$, only occurs for a unique pair of positive integers (p,q). This ensures that, in (A.1) for $j = p(K+1) + q(J+1) \leq (n-1)(J+1)$, the only Lie derivatives that contribute are the ones of order $p+q$, with p times f^1 and q times f^2. As a result, for any p and q with $p+q \leq n-1$, we have available one equation (that is, (A.1) for $j = p(K+1) + q(J+1)$) that only depends on the Lie derivatives of order $p+q$, with p times f^1 and q times f^2.

For instance, for $p = 2$ and $q = 3$, (A.1) for $j = 2(K+1) + 3(J+1)$ only depends on $\binom{5}{2} = 10$ Lie derivatives: L_{11222}^5, L_{12122}^5, L_{12212}^5, L_{12221}^5, L_{21122}^5, L_{21212}^5, L_{21221}^5, L_{22112}^5, L_{22121}^5, L_{22211}^5. Note that the dependence is linear through the coefficients $C_{k_1,\ldots,k_p}^{j,p}$.

The existence of such K and J is guaranteed by Lemma A.1, in particular by applying Lemma A.1 with $m = 2$ and $l = n - 1$ and by setting $K + 1 = z_1$ and $J + 1 = z_2$.

However, this does not allow us to obtain each single Lie derivative, since we have a single equation in $\binom{p+q}{p}$ Lie derivatives. On the other hand, Lemma A.1 ensures that we can obtain as many equations as we want. In particular, for any positive integer m, we can find m positive integers $z_1, \ldots, z_m$ for which the decomposition given in (A.3) is unique for all $z = j \leq (n-1)z_m$. Let us suppose that m is an even integer, and let us set

$$u_1^{(z_{2i})}(0) = 1, \quad u_2^{(z_{2i})}(0) = 0,$$
$$u_1^{(z_{2i-1})}(0) = 0, \quad u_2^{(z_{2i-1})}(0) = 1$$

for $i = 1, 2, \ldots, m/2$. Since there is no restriction on m, by setting its value large enough we can ensure that for each pair of nonnegative integers α, β (with $\alpha + \beta \leq n-1$) we have at least $\binom{\alpha+\beta}{\alpha}$ equations, which are linear in the Lie derivatives of order $\alpha + \beta$, with α times f^1 and β times f^2. It is possible to show that the matrix that characterizes this linear system is nonsingular. This is a consequence of the properties of the coefficients $C_{k_1,\ldots,k_p}^{j,p}$ that can be obtained from (A.2) (in particular that they variate by changing the degree of the derivative j and the order of $k_1, \ldots, k_p$). As a result, we can obtain all the Lie derivatives up to the order $n - 1$. These Lie derivatives certainly include the generators $l_1(x_0), \ldots, l_s(x_0)$.

Lemma A.1. *Given two positive integers m and l, there exist m positive integers $z_1 < z_2 \cdots < z_m$ such that all the integers z smaller than lz_m, generated by $z_1, \ldots, z_m$, i.e., such that*

$$z = \sum_{i=1}^{m} k_i z_i, \qquad (A.3)$$

where k_i are nonnegative integers, admit a unique decomposition of type (A.3).

We start by proving a simpler result (Lemma A.2 below) before giving the proof of Lemma A.1. We introduce the following terminology:

- Given m positive integers $z_1, \ldots, z_m$, we say that an integer z is generated by $z_1, \ldots, z_m$ with degree k if $z = \sum_{i=1}^{m} k_i z_i$ where k_i are nonnegative integers such that $\sum_{i=1}^{m} k_i = k$.
- We call $z = \sum_{i=1}^{m} k_i z_i$ a decomposition of z of degree k in terms of $z_1, \ldots, z_m$.

Lemma A.2. *Given two positive integers m and l, there exist m positive integers $z_1, \ldots, z_m$ such that all the integers generated by $z_1, \ldots, z_m$ with degree less or equal than l admit a unique decomposition in terms of $z_1, \ldots, z_m$.*

Proof. The proof is constructive and by induction on m. Trivially, it holds for $m = 1$ and for any integer z_1 (it is a trivial consequence of the fundamental theorem of arithmetic). Now, let us suppose that $z_1, \ldots, z_{m-1}$ are $m - 1$ integers that satisfy the property for a given l. In other words,

$$z = \sum_{i=1}^{m-1} k_i z_i \neq \sum_{i=1}^{m-1} j_i z_i, \tag{A.4}$$

where $k_1, \ldots, k_{m-1}$ and $j_1, \ldots, j_{m-1}$ are nonnegative integers such that $\sum_{i=1}^{m-1} k_i \leq l$, $\sum_{i=1}^{m-1} j_i \leq l$ and when $k_i \neq j_i$ for at least one i.

We want to build m integers satisfying the same property starting from them. We define

$$z_i' = z_0' + z_i, \qquad i = 1, \ldots, m-1,$$

where z_0' is the new integer to be defined in order to have

$$z = \sum_{i=1}^{m-1} k_i z_i' + k_0 z_0' \neq \sum_{i=1}^{m-1} j_i z_i' + j_0 z_0', \tag{A.5}$$

where $k_0, k_1, \ldots, k_{m-1}$ and $j_0, j_1, \ldots, j_{m-1}$ are nonnegative integers such that $\sum_{i=0}^{m-1} k_i \leq l$, $\sum_{i=0}^{m-1} j_i \leq l$ and when $k_i \neq j_i$ for at least one i.

From (A.4) and (A.5) we obtain

$$\sum_{i=1}^{m-1} k_i z_i + k z_0' \neq \sum_{i=1}^{m-1} j_i z_i + j z_0', \tag{A.6}$$

where $k \triangleq \sum_{i=0}^{m-1} k_i$ and $j \triangleq \sum_{i=0}^{m-1} j_i$. Now, if $k = j$, the previous equation holds for the inductive assumption for any choice of z_0'. Let us assume that $k \neq j$. We have that $|(k-j)z_0'| \geq z_0'$. On the other hand, the difference

$$\sum_{i=1}^{m-1} k_i z_i - \sum_{i=1}^{m-1} j_i z_i$$

cannot exceed $l z_m - z_1$. Therefore, it suffices to set z_0' larger than $l z_m - z_1$ to ensure the validity of (A.6). ◄

Now we are ready to prove Lemma A.1.

Proof. Let us suppose that $z_1, \ldots, z_m$ are m integers that satisfy Lemma A.2 for a given l. In other words,

$$z = \sum_{i=1}^{m} k_i z_i \neq \sum_{i=1}^{m} j_i z_i,$$

where $k_1, \ldots, k_m$ and $j_1, \ldots, j_m$ are nonnegative integers such that $\sum_{i=1}^{m} k_i \leq l$, $\sum_{i=1}^{m} j_i \leq l$ and when $k_i \neq j_i$ for at least one i.

Appendix A. Proof of Theorem 4.5

We define
$$z'_i = N + z_i, \qquad i = 1, \ldots, m-1, \tag{A.7}$$
where N must be defined in order to have
$$z = \sum_{i=1}^{m} k_i z'_i \neq \sum_{i=1}^{m} j_i z'_i,$$
where $k_1, \ldots, k_{m-1}$ and $j_1, \ldots, j_{m-1}$ are nonnegative integers such that $z \leq l z'_m$ and when $k_i \neq j_i$ for at least one i.

First of all, we require that
$$(l+1) z'_1 > l z'_m.$$
This condition ensures that any $z = \sum_{i=1}^{m} k_i z'_i \leq l z'_m$ admit decomposition in terms of $z'_1, \ldots, z'_m$ of a degree that cannot exceed l. The above condition is honored by requiring
$$N > l z_m - l z_1 - z_1.$$

Thanks to this condition we require that
$$\sum_{i=1}^{m} k_i z'_i \neq \sum_{i=1}^{m} j_i z'_i \tag{A.8}$$

holds with the restriction on k_i and j_i that the degree of the decomposition cannot exceed l, i.e., $k \triangleq \sum_{i=1}^{m} k_i \leq l, j \triangleq \sum_{i=1}^{m} d_i \leq l$.

From (A.7) and (A.8) we must require
$$kN + \sum_{i=1}^{m} k_i z_i \neq jN + \sum_{i=1}^{m} j_i z_i.$$

If $k = j$, the above inequality holds. Let us assume that $k \neq j$. We have that $|(k-j)N| \geq N$. On the other hand, the difference
$$\sum_{i=1}^{m} k_i z_i - \sum_{i=1}^{m} j_i z_i$$
cannot exceed $l z_m - z_1$. Therefore, it suffices to set $N > l z_m - z_1$. ◂

Appendix B
Reminders on Quaternions and Rotations

We provide a very concise summary on the quaternions and their use in characterizing rotations in 3D. The interested reader can find more material in the literature (e.g., in [29]).

A quaternion q consists of four real numbers. Let us denote them by q_t, q_x, q_y, q_z. In this book, we represent a quaternion in two manners. The first is by using a column vector, i.e.,

$$q = \begin{bmatrix} q_t \\ q_x \\ q_y \\ q_z \end{bmatrix}.$$

The second uses the symbols i, j, k, which extend the imaginary unit i and are called the fundamental quaternion units:

$$q = q_t + q_x\, i + q_y\, j + q_z\, k.$$

The sum of two quaternions is obtained by summing separately the four components, i.e.,

$$q = q_t + q_x\, i + q_y\, j + q_z\, k,$$

$$p = p_t + p_x\, i + p_y\, j + p_z\, k,$$

$$q + p = (q_t + p_t) + (q_x + p_x)\, i + (q_y + p_y)\, j + (q_z + p_z)\, k$$

or

$$q + p = \begin{bmatrix} q_t + p_t \\ q_x + p_x \\ q_y + p_y \\ q_z + p_z \end{bmatrix}.$$

Hence, the sum of quaternions is a commutative operation. The product of two quaternions is obtained by defining the product of the fundamental quaternion units as follows:

$$\begin{aligned} i^2 &= j^2 &&= k^2 = -1, \\ ij &= -ji &&= k, \\ jk &= -kj &&= i, \\ ki &= -ik &&= j. \end{aligned}$$

From them we obtain

$$qp = (q_t p_t - q_x p_x - q_y p_y - q_z p_z) + (q_t p_x + q_x p_t + q_y p_z - q_z p_y) \, i$$
$$+ (q_t p_y + q_y p_t + q_z p_x - q_x p_z) \, j + (q_t p_z + q_z p_t + q_x p_y - q_y p_x) \, k$$

or

$$qp = \begin{bmatrix} q_t p_t - q_x p_x - q_y p_y - q_z p_z \\ q_t p_x + q_x p_t + q_y p_z - q_z p_y \\ q_t p_y + q_y p_t + q_z p_x - q_x p_z \\ q_t p_z + q_z p_t + q_x p_y - q_y p_x \end{bmatrix}.$$

Hence, it is not commutative.

Given a quaternion $q = q_t + q_x \, i + q_y \, j + q_z \, k$ we define its conjugate as follows:

$$q^* = q_t - q_x \, i - q_y \, j - q_z \, k = \begin{bmatrix} q_t \\ -q_x \\ -q_y \\ -q_z \end{bmatrix}.$$

It is immediate to verify that

$$qq^* = q^* q = q_t^2 + q_x^2 + q_y^2 + q_z^2,$$

which is a real and nonnegative number. Its square root is the quaternion *norm*. A quaternion q such that $qq^* = 1$ is a *unit* quaternion. A quaternion such that the first component (q_t) vanishes is an *imaginary* quaternion.

A unit quaternion can be used to characterize a rotation in 3D. We remind the reader that, in 3D, any rotation can be obtained by defining a rotation axis (τ) and an angle of rotation (θ). A rotation axis is a unit vector that can be characterized by two angles:

$$\tau = \begin{bmatrix} \sin \psi \cos \phi \\ \sin \psi \sin \phi \\ \cos \psi \end{bmatrix}.$$

See Figure B.1 for an illustration. The existence of a rotation axis only holds in three dimensions.[36]

Let us consider two frames that share the same origin. Let us suppose that the second frame can be obtained by rotating the first frame by an angle θ about the axis τ. The unit vector τ has the above expression in the first frame.

We introduce the following unit quaternion:

$$q = \cos\left(\frac{\theta}{2}\right) + \sin\left(\frac{\theta}{2}\right) (\tau_x \, i + \tau_y \, j + \tau_z \, k), \tag{B.1}$$

where we denoted by τ_x, τ_y, τ_z the three components of the unit vector τ.

---

[36] In n dimensions a rotation is characterized by $\frac{n(n-1)}{2}$ parameters (i.e., the number of free parameters of a skew-symmetric matrix) and the equality

$$\frac{n(n-1)}{2} = n$$

only holds for $n = 3$. Hence, for $n \neq 3$ it is not possible to define a rotation axis.

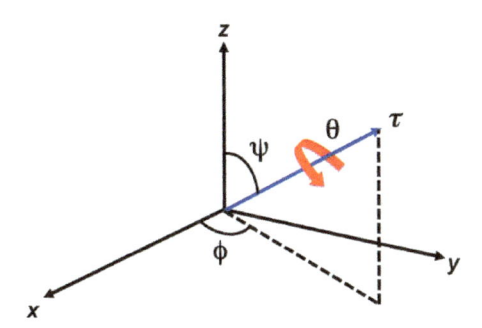

Appendix B. Reminders on Quaternions and Rotations

Figure B.1. *A rotation in 3D can be characterized by a rotation axis (τ) and an angle of rotation (θ). The rotation axis τ can be characterized by two angles (ϕ, ψ).*

Let us consider a given vector, and let us denote by

$$w = \begin{bmatrix} w_x \\ w_y \\ w_z \end{bmatrix}$$

its coordinates in the first frame and by

$$W = \begin{bmatrix} W_x \\ W_y \\ W_z \end{bmatrix}$$

its coordinates in the second frame. We provide a simple method to obtain W in terms of w and the reverse, by using the unit quaternion q.

First of all, we associate to w and W the following two imaginary quaternions:

$$w \to w_q = w_x\,i + w_y\,j + w_z\,k = \begin{bmatrix} 0 \\ w_x \\ w_y \\ w_z \end{bmatrix}, \tag{B.2}$$

$$W \to W_q = W_x\,i + W_y\,j + W_z\,k = \begin{bmatrix} 0 \\ W_x \\ W_y \\ W_z \end{bmatrix}. \tag{B.3}$$

The following fundamental equality holds:

$$W_q = q^* w_q q, \tag{B.4}$$

from which it is immediate to obtain the reverse:

$$w_q = q W_q q^*. \tag{B.5}$$

The two above equalities allow us to obtain the new coordinates of a given vector with respect to the second reference frame and vice versa.

We conclude by summarizing the overall procedure to obtain the coordinates of a given vector after a rotation. It consists of the following four steps:

1. Determine the rotation axis τ and the rotation angle θ and build the unit quaternion by using (B.1).

2. Use (B.2) (or (B.3)) to build the imaginary quaternion associated to w (or W).

3. Use (B.4) (or (B.5)) to compute the imaginary quaternion associated to W (or w).

4. Use (B.3) (or (B.2)) to obtain W (or w) from the associated imaginary quaternion.

Appendix C
Canonic Form with Respect to the Unknown Inputs

We refer to systems characterized by the following equations:

$$\begin{cases} \dot{x} = g^0(x) + \sum_{i=1}^{m_u} f^i(x)u_i + \sum_{j=1}^{m_w} g^j(x)w_j, \\ y = [h_1(x), \ldots, h_p(x)]^T. \end{cases} \quad \text{(C.1)}$$

We use the following notation:

- $\mathcal{F}$ is defined by Definition 6.13. According to Remark 6.14 it includes functions which are weakly observable.

- $\mathcal{DF}$ is the codistribution generated by the gradients, with respect to the state x, of the functions in $\mathcal{F}$.

- $\mathcal{L}_G\mathcal{F}$ is the space of functions that consists of the Lie derivatives along G of the functions in $\mathcal{F}$ (we remind the reader that G is the vector defined in (6.6)).

- $\mathcal{D}_w\mathcal{L}_G\mathcal{F}$ is the codistribution generated by the gradients, with respect to the unknown input vector w, of the functions in $\mathcal{L}_G\mathcal{F}$.

It is immediate to prove the following properties:

- From $\mathcal{F}$ we can select a set of functions such that their gradients generate $\mathcal{DF}$.

- There exists an integer m such that, by running m recursive steps of Algorithm 6.1, we obtain a codistribution $\overline{\Omega}_m$ that contains $[\mathcal{DF}, 0_{mm_w}]$ (i.e., the codistribution $\mathcal{DF}$ once embedded in the extended space).

- In general, the functions belonging to $\mathcal{L}_G\mathcal{F}$ are functions of x and $w_1, \ldots, w_{m_w}$.

- The dimension of the codistribution $\mathcal{D}_w\mathcal{L}_G\mathcal{F}$ cannot exceed the dimension of w (i.e., it cannot exceed m_w).

We introduce the following definition.

Definition C.1 (Canonic Form). *The system in* (C.1) *is in canonic form with respect to the unknown inputs if the dimension of* $\mathcal{D}_w \mathcal{L}_G \mathcal{F}$ *is* m_w.

We remark that if the system is in canonic form with respect to the unknown inputs, we can select m_w functions from $\mathcal{L}_G \mathcal{F}$ whose gradients with respect to w are independent. Therefore, we can select m_w functions from $\mathcal{F}$ such that, by using these functions for the selection of $\tilde{h}_1, \ldots, \tilde{h}_{m_w}$ needed to run Algorithm 8.3, the matrix in (8.2) is nonsingular. In the case $m_w = 1$, this means that we can select a function from $\mathcal{F}$ such that the quantity L_g^1 defined in (7.2) does not vanish. Note that we are allowed to use the functions in $\mathcal{F}$ because these functions are weakly observable (see Remark 6.14).

In the rest of this appendix, we show that there exists a finite number of transformations such that any system characterized by (C.1) can be either reduced in canonic form, or part of its unknown inputs is spurious (i.e., they do not affect the observability properties of the state). We will call the procedure that consists of these transformations *system canonization*. For clarity's sake, we distinguish the case of $m_w = 1$ from the general case. In particular, in section C.1 we discuss the case $m_w = 1$ (both in the driftless case (section C.1.1) and in the case with a drift (section C.1.2)) and in section C.2 the general case.

C.1 ▪ System canonization in the case of a single unknown input

Let us suppose that the system is not in canonic form. From Definition C.1, we know that the dimension of the codistribution $\mathcal{D}_w \mathcal{L}_G \mathcal{F}$ is 0. This means that the functions in $\mathcal{L}_G \mathcal{F}$ are independent of w. We discuss separately the cases without and with a drift.

C.1.1 ▪ Driftless case

In this case, $\mathcal{L}_G \mathcal{F}$ only contains the zero function (the Lie derivative along G of any function in $\mathcal{F}$ vanishes). As a result, any order Lie derivative of any outputs, computed along $F^1, \ldots, F^{m_u}$ (which are the vector fields defined in (6.6)) and at least once along G, vanishes. This means that, in Algorithm 6.1, we can ignore the contribution due to the Lie derivative along G. But this trivially means that the observable codistribution is precisely $\mathcal{D}\mathcal{F}$ and the unknown input is spurious.

C.1.2 ▪ The case with a drift

We perform a recursive procedure that consists of a finite number of steps. For the initialization, we set

$$\mathcal{F}^0 \triangleq \mathcal{F}, \quad \mathcal{D}\mathcal{F}^0 \triangleq \mathcal{D}\mathcal{F}, \quad \mathcal{D}_w \mathcal{L}_G \mathcal{F}^0 \triangleq \mathcal{D}_w \mathcal{L}_G \mathcal{F}.$$

The $(l+1)$th step of the procedure ($l \geq 0$) consists of the following operations:

1. Define the space of functions $\mathcal{F}^{l+1}$ as the space that contains all the functions in $\mathcal{F}^l \oplus \mathcal{L}_G \mathcal{F}^l$ and their Lie derivatives along the vector fields $f^1, \ldots, f^{m_u}$ up to any order.

2. Build the codistribution $\mathcal{D}\mathcal{F}^{l+1}$ defined as the codistribution that contains all the gradients, with respect to the state x, of the functions in $\mathcal{F}^{l+1}$ (note that $\mathcal{D}\mathcal{F}^{l+1}$ is included in the observable codistribution of the system defined by (C.1), as long as there exists an integer

C.2. System canonization in the general case

m such that, by running m recursive steps of Algorithm 6.1, we obtain a codistribution $\overline{\Omega}_m$ that contains $[\mathcal{DF}^{l+1}, 0_{mm_w}]$.

3. Build the codistribution $\mathcal{D}_w\mathcal{L}_G\mathcal{F}^{l+1}$ as the codistribution generated by the gradients with respect to w of the functions that belong to $\mathcal{L}_G\mathcal{F}^{l+1}$.

We remark that, at each step, we can have two distinct results:

- The dimension of $\mathcal{D}_w\mathcal{L}_G\mathcal{F}^{l+1}$ is 0 ($\dim(\mathcal{D}_w\mathcal{L}_G\mathcal{F}^{l+1}) = 0$).
- The dimension of $\mathcal{D}_w\mathcal{L}_G\mathcal{F}^{l+1}$ is 1 ($\dim(\mathcal{D}_w\mathcal{L}_G\mathcal{F}^{l+1}) = 1$).

In the latter case we stop the procedure. We select a scalar function in $\mathcal{F}^{l+1}$ such that its Lie derivative along $g \triangleq g^1$ does not vanish (it exists because $\dim(\mathcal{D}_w\mathcal{L}_G\mathcal{F}^{l+1}) = 1$). The system is in canonic form by using this scalar function as a system output (we are allowed to use this function as an output since its gradient belongs to $\mathcal{DF}^{l+1}$, which is included in the observable codistribution of the system defined by (C.1)).

Let us consider the former case (i.e., $\dim(\mathcal{D}_w\mathcal{L}_G\mathcal{F}^{l+1}) = 0$). We easily have $\mathcal{L}_g\mathcal{F}^{l+1} = \{0\}$. Hence, $\mathcal{F}^{l+2} = \mathcal{F}^{l+1} \oplus \mathcal{L}_G\mathcal{F}^{l+1} = \mathcal{F}^{l+1} \oplus \mathcal{L}_{g^0}\mathcal{F}^{l+1}$ We proceed as follows. We check whether $\mathcal{L}_{g^0}\mathcal{DF}^{l+1} \subseteq \mathcal{DF}^{l+1}$. Note that this condition will be satisfied in at least $n-1$ steps. Once this condition is satisfied, it means that $\mathcal{L}_g\mathcal{L}_{g^0}\mathcal{F}^{l+1} = \{0\}$. Hence, $\mathcal{L}_g\mathcal{L}_G\mathcal{F}^{l+1} = \{0\}$ and $\mathcal{L}_g\mathcal{F}^{l+2} = \{0\}$. Therefore, $\dim(\mathcal{D}_w\mathcal{L}_G\mathcal{F}^{l+2}) = 0$. By induction, this means that $\dim(\mathcal{D}_w\mathcal{L}_G\mathcal{F}^{l+p}) = 0$ for any integer p. In this case, the unknown input is spurious and the observable codistribution is $\mathcal{DF}^{l+1}$.

Algorithm C.1 provides the pseudocode for the system canonization.

ALGORITHM C.1. Canonization for driftless systems with a single unknown input.

Set $\mathcal{F}, \mathcal{DF}, \mathcal{L}_G\mathcal{F}$, and $\mathcal{D}_w\mathcal{L}_G\mathcal{F}$ as explained above
if $\dim(\mathcal{D}_w\mathcal{L}_G\mathcal{F}) == 1$ **then**
    Set h as one of the functions in $\mathcal{L}_G\mathcal{F}$ s.t. $\mathcal{L}_g h \neq 0$. The system is in canonic form. Algorithm 7.3 can be implemented by using this function.
    RETURN
end if
while $\mathcal{L}_{g^0}\mathcal{DF} \not\subseteq \mathcal{DF}$ **do**
    Set $\mathcal{F} := \mathcal{F} \oplus \mathcal{L}_G\mathcal{F}$ and close with respect to $\mathcal{L}_{f^1}, \ldots, \mathcal{L}_{f^{m_u}}$
    if $\dim(\mathcal{D}_w\mathcal{L}_G\mathcal{F}) == 1$ **then**
        Set h as one of the functions in $\mathcal{L}_G\mathcal{F}$ s.t. $\mathcal{L}_g h \neq 0$. The system is in canonic form. Algorithm 7.3 can be implemented by using this function.
        RETURN
    end if
end while
The unknown input is spurious and the observable codistribution is $\mathcal{DF}$

C.2 • System canonization in the general case

Let us suppose that the system is not in canonic form. From Definition C.1, we know that the dimension of the codistribution $\mathcal{D}_w\mathcal{L}_G\mathcal{F}$ is smaller than m_w. As in the case with a drift and $m_w = 1$, we introduce a recursive procedure.

C.2.1 ▪ The recursive procedure to perform the system canonization in the general case

For the initialization, we set

$$\mathcal{F}^0 \triangleq \mathcal{F}, \quad \mathcal{D}\mathcal{F}^0 \triangleq \mathcal{D}\mathcal{F}, \quad {}^0w \triangleq w, \quad \mathcal{D}_w\mathcal{L}_G\mathcal{F}^0 \triangleq \mathcal{D}_w\mathcal{L}_G\mathcal{F}.$$

In addition, we denote by d^0 the dimension of $\mathcal{D}_w\mathcal{L}_G\mathcal{F}^0$ ($d^0 < m_w$).

The $(l+1)$th step of this procedure ($l \geq 0$) consists of six operations. In particular, the last three operations are the three operations that characterize the $(l+1)$th step of the recursive procedure introduced in the case with a drift and $m_w = 1$. We provide first the list of the operations and after we detail them. Note that, at each step, the state will be in general augmented and the vector fields, $f^1, \ldots, f^{m_u}$, will be modified accordingly (i.e., the modified vectors will characterize the dynamics of the augmented state). The six operations are as follows:

1. Redefine the unknown inputs in such a way that all the functions in $\mathcal{L}_G\mathcal{F}^l$ depend only on x and d^l unknown inputs (i.e., they are independent of the remaining $m_w - d^l$ unknown inputs). In this section we detail the method needed to obtain this result. The new unknown input vector will be denoted by $\tilde{w}$. The functions in $\mathcal{L}_G\mathcal{F}^l$ depend only on x and the first d^l entries of $\tilde{w}$.

2. Augment the state by including in it the first d^l entries of $\tilde{w}$. Now the functions in $\mathcal{L}_G\mathcal{F}^l$ depend only on the augmented state.

3. Define the new vector of unknown inputs (^{l+1}w) as follows. Its first d^l entries are the first-order time derivatives of the entries of $\tilde{w}$. The last $m_w - d^l$ coincide with the last $m_w - d^l$ entries of $\tilde{w}$. In other words, $^{l+1}w \triangleq [\dot{\tilde{w}}_1, \ldots, \dot{\tilde{w}}_{d^l}, \tilde{w}_{d^l+1}, \ldots, \tilde{w}_{m_w}]$.

4. Define the space of functions $\mathcal{F}^{l+1}$ as the space that contains all the functions in $\mathcal{F}^l \oplus \mathcal{L}_G\mathcal{F}^l$ and their Lie derivatives along the vector fields $f^1, \ldots, f^{m_u}$ up to any order. Note that the vector fields $f^1, \ldots, f^{m_u}$ are now the ones that characterize the dynamics of the augmented state.

5. Build the codistribution $\mathcal{D}\mathcal{F}^{l+1}$ defined as the smallest codistribution that contains the gradients with respect to the augmented state of the functions in $\mathcal{F}^l \oplus \mathcal{L}_G\mathcal{F}^l$ and is invariant with respect to the Lie derivatives along the vector fields $f^1, \ldots, f^{m_u}$. In other words, $\mathcal{D}\mathcal{F}^{l+1}$ is the codistribution generated by the gradients with respect to the augmented state of the functions in $\mathcal{F}^{l+1}$ (note that $\mathcal{D}\mathcal{F}^{l+1}$ is included in the observable codistribution of the system defined by (C.1)).

6. Build the codistribution $\mathcal{D}_w\mathcal{L}_G\mathcal{F}^{l+1}$ as the codistribution generated by the gradients with respect to ^{l+1}w of the functions that belong to $\mathcal{L}_G\mathcal{F}^{l+1}$.

First operation

Since the dimension of $\mathcal{D}_w\mathcal{L}_G\mathcal{F}^l$ is d^l, we can select d^l functions in $\mathcal{L}_G\mathcal{F}^l$ whose gradients with respect to ^lw are independent. Let us denote them by

$$^l\gamma_1, \ldots, {}^l\gamma_{d^l}.$$

We also denote by $^l\tilde{h}_i$ ($i = 1, \ldots, d^l$) the function in $\mathcal{F}^l$ such that $^l\gamma_i = \mathcal{L}_G \, {}^l\tilde{h}_i$. We have

$$^l\gamma_i = \mathcal{L}_G \, {}^l\tilde{h}_i = \mathcal{L}_{g^0} \, {}^l\tilde{h}_i + \mathcal{L}_{g^j} \, {}^l\tilde{h}_i \, {}^lw_j,$$

C.2. System canonization in the general case

where, in accordance with the Einstein notation, the Latin index j is a dummy index and is summed over $j = 1, \ldots, m_w$. By denoting ${}^l\gamma_i^j \triangleq \mathcal{L}_{g^j}\,{}^l\tilde{h}_i$ we have

$${}^l\gamma_i = \mathcal{L}_{g^0}\,{}^l\tilde{h}_i + {}^l\gamma_i^j\,{}^lw_j. \tag{C.2}$$

Since the gradients with respect to ${}^lw$ of ${}^l\gamma_1, \ldots, {}^l\gamma_{d^l}$ are independent, we can extract from ${}^l\gamma_i^j$ a nonsingular two-index tensor, whose indices take the values $1, \ldots, d^l$. Additionally, let us reorder the unknown inputs in such a way that this tensor coincides with ${}^l\gamma_i^j$ for $j = 1, \ldots, d^l$. We denote this tensor by ${}^l\mu_i^j$. We can write (C.2) as follows:

$${}^l\gamma_i = \mathcal{L}_{g^0}\,{}^l\tilde{h}_i + \sum_{j=1}^{d^l} {}^l\mu_i^j\,{}^lw_j + \sum_{j=d^l+1}^{m_w} {}^l\gamma_i^j\,{}^lw_j. \tag{C.3}$$

We denote by ${}^l\nu$ the inverse of ${}^l\mu$. We introduce the following coordinate change (in the unknown inputs):

$$\begin{array}{ll} {}^lw_j \to \tilde{w}_j \triangleq {}^lw_j + \sum_{i=1}^{d^l}\sum_{k=d^l+1}^{m_w} {}^l\nu_j^i\,{}^l\gamma_i^k\,{}^lw_k, & j = 1, \ldots, d^l, \\ {}^lw_j \to \tilde{w}_j \triangleq {}^lw_j, & j = d^l+1, \ldots, m_w. \end{array} \tag{C.4}$$

Note that this coordinate change corresponds to a redefinition of the vector fields g^j. Specifically we have

$$\begin{array}{ll} g^j \to g^j, & j = 1, \ldots, d^l, \\ g^j \to g^j - \sum_{i=1}^{d^l}\sum_{k=1}^{d^l} {}^l\nu_k^i\,{}^l\gamma_i^j g^k, & j = d^l+1, \ldots, m_w. \end{array} \tag{C.5}$$

By an explicit computation it is possible to verify that, after this change, the functions in $\mathcal{L}_G\mathcal{F}^l$ only depend on $\tilde{w}_1, \ldots, \tilde{w}_{d^l}$, namely on the first d^l entries of the new unknown input vector $\tilde{w}$.

Second operation

We include the first d^l entries of $\tilde{w}$ in the state, i.e.,

$$x \to [x^T, \tilde{w}_1, \ldots, \tilde{w}_{d^l}]^T. \tag{C.6}$$

Third operation

We define the new unknown input vector:

$$^{l+1}w \triangleq [\dot{\tilde{w}}_1, \ldots, \dot{\tilde{w}}_{d^l}, \tilde{w}_{d^l+1}, \ldots, \tilde{w}_{m_w}]. \tag{C.7}$$

Fourth operation

We define the space of functions $\mathcal{F}^{l+1}$ as follows. This space contains $\mathcal{F}^l \oplus \mathcal{L}_G\mathcal{F}^l$ and their Lie derivative along the vector fields $f^1, \ldots, f^{m_u}$ up to any order.

Fifth operation

We denote by $\mathcal{DF}^{l+1}$ the codistribution generated by the gradients with respect to the new state in (C.6) of the functions in $\mathcal{F}^{l+1}$. $\mathcal{DF}^{l+1}$ is the smallest codistribution that contains the gradients of $\mathcal{F}^l \oplus \mathcal{L}_G\mathcal{F}^l$, and it is invariant with respect to the Lie derivative along the vector fields $f^1, \ldots, f^{m_u}$. Note that $\mathcal{DF}^{l+1}$ is included in the observable codistribution of the system defined by (C.1).

Sixth operation

We build the codistribution $\mathcal{D}_w\mathcal{L}_G\mathcal{F}^{l+1}$ as the codistribution generated by the gradients with respect to ^{l+1}w of the functions that belong to $\mathcal{L}_G\mathcal{F}^{l+1}$. The dimension of this codistribution is larger than or equal to d^l. On the other hand, it cannot exceed the dimension of ^{l+1}w, i.e., m_w. We denote this dimension by d^{l+1}, and we have $d^l \leq d^{l+1} \leq m_w$.

C.2.2 ▪ Convergence of the recursive procedure

We start by remarking that, in the case $d^{l+1} = m_w$, the canonization has been completed. Indeed, we can select m_w functions in $\mathcal{L}_G\mathcal{F}^{l+1}$ whose gradients with respect to ^{l+1}w are independent. Let us denote these functions by $\mathcal{L}_G\tilde{h}_1, \ldots, \mathcal{L}_G\tilde{h}_{m_w}$. The system is in canonic form by using $\tilde{h}_1, \ldots, \tilde{h}_{m_w}$ as system outputs (we are allowed to use these functions as outputs since their gradients belong to $\mathcal{D}\mathcal{F}^{l+1}$, which is included in the observable codistribution of the system defined by (C.1)).

In general, at each step, we can have two distinct results:

- $d^{l+1} > d^l$;

- $d^{l+1} = d^l$.

We remark that the first case cannot occur indefinitely. Hence, each time that the second case occurs, we apply Algorithm 8.3 to the current system by only considering the first d^{l+1} unknown inputs. Let us denote by Ω^* the codistribution provided by this algorithm, once converged. We then continue to proceed with the above procedure. At each step, denoted by $l+s$, $s \geq 2$, we check, first of all, whether $d^{l+s} = d^{l+1}$. If it is larger, we do not check anything else and we start again with the steps above until at a given step l' we obtain again that $d^{l'+1} = d^{l'}$. If $d^{l+s} = d^{l+1}$, we check whether $\Omega^* \subseteq \mathcal{D}\mathcal{F}^{l+s}$. If this is the case we conclude that only d^{l+1} unknown inputs affect the observability properties and the remaining $m_w - d^{l+1}$ unknown inputs are spurious and can be ignored. Additionally, we conclude that Ω^* is the observable codistribution.

Note that the final system is characterized by a new state, according to the change described by (C.6), which is performed at each step. Additionally, this final system is in canonic form with respect to the d^{l+1} unknown inputs, which are related to the original unknown inputs by the change in (C.4) and (C.7).

Algorithm C.2 provides the pseudocode for the system canonization in the general case.

ALGORITHM C.2. Canonization in the general case.

> Set $\mathcal{F}, \mathcal{D}\mathcal{F}, \mathcal{L}_G\mathcal{F}$, and $\mathcal{D}_w\mathcal{L}_G\mathcal{F}$ as explained above
> $d := \dim(\mathcal{D}_w\mathcal{L}_G\mathcal{F})$
> **if** $d == m_w$ **then**
>   The system is in canonic form. Select m_w functions in $\mathcal{F}$ such that their gradients (with respect to w) of their Lie derivatives along G are independent.
>   Use these m_w functions for the functions $\tilde{h}_1, \ldots, \tilde{h}_{m_w}$ to implement Algorithm 8.3.
>   RETURN
> **end if**
> Redefine w according to (C.4)
> Augment the state according to (C.6)
> Redefine w according to (C.7)
> Set $\mathcal{F} := \mathcal{F} \oplus \mathcal{L}_G\mathcal{F}$ and close with respect to $\mathcal{L}_{f^1}, \ldots, \mathcal{L}_{f^{m_u}}$

C.2. System canonization in the general case

$d_{old} := d$
$d := \dim(\mathcal{D}_w \mathcal{L}_G \mathcal{F})$
if $d == m_w$ **then**
    The system is in canonic form. Select m_w functions in $\mathcal{F}$ such that their gradients (with respect to w) of their Lie derivatives along G are independent.
    Use these m_w functions for the functions $\tilde{h}_1, \ldots, \tilde{h}_{m_w}$ to implement Algorithm 8.3.
    RETURN
end if
loop
    if $d == d_{old}$ **then**
        Select d functions in $\mathcal{F}$ such that their gradients (with respect to w) of their Lie derivatives along G are independent.
        Implement Algorithm 8.3 by ignoring the remaining $m_w - d$ unknown inputs. Specifically, use the selected d functions for the functions $\tilde{h}_1, \ldots, \tilde{h}_d$ to implement Algorithm 8.3 with d unknown inputs. Denote by Ω^* the codistribution provided by the algorithm, once converged.
    end if
    while $d == d_{old}$ **do**
        if $\Omega^* \subseteq \mathcal{DF}$ **then**
            The system is in canonic form only with respect to the first d unknown inputs. The remaining $m_w - d$ inputs are spurious. The observable codistribution is $\mathcal{DF}$.
            RETURN
        end if
        Redefine w according to (C.4)
        Augment the state according to (C.6)
        Redefine w according to (C.7)
        Set $\mathcal{F} := \mathcal{F} \oplus \mathcal{L}_G \mathcal{F}$ and close with respect to $\mathcal{L}_{f^1}, \ldots, \mathcal{L}_{f^{m_u}}$
        $d := \dim(\mathcal{D}_w \mathcal{L}_G \mathcal{F})$
        if $d == m_w$ **then**
            The system is in canonic form. Select m_w functions in $\mathcal{F}$ such that their gradients (with respect to w) of their Lie derivatives along G are independent.
            Use these m_w functions for the functions $\tilde{h}_1, \ldots, \tilde{h}_{m_w}$ to implement Algorithm 8.3.
            RETURN
        end if
    end while
    Redefine w according to (C.4)
    Augment the state according to (C.6)
    Redefine w according to (C.7)
    Set $\mathcal{F} := \mathcal{F} \oplus \mathcal{L}_G \mathcal{F}$ and close with respect to $\mathcal{L}_{f^1}, \ldots, \mathcal{L}_{f^{m_u}}$
    $d_{old} := d$
    $d := \dim(\mathcal{D}_w \mathcal{L}_G \mathcal{F})$
    if $d == m_w$ **then**
        The system is in canonic form. Select m_w functions in $\mathcal{F}$ such that their gradients (with respect to w) of their Lie derivatives along G are independent.
        Use these m_w functions for the functions $\tilde{h}_1, \ldots, \tilde{h}_{m_w}$ to implement Algorithm 8.3.
        RETURN
    end if
end loop

C.2.3 • Nonsingularity and canonic form

On the basis of the above discussion, we have obtained the following two results:

- If the codistribution $\overline{\Omega}_k$ is nonsingular at a given extended state

$$^k x = \begin{bmatrix} x \\ w_1^{(0)}(0) \\ \cdots \\ w_{m_w}^{(0)}(0) \\ w_1^{(1)}(0) \\ \cdots \\ w_{m_w}^{(1)}(0) \\ \cdots \\ w_1^{(k-1)}(0) \\ \cdots \\ w_{m_w}^{(k-1)}(0) \end{bmatrix},$$

then $w_1^{(0)}(0), \ldots, w_k^{(0)}(0)$ do not vanish (necessary condition).

- If the codistribution $\overline{\Omega}_k$ is nonsingular at a given extended state

$$^k x = \begin{bmatrix} x \\ w_1^{(0)}(0) \\ \cdots \\ w_{m_w}^{(0)}(0) \\ w_1^{(1)}(0) \\ \cdots \\ w_{m_w}^{(1)}(0) \\ \cdots \\ w_1^{(k-1)}(0) \\ \cdots \\ w_{m_w}^{(k-1)}(0) \end{bmatrix},$$

then it is also nonsingular at any extended state:

$$^k x' = \begin{bmatrix} x \\ w_1'^{(0)}(0) \\ \cdots \\ w_{m_w}'^{(0)}(0) \\ w_1^{(1)}(0) \\ \cdots \\ w_{m_w}'^{(1)}(0) \\ \cdots \\ w_1'^{(k-1)}(0) \\ \cdots \\ w_{m_w}'^{(k-1)}(0) \end{bmatrix},$$

such that $w_1'^{(0)}(0), \ldots, w_k'^{(0)}(0)$ do not vanish.

C.2.4 ▪ Remarks to reduce the computation

We conclude this section by adding two important remarks that can significantly reduce the computational burden to perform the observability analysis of a system once it has been set in canonic form, by using the procedure above.

The first remark is the following. For each step (m) of the above procedure, we remark that the gradients of all the scalar functions that belong to $\mathcal{F}^m$ belong to the observable codistribution. Hence, in order to perform an observability analysis, we can consider any functions in $\mathcal{F}^m$ as a system output. Therefore, we have the following result.

Remark C.2. *Let us suppose that the system is set in canonic form after m steps of the above procedure. Let us suppose that we can find m_w scalar functions, $\tilde{h}_1, \ldots, \tilde{h}_{m_w}$, in $\mathcal{F}^m$ that only depend on the state after $k < m$ steps. Additionally, let us suppose that the system obtained after k steps is in canonic form, by using these functions as outputs. Instead of performing the observability analysis by implementing Algorithm 8.3 on the system after m steps, we can implement Algorithm 8.3 on the system after k steps, by using the outputs $\tilde{h}_1, \ldots, \tilde{h}_{m_w}$.*

Similarly, once we know that our system can be set in canonic form, we can run Algorithm 6.1 and try to find functions of the original state, whose gradient belongs to the codistribution generated by this algorithm. We remark that the gradients of all these functions belong to the observable codistribution. Hence, in order to perform an observability analysis, we can consider these functions as system outputs. Therefore, we have the following result.

Remark C.3. *Let us consider m_w scalar functions that only depend on the original state. Let us suppose that their gradients belong to the codistribution generated by running Algorithm 6.1 for a given number of steps. Additionally, let us suppose that, by using these functions as outputs, the original system is in canonic form. We are allowed to perform the observability analysis by applying Algorithm 8.3 to the original system and by selecting these functions to run Algorithm 8.3.*

Bibliography

[1] J.-P. Barbot, D. Boutat, and T. Floquet. An Observation Algorithm for Nonlinear Systems with Unknown Inputs. Automatica, vol. 45, no. 8, pp. 1970–1974, August 2009 (Cited on p. 14)

[2] G. Basile and G. Marro. On the Observability of Linear, Time Invariant Systems with Unknown Inputs. Journal of Optimization Theory and Applications, vol. 3, pp. 410–415, 1969 (Cited on p. 14)

[3] F. A. W. Belo, P. Salaris, and A. Bicchi, 3 Known Landmarks are Enough for Solving Planar Bearing SLAM and Fully Reconstruct Unknown Inputs, IEEE/RSJ International Conference on Intelligent Robots and Systems (IROS), October 2010, Taipei, Taiwan (Cited on p. 139)

[4] A. Berthoz, B. Pavard, and L. R. Young. Perception of Linear Horizontal Self-Motion Induced by Peripheral Vision (Linearvection) Basic Characteristics and Visual-Vestibular Interactions, Experimental Brain Research, vol. 23, pp. 471–489, 1975 (Cited on p. 105)

[5] S. P. Bhattacharyya. Observer Design for Linear Systems with Unknown Inputs, IEEE Transactions on Automatic Control, vol. 23, pp. 483–484, 1978 (Cited on p. 14)

[6] J. Y. Bouguet. Camera Calibration Toolbox for Matlab. www.vision.caltech.edu/bouguetj (Cited on p. 126)

[7] J. L. Casti. Recent Developments and Future Perspectives in Nonlinear System Theory, SIAM Review, vol. 24, no. 2, pp. 301–331, July 1982 (Cited on pp. xi, 7)

[8] M. Darouach, M. Zasadzinski, and S. J. Xu. Full-order Observers for Linear Systems with Unknown Inputs, IEEE Transactions on Automatic Control, vol. 39, no. 3, 1994 (Cited on p. 14)

[9] C. De Persis and A. Isidori. A Geometric Approach to Nonlinear Fault Detection and Isolation, IEEE Transactions on Automatic Control, vol. 46, no. 6, pp. 853–865, 2001 (Cited on p. 14)

[10] K. Dokka, P. R. MacNeilage, G. C. De Angelis, and D. E. Angelaki. Estimating Distance During Self-motion: A Role for Visual-vestibular Interactions, Journal of Vision, vol. 11, no. 13, pp. 1–16, 2011 (Cited on p. 105)

[11] C. R. Fetsch, G. C. De Angelis, and D. E. Angelaki. Visual-Vestibular Cue Integration for Heading Perception: Applications of Optimal Cue Integration Theory, European Journal of Neuroscience, vol. 31, no. 10, pp. 1721–1729, May 2010 (Cited on p. 105)

[12] T. Floquet, C. Edwards, and S. Spurgeon. On Sliding Mode Observers for Systems with Unknown Inputs. International Journal of Adaptive Control and Signal Processing, vol. 21, no. 89, pp. 638–656, 2007 (Cited on p. 14)

[13] C. Forster, L. Carlone, F. Dellaert, and D. Scaramuzza. On-manifold Preintegration for Real-time Visual-inertial Odometry, IEEE Transactions Robotics, vol. 33, no. 1, pp. 1–21, February 2017 (Cited on p. 105)

[14] E. W. Griffith and K. S. P. Kumar. On the Observability of Nonlinear Systems, Journal of Mathematical Analysis and Applications, vol. 35, no. 1, pp. 135–147, 1971 (Cited on p. 67)

[15] Y. Guan and M. Saif. A Novel Approach to the Design of Unknown Input Observers. IEEE Transactions on Automatic Control, vol. 36, no. 5, pp. 632–635, 1991 (Cited on p. 14)

[16] R. Guidorzi and G. Marro. On Wonham Stabilizability Condition in the Synthesis of Observers for Unknown-Input Systems, IEEE Transactions on Automatic Control, vol. 16, no. 5, pp. 499–500, October 1971 (Cited on p. 14)

[17] Q. P. Ha and H. Trinh. State and Input Simultaneous Estimation for a Class of Nonlinear Systems, Automatica, vol. 40, pp. 1779–1785, 2004 (Cited on p. 14)

[18] H. Hammouri and Z. Tmar. Unknown Input Observer for State Affine Systems: A Necessary and Sufficient Condition, Automatica, vol. 46, no. 2, pp. 271–278, 2010 (Cited on p. 14)

[19] R. Hermann and A. J. Krener. Nonlinear Controllability and Observability, IEEE Transactions on Automatic Control, vol. 22, no. 5, pp. 728–740, 1977 (Cited on pp. xi, 7, 67)

[20] M. Hou and P. C. Müller. Design of Observers for Linear Systems with Unknown Inputs. IEEE Transactions on Automatic Control, vol. 37, no. 6, pp. 871–875, 1992 (Cited on p. 14)

[21] G. Huang, A. Mourikis, and S. Roumeliotis. An Observability-constrained Sliding Window Filter for Slam, IEEE/RSJ International Conference on Intelligent Robots and Systems, San Francisco, 2011 (Cited on p. 105)

[22] B. R. Hunt, T. Sauer, and J. A. Yorke. Prevalence: A Translation-invariant "Almost Every" on Infinite-Dimensional Spaces, Bulletin of the American Mathematical Society, vol. 27, no. 2, October 1992 (Cited on p. 137)

[23] V. Indelman, S. Williams, M. Kaess, and F. Dellaert. Information Fusion in Navigation Systems via Factor Graph Based Incremental Smoothing, Robotics and Autonomous Systems, vol. 61, pp. 721–738, 2013 (Cited on p. 105)

[24] A. Isidori. *Nonlinear Control Systems*, 3rd ed., London, Springer-Verlag, 1995 (Cited on pp. 181, 226)

[25] R. E. Kalman. On the General Theory of Control Systems, Proceedings of the 1st International Congress of IFAC, Moscow, 1960, Butterworth, London, 1961 (Cited on p. 67)

[26] R. E. Kalman. Mathematical Description of Linear Dynamical Systems, Journal of the Society for Industrial and Applied Mathematics Series A Control, vol. 1, pp. 152–192, 1963 (Cited on p. 67)

[27] D. Koening, B. Marx, and D. Jacquet. Unknown Input Observers for Switched Nonlinear Discrete Time Descriptor System, IEEE Transactions on Automatic Control, vol. 53, no. 1, 2008 (Cited on p. 14)

[28] S. R. Kou, D. L. Elliott, and T. J. Tarn. Observability of Nonlinear Systems, Information and Control, vol. 22, pp. 89–99, 1973 (Cited on p. 67)

[29] J. B. Kuipers. Quaternions and Rotation Sequences: A Primer with Applications to Orbits, Aerospace, and Virtual Reality. Princeton University Press, 1999 (Cited on p. 243)

[30] S. Leutenegger, P. Furgale, V. Rabaud, M. Chli, K. Konolige, and R. Siegwart. Keyframe-based Visual-Inertial Odometry Using Nonlinear Optimization, International Journal of Robotics Research, vol. 34, pp. 314–334, 2015 (Cited on p. 105)

Bibliography

[31] T. Lupton and S. Sukkarieh. Visual-inertial-aided Navigation for High-dynamic Motion in Built Environments without Initial Conditions, IEEE Transactions on Robotics, vol. 28, pp. 61–76, 2012 (Cited on p. 105)

[32] A. Martinelli. State Estimation Based on the Concept of Continuous Symmetry and Observability Analysis: The Case of Calibration, IEEE Transactions on Robotics, vol. 27, no. 2, pp. 239–255, 2011 (Cited on pp. 5, 105)

[33] A. Martinelli. Closed-form Solution of Visual-inertial Structure From Motion, International Journal of Computer Vision, vol. 106, no. 2, pp. 138–152, 2014 (Cited on p. 111)

[34] A. Martinelli. The Unicycle in Presence of a Single Disturbances: Observability Properties, in 2017 Proceedings of the Conference on Control and its Applications, Pittsburgh, July 2017 (Cited on pp. 165, 171)

[35] A. Martinelli. State Observability in Presence of Disturbances: The Analytic Solution and its Application in Robotics, International Conference on Intelligent Robots and Systems (IROS), 2017 IEEE/RSJ, Vancouver, BC, Canada, September 2017 (Cited on pp. 125, 126, 171, 172, 175)

[36] A. Martinelli, A. Oliva, and B. Mourain. Cooperative Visual-Inertial Sensor Fusion: The Analytic Solution, IEEE Robotics and Automation Letters, vol. 4, no. 2, pp. 453–460, 2019 (Cited on p. 124)

[37] A. Martinelli. Cooperative Visual-Inertial Odometry: Analysis of Singularities, Degeneracies and Minimal Cases, IEEE Robotics and Automation Letters, vol. 5, no. 2, pp. 668–675, 2020 (Cited on p. 124)

[38] A. Martinelli. Nonlinear Unknown Input Observability: Extension of the Observability Rank Condition, IEEE Transactions on Automatic Control, vol. 64, no. 1, pp. 222–237, 2019 (Cited on pp. 16, 153, 158, 175)

[39] A. Martinelli. Rank Conditions for Observability and Controllability for Time-Varying Nonlinear Systems, arXiv:2003.09721 [math.OC] (Cited on p. 93)

[40] A. Mourikis and S. Roumeliotis. A Dual-Layer Estimator Architecture for Long-term Localization, IEEE Computer Society Conference on Computer Vision and Pattern Recognition, Anchorage, Alaska, 2008 (Cited on p. 105)

[41] L. M. Silverman and H. E. Meadows. Controllability and Observability in Time-Variable Linear Systems, Journal of the Society for Industrial and Applied Mathematics Series A Control, vol. 5, pp. 64–73, 1967 (Cited on p. 93)

[42] S. N. Singh. Observability in Non-linear Systems with Immeasurable Inputs, International Journal of Systems Science, vol. 6, pp. 723–732, 1975 (Cited on p. 67)

[43] S. Telen, B. Mourrain, and M. Van Barel. Solving Polynomial Systems via Truncated Normal Forms, SIAM Journal on Matrix Analysis and Applications, vol. 39, pp. 1421–1447, 2018 (Cited on p. 124)

[44] C. Troiani and A. Martinelli. Vision-aided Inertial Navigation Using Virtual Features, IEEE/RSJ International Conference on Intelligent Robots and Systems, Vilamoura, Portugal, 7–12 October 2012 (Cited on p. 132)

[45] S .H. Wang, E. J. Davison, and P. Dorato. Observing the States of Systems with Unmeasurable Disturbance IEEE Transactions on Automatic Control, vol. 20, no. 5, pp. 716–717, 1975 (Cited on p. 14)

[46] F. W. Warner. *Foundations of Differentiable Manifolds and Lie Groups*, Springer, New York, 1983 (Cited on p. 21)

[47] F. Yang and R. W. Wilde. Observer for Linear Systems With Unknown Inputs. IEEE Transactions on Automatic Control, vol. 33, no. 7, 1988 (Cited on p. 14)

Index

Abelian group, *see* group
analytic function, 11
atlas, 23–25
augmented space, *see* extended space
augmented state, *see* state
autobracket, 191

Baker–Hausdorff formula, 50

calibration, 98–105, 126–128
   extrinsic calibration, 98, 111
canonic form, 248
canonization, 248–253
chart, 23–25
chronospace, 58–61, 91, 189–193, 235
chronostate, *see* state
codistribution, 39
   integrable codistribution, 39–42
   invariant codistribution, 40
   nonsingular codistribution, 39
   observable codistribution, 71–74, 77–78, 90–93, 95, 143–148, 154–156, 188–195, 219–226, 235–236
   orthogonal codistribution, 42
continuous symmetry, 8, 88–89
contraction, 30–31
coordinates' change, 24
cotangent space, 27
covariant vector, *see* covector
covector, 26
covector field, *see* tensor field

directional derivative, 27
distribution, 36

involutive distribution, 37
nonsingular distribution, 37
orthogonal distribution, 42, 88–89, 95
drift, 2, 13, 57–65
driftless, *see* system

Einstein notation, 26–27
extended space, 143
extended state, *see* state

Frobenius theorem, 41–45

group, 45
   Abelian group, 15–16, 45, 48, 54, 62–65
   coordinates transformations' group, 9
   $GL(n, \mathbb{R})$, 46
   Lie group, 45–53
   scale group, 45
   shift group, 46
   simply connected group, 50
   $SO(n)$, 46, 51–53
   $SU(n)$, 47, 52–53
group of invariance of observability, 9, 57
   input transformations' group, 10, 62–63
   output transformations' group, 10, 61–63
   Simultaneous Unknown Input-Output transformations' group, 16, 64–65, 187–195
group representation, 47–48

indistinguishability, 85–89, 148–151, 224
indistinguishable set, 89

inertial sensors, 105, 196, 206
input
   known input, 11, 13, 61
   system input, xi, 2, 59
   unknown input, 11, 13, 63
input transformations' group, *see* group of invariance of observability
input-output map, 74–78
inverse function
   inverse function problem, 68, 71–72, 74
   inverse function theorem, 8, 68, 71–72, 74

Lie algebra, 49
   infinitesimal generators of a Lie algebra, 50
   Lie algebra representation, 49
   structure constants of a Lie algebra, 49
Lie brackets, 35
Lie derivative, 32–36
Lie group, *see* group
local observable subsystem, 87–88, 95

manifold, 21–23
   differential manifold, 23
monocular camera, *see* monocular vision
monocular vision, 107

observability
   constructive definition of observability, 7, 68
   standard definition of observability, 85
   theory of observability, 2

261

weak observability, 8, 68, 85
observability rank condition, 8, 67–81
   extension of the observability rank condition, 14, 90–93, 153–157, 187–196, 234–236
observable codistribution, *see* codistribution
observable function, 8, 82–85
   weak observable function, 82–85
one-to-one correspondence, 22
open set, 21
output
   system output, xi, 2
output transformations' group, *see* group of invariance of observability

prevalent set, 137

quaternion, 243
   quaternions and rotations, 243–246
   unit quaternion, 244

regular point
   regular point of a codistribution, 39, 41, 72, 76–78

regular point of a distribution, 37

scalar, *see* tensor
Schur's lemma, 15, 48, 58
shy set, 137
Simultaneous Unknown Input-Output transformations' group, *see* group of invariance of observability
state, 2
   chronostate, 58–61, 189–190
   extended state, 139–145, 254
state extension, 140, 144
symmetry, *see* continuous symmetry
system
   driftless system, 15–16, 63–65, 153–186
   input-output system, 1, 2
   time-variant system, 16–17, 90–93, 234–236

tangent space, 27
Taylor theorem, 68–69
tensor, 25, 30–32
   tensor associated with a group of transformations, 53

tensor field, 31
tensor rank, 30
tensor type, 30
time
   chronological time, 12, 57
   ordinary time, 12, 57
   twofold role of time, 12
topological space, 21–22

unicycle, 1–2, 78–81, 96–97, 145–148, 158–171
unknown input observability, 12
   constructive definition of unknown input observability, 136–138
   standard definition of unknown input observability, 149
   w-observability, 136–137
   weak unknown input observability, 149
unobservability, 8, 87–89

vector
   contravariant vector, 25
vector field, *see* tensor field
visual inertial sensor fusion, 105–132, 196–218
visual-vestibular integration, 217–218